ORIGINS OF
Molecular Biology

A Tribute to Jacques Monod

ORIGINS OF
Molecular Biology

A Tribute to Jacques Monod

REVISED EDITION

EDITED BY

Agnes Ullmann

**ASM
PRESS**

Copyright © 2003   ASM Press
American Society for Microbiology
1752 N Street, N.W.
Washington, DC 20036-2904

Library of Congress Cataloging-in-Publication Data

Origins of molecular biology : a tribute to Jacques Monod / edited by
Agnes Ullmann.—Rev. ed.
    p. ; cm.
Includes bibliographical references and index.
ISBN 1-55581-281-3 (alk. paper)
    1. Monod, Jacques.   2. Molecular biologists—France—Biography.
3. Molecular biology—History.   4. Molecular microbiology—History.
[DNLM: 1. Monod, Jacques.   2. Molecular Biology—Biography.
3. Molecular Biology—history. WZ 100 M7510 2004]   I. Ullmann, Agnes.
II. Title.
    QH31.M544O74   2004
    572.8′092—dc22   2003021494

ISBN 1-55581-281-3

10 9 8 7 6 5 4 3 2 1
Address editorial correspondence to: ASM Press, 1752 N St., N.W.,
Washington, DC
20036-2904, U.S.A.

Send orders to: ASM Press, P.O. Box 605, Herndon, VA 20172, U.S.A.
Phone: 800-546-2416; 703-661-1593
Fax: 703-661-1501
Email: books@asmusa.org
Online: www.asmpress.org

Contents

Preface to the New Edition

A quarter of a century after the first edition of this book, we can still feel the impact of Jacques Monod's contribution on modern biology. Jacques Monod was undoubtedly one of the most creative minds in 20th century science. His time was that of the "golden age" of molecular biology, which then offered numerous opportunities for making models, proposing new concepts, and performing hypothesis-driven research. This science then underwent deep structural transformations, mainly after it shifted its interests from bacteria to multicellular organisms and started to deal with the study of complex problems such as cell differentiation, the immune response, or the functioning of the brain. New technologies have emerged, which have led to mega-projects such as the sequencing of the human genome. In parallel, the interactions within the scientific community became different, perhaps because there were fewer great personalities, imposing figures similar to Jacques Monod who, at the time, set the standards of scientific conduct and created schools of thought around them.

The late André Lwoff, my coeditor of the first edition of this book, used to say: "For many young scientists the future is more important than the past and the history of science begins tomorrow." However, to be prepared for the future, one should know something about the past. This book relates historical rather than contemporary events; it recalls how science was made during the post-World War II period; it could be considered as a journey in the past, but also as a reminder of the contributions of this epoch-making period. In reading it, one will realize that the concepts that Jacques Monod developed are absolutely central to modern biology. The concept of the regulation of gene expression—essentially the Jacob-Monod operon model—was a main forerunner of the biotechnological revolution and proved to constitute the basis of the control mechanisms of gene expression in eukaryotic systems. Messenger RNA, a concept derived from the operon model, became a tool in the modern microarray techniques of hybridization. Even β-galactosidase, used by Jacques Monod very early as a paradigm for enzyme induction, has become a classical tool in modern biological research: *lacZ* is the most commonly used reporter gene in the analysis of developmentally regulated systems and tissue-specific expression. Another new and fertile concept put forward by Jacques Monod dealt with the existence of lactose permease, a membrane-associated protein believed to allow bacterial cells to pump β-galactosides from the medium. The importance of such membrane-associated pumps in biological phenomena is well recognized today.

One of the major interests of Jacques Monod was to understand how proteins recognize chemical signals. This led him to develop one of the most elegant concepts of molecular biology, the theory of allostery. This concept was one of the most important ideas to emerge from the study of bacterial regulatory mechanisms. We know today that most mechanisms of cell signaling involve allosteric interactions. The theory of allostery even inspired the prion theory, which implies a transmission of a conformational change between identical protein molecules, as was postulated to happen between protomers in an allosteric protein.

By his theories and experiments, and also because of the strong impact he had as a person on the scientific community, Jacques Monod profoundly altered the thinking of biologists. His logically designed and beautifully written papers have remained highlights in the scientific literature. He often expressed his belief in the aesthetics of science by his saying, "A beautiful model or theory may not be right, but an ugly one must be wrong."

The relevance of Jacques Monod's work for modern biology as well as that related by the contributors of this book justified this new edition. The book will thus be made available to a new generation of scientists who might discover how science was made in those not-so-ancient times. This new edition includes an additional introductory chapter by Edmond H. Fischer and short biographies of the different contributors to this book, who all worked at some point with Jacques Monod or interacted with him and who all played their role in the development of molecular biology. It also includes the text of Jacques Monod's Nobel Lecture and a complete bibliography of his scientific publications.

I am grateful to Stanley Maloy, who had the idea for this second edition and put me in contact with Jeff J. Holtmeier, Director of ASM Press, to whom, and to the staff of ASM Press, especially to Ellie Tupper, I wish to express my gratitude for their friendly cooperation. I sorely missed André

Lwoff, with whom coediting the first edition was a pleasure. I particularly thank Maxime Schwartz, who provided me support during this editorial work.

The royalties from this book will be deposited in a Jacques Monod Memorial Fund, administered by the Institut Pasteur.

Agnes Ullmann

Preface to the First Edition

Occasionally the career of a scientist is marked by an important discovery. It is most unusual that it be illuminated by an uninterrupted series of great discoveries and still more unusual when each discovery gives rise to new concepts and opens new vistas.

Sometimes a scientist by his work or personality influences his contemporaries. It is rare that he establishes a school. The founder of a school must dominate a field. He must have enought insight to foresee the direction research has to assume in order to achieve his goal. He should be able to judge the potential of young scientists and to assess the manifold aspects of their personalities so that he can provide them with projects in harmony with their interests and talents. He should be able to propose projects that can be solved or be channeled in a productive manner. He should love his students and collaborators and be generous. Jacques Monod possessed all these qualities; therefore, he was not only a brilliant scientist but also the founder of a renowned school.

During the first phase of his career, Jacques Monod worked alone. In the Institut Pasteur, he attracted a number of scientists. Some were students at various phases of their scientific careers, often at onset; others were mature, accomplished scientists. They worked with him for a few months or a few years; some stayed at the Institut Pasteur. They had different personalities.

These scientists were asked to narrate their adventure, to relate their experiences with Jacques Monod. Almost all responded with enthusiasm, and most provided the contributions they promised. The result is fascinating. One sees Jacques Monod through the eyes of his technician, his secretary, his peers, his friends, and also of his enemies—love, friendship, and hate. The portraits of Jacques Monod—or, better still, images of the manifold aspects of his personality—are often painted with talent. Necessarily, the personality of the contributor appears as a watermark.

More important, the history of various discoveries is unfolded. This unique document illustrates the birth and development of concepts. It also shows the importance of a close, friendly, confident cooperation between different types of minds, the importance of interactions. One learns how a great scientist receives, discusses, rejects, accepts, assimilates, and creates ideas; how ideas are turned into experiments; how experimental results are interpreted and how concepts are born—in short, how science is constructed. The reader participates in the formulation of problems, in the conquest of knowledge, and in the building of a discipline—a unique contribution to the life of a laboratory and to the dynamic history of science.

It will be noted that the depiction of the same discovery may be told differently by different scientists who worked at the same time in the same laboratory. Obviously, the personality of the narrator has sometimes influenced the narration. Certainly each one perceives the importance of his own contribution better than that of others. Where is truth? Does truth exist? The métier of detective, or coroner, is difficult.

Some aspects of Jacques Monod's activities have been omitted: mountain climbing, music, the Underground, human rights, philosophy. Our goal was to depict the scientist. An intense light has been projected on one of the founders of molecular biology. Light engenders shades, and the contrast contributes to the relief. The image of Jacques Monod has been shaped step by step, and a portrait has emerged. "One becomes and one remains as others have seen you."

The royalties of this book will be deposited in a Jacques Monod Memorial Fund, which will be administered by the Institut Pasteur. We wish to express our gratitude to the staff of Academic Press for their friendly cooperation.

André Lwoff
Agnes Ullmann

Foreword to the New Edition

Fraternellement, Jacques

When Jacques gave me *Le hasard et la nécessité*, he simply wrote: "À Ed–fraternellement, Jacques." Just one word, nothing more, but for me, that word said it all. It is amazing that I felt so close to him considering the fact that unlike most of his other friends, I never was a student of his, nor did we ever collaborate or work or publish together. And while we clearly loved the same things: music, mountain climbing, etc., I never had a chance to do any of this with him; I never accompanied him on the cello and we never went rock climbing together, not even at "Bleau." If anything, regarding our common interests in enzyme regulation, we were looking at opposite faces of the coin, so to speak; I was on the other side of the fence. Indeed, when I first met him, he was totally committed to allosteric interactions as a means to regulate cellular processes, while I was almost exclusively concerned with the role covalent modifications might play in these events. At that time, of course, we knew of only a few such examples: no more than three or four enzymes had been shown to be regulated by reversible protein phosphorylation and we had absolutely no inkling that this kind of reaction would turn out to be so prevalent and ubiquitous.

While I had met Jacques fleetingly once before, I really got to know him well only during my first sabbatical leave of absence from the University of Washington in 1963–1964. My wife Bev and I had decided to spend the year with the kids in Europe and Israel: three months at the Lister Institute in London with Bill Whelan, another three at the Weizmann Institute with Efraim Katchalski-Katzir and Michael Sela who were already good friends, before ending the year at the Institut Pasteur. I had written to Elie Wollman to this effect, but Elie thought I would feel more at home if I joined Jacques's group and Jacques, generously, accepted me in his lab.

I will never forget our first encounter. It was in his office, in the late afternoon, and Agi Ullmann was also present. She had arrived from Hungary four years earlier, and I already knew of her from the literature. In fact, we had probably corresponded together because of our previous involvement with α-amylases. She had worked on the biosynthesis of the pancreatic enzyme with Bruno Straub in Budapest, while in Geneva, in Kurt H. Meyer's lab, I had crystallized and characterized several of these enzymes from a variety of sources. Anyway, after the usual civilities, Jacques all of a sudden produced a sort of a tubular contraption and, beaming from here to there, started to twist it around while proclaiming triumphantly: "This—is—an—allosteric—protein." I was speechless, totally dumbfounded, having absolutely no idea of what the hell he was talking about. It started to make sense only when, a few days later, he handed me a couple of papers including the preprint of his remarkable article with Jeffries Wyman and Jean-Pierre Changeux. But from this point on, we had some common grounds for scientific discussions because my knowledge of molecular biology at that time was essentially nonexistent.

In a way, those few months spent in Paris in the spring of 1964 were extraordinary: absolutely wonderful on the one hand but also strange if not even slightly unsettling. We had initially rented an apartment in Versailles but this turned out to be too inconvenient because it was so far away from the lab. Fortunately, Madeleine Brunerie succeeded in digging up for us a tiny house in Vanves, just outside the périphérique, within a forty minute walk from the institute.

Jacques housed me in the big lab next to his office, placing me in a sort of telephone booth from which Jean-Pierre Changeux had carried out his thesis research on threonine deaminase. It is partly on the basis of this study that the two of them formulated their theory of allosteric interactions in protein. Agi Ullmann and David Perrin were also in the lab, and I'm not sure who else. Jacques also gave me a bench but, frankly, I had little intention of doing much work. I had brought with me some muscle phosphorylase and a few related enzymes, and some cyclic AMP, to tinker a little with the system.

But Paris was Paris and we loved all it had to offer: exhibits, concerts, and museums; the bookstalls on the Left Bank; Notre Dame and Ste. Chapelle; St. Denis and its recumbent statues and so much more. One day, Jean-Pierre offered to give me a guided tour of the Marais and it was a thrilling experience. But I was missing terribly the Pacific Northwest and its mountains and lakes, its huge forests and boundless wilderness.

I remember so fondly the lunches we would eat together in the "ver-rière." We would pick up our food at a small but savory delicatessen store a few paces down the block. Regularly, the jovial owner would greet us by asking: "Are you working on the messenger?" And I, somewhat embarrassed, would sheepishly answer no. We would then sit around the large table and Jacques, always exuberant and stimulating, would keep the conversation rolling by telling us a bunch of stories. We would often have exciting seminars by various luminaries who would not pass through Paris without stopping at the Pasteur.

I never really understood why Jacques and François Jacob, who had known each other for fifteen years and worked closely together for six, continued to say "vous" to one another. Coming from Switzerland, it sounded incredibly "Vieille France." Indeed, in Switzerland as in Belgium or Canada, one would rapidly use the familiar "tu" with just about anybody. In fact, it was commonly said that in French-speaking Canada, one recognized immediately a professor coming from abroad because he would say "vous" to his students while they said "tu" to him. Anyway, in a very short time, I would say "tu" to both of them.

I remember with enormous fondness those ends of days when, being through with the lab, we would congregate in Jacques's office. Agi would dig up a bottle of Scotch from the ample supply she kept under her bench; we would flop down on the well-worn couch that stood behind the coffee table and start speaking of anything and everything: science, politics, mountain climbing, music, the arts and literature, whatever. It was the time when both allostery and Dan Koshland's induced-fit theory were starting to take shape, and Jacques would discuss these at the blackboard, chalk in hand, with absolute confidence. He had an amazing talent of exposing and analyzing complicated facts and then drawing rational deductions from them. His logic was implacable; there was never any doubt in his mind that the interpretation he was proposing was correct. Time would fly and we would stay there for an hour or more, and every time I would walk home absolutely dazzled.

Jacques was a totally committed person. I remember a Gordon Conference on Proteins that Dan Koshland and I had organized and to which both Jacques and Agi had been invited. It was in June 1968, soon after the huge riots of May that had paralyzed Paris and just about crippled the rest of the country. As usual, the Thursday evening session was reserved for a variety of nonscientific subjects, and I had asked Jacques to tell us something about those extraordinary events and the conditions that had brought them along, a situation totally incomprehensible to any foreigner.

He hesitated for several days, obviously would have liked to do so, but finally told me he just couldn't speak about that, so shaken had he been by what had taken place.

I have two other memories from that conference. The first concerns Agi; we hadn't been able to pay for her trip because of lack of funds. One afternoon, we were all at the nearby lake where the Anfinsens, Sobers, and Harringtons always stayed, and where there was a ping-pong table. Now, Dan Koshland also had a ping-pong table at home and he played very well. At one point, he said:

"Hey, Agi, how about a game ?"

"Sure," she said, "but what do I get if I win?"

"Whatever you want," said Dan, quite confident of the outcome.

"Will you guys pay for my trip if I win?" asked Agi.

"Of course," he replied. "If you win, we'll take care of your ticket."

Little did Dan know that Agi had been a champion ping-pong player in her native country and had participated in several tournaments. You can guess the rest.

The last souvenir I have of that conference comes from an evening session that was held a day or two later. I'm not really sure what brought this about. It had been a glorious, resplendent day and, once again, we had spent the whole afternoon at the lake. It was very warm and, naturally, we had quite a bit to drink. Whatever, in the middle of the session, Jacques jumped out of his seat all of a sudden and proclaimed with absolute finality, "Covalent bonds don't exist!" and sat down again. Everybody was stunned. Of course, it was at the time when the controversy between the allosteric and induced-fit theories was on everybody's mind, further enhanced by the presence of the two main protagonists. As a consequence, the notion that covalent regulation might also be of importance was largely dismissed. Until someone (it could have been Saya Pocker) remarked that insofar as he knew, the only person who was not held together by covalent bonds was the wife of Lot after she had been transformed into a pillar of salt.

Of Jacques, I keep another touching memory. During one of his visits to Seattle, I had shown some slides taken from the air of the massif of Mont Blanc and of the Swiss Alps all the way to the Matterhorn. A good friend of mine from Geneva had taken me in his Piper Cub, and I had taken many shots of the mountains and peaks I had climbed and the routes I had taken. Jacques, of course, knew them all. So we spent hours arguing the advantages or disadvantages of climbing by this way or that, which approach was more easily manageable and which was tricky, and recalling all the things that had happened to us on one occasion or the other. For days if not weeks after that, Jacques was still dreaming of those faces, crags, and ridges and of the times gone by.

In the fall of 1970, we came back to Europe for a six-month sabbatical leave of absence. The first half was spent in Würzburg with Ernst Helmreich; the second half was in Geneva with my former student and collaborator Eric Stein. One day in early January 1971, I received a call from Jacques asking if I would come to Paris to present a lecture within a course he was giving at the Collège de France in which he wanted to include enzyme regulation. I was to cover covalent regulation by protein phosphorylation while he would take up allosteric reactions.

"You'll see," he said, "it is quite an experience. These lectures are open to the public and anybody can attend. You'll find seated in the front row those ladies in fur coats. In the back, you'll see a row of 'clochards,' tramps and winos sitting there to take refuge from the cold. But don't worry: they don't bother anybody and stay perfectly quiet whether they listen or not." And so it was. The rest of the audience was made up of students from the IP and elsewhere, with quite a few colleagues and friends.

There is little doubt in my mind that "ces dames aux manteaux de fourrure" came less for the intellectual challenge of the lectures than for the thrill of listening to a handsome, polished Nobel Laureate. It is true that Jacques spoke extremely well and easily, with a supreme elegance and distinction. He and his brother Philo always represented for me the epitome of the perfect scholars, the finest examples of what the intellectual aristocracy of France had to offer.

Admittedly, seven years had passed since my first visit to Paris and the field of protein phosphorylation had come a long way during that period of time. In fact, it was ready to explode. Many new enzymatic systems and physiological events such as muscle contraction had been uncovered and shown to be modulated by protein phosphorylation. It was already evident that such a process would play a role just as fundamental and widespread as allostery. Furthermore, it was obvious that one system did not exclude the other. On the contrary, they had to complement one another, providing an extra dimension or higher level of sophistication to the control of cellular processes.

I recounted the story that, in fact, we had uncovered the regulation of phosphorylase by phosphorylation/dephosphorylation with my colleague and friend Ed Krebs when we were trying to understand how this enzyme was activated by AMP. Indeed, Carl and Gertie Cori in St. Louis had shown that one form of the muscle enzyme (phosphorylase b) had an absolute requirement for this nucleotide, but neither they nor anybody else knew why. So, with Ed, we decided to take a crack at this problem, but we never got the answer either. For that, we had to wait five or six years for the Pasteur group to come up with their allosteric model of enzyme regulation. In retrospect, this type of regulation made a lot of sense. Phosphorylase catalyzed the first step in the degradation of glycogen and, hence, of carbohydrate metabolism. According to the rule that most enzymes are subjected to end-product or feedback inhibition, it would be expected to be inhibited by glucose 6-phosphate, rapidly generated in the course of its reaction, and by ATP, the ultimate end product of carbohydrate metabolism. By the same token, it would be activated by AMP, which indeed it was. But then, if one had to rely solely on this type of regulation, all the enzymes susceptible to the same allosteric effector would be similarly affected. Unless one had strict intracellular compartmentalization, one would have to simultaneously open many doors or close many others. In contrast, covalent modification such as protein phosphorylation was mediated by regulatory enzymes (kinases and phosphatases) that were highly specific. It provided the opportunity of controlling a single step, of opening or closing a single door without having to touch any other. But far more importantly, it was becoming increasingly evident that covalent regulation responded mainly to a variety of external signals reaching the cell surface in the form of hormones, drugs, growth factors, neuromediators, light, whatnot. And these signals would then be transduced by a cascade of phosphorylation reactions. Anyway, those were the topics that I brought up, followed by Jacques, who presented with his usual brilliance an overview of the characteristics of allosteric systems. Nonetheless, I had the feeling that he still remained somewhat skeptical of the arguments I had presented.

The last time I spoke with Jacques in person was during the summer before he died. He had told me: "You know, next summer, we'll go sailing through the Greek Islands." I was so enthusiastic, so enthralled about the idea that I said, "Terrific, but on one condition only: I'll take care of all of the booze." He shook his head, looked at me sadly, and said, "Mon pauvre, tu n'est pas assez riche."

But perhaps the most moving memory I have of Jacques is from the day we were listening to some music with Bev and Agi in our home in Seattle. At one point, he spotted Bev's cello lying under the piano and picked it up. And for a long long while, he held it in his arms, tenderly, bow in hand, reading without playing the suites of Bach. It is this picture of Jacques, listening to the past, that I will never forget.

Edmond H. Fischer

Edmond H. Fischer

Edmond Fischer was born in Shanghai, China, on 6 April 1920. He carried out all his studies in Switzerland, where he earned a Ph.D. degree in Chemistry at the University of Geneva. After spending a year at the California Institute of Technology, he joined the Department of Biochemistry at the University of Washington, Seattle, where he is now Professor Emeritus. It was in Seattle that he initiated his collaborative studies with Edwin G. Krebs on the regulation of cellular processes by reversible protein phosphorylation.

In recent years, Dr. Fischer has been particularly interested in the role played by cytoplasmic and receptor protein tyrosine phosphatases in signal transduction and cell transformation. He has served on various Scientific Advisory Boards and is on the Board of Scientific Governors of the Scripps Research Institute in La Jolla, Calif., and of the Weizmann Institute of Sciences in Rehovot, Israel. He is a member of the National Academy of Science and of several science academies abroad and was awarded honorary degrees from a number of U.S. and international universities. In 1992, he and Edwin G. Krebs were awarded the Nobel Prize in Physiology or Medicine for their discoveries concerning reversible protein phosphorylation as a biological regulatory mechanism.

Jacques Lucien Monod: 1910–1976*

A. M. Lwoff, For. Mem. R. S.

Childhood

Jacques Lucien Monod was born in Paris on 9 February 1910. When he was seven, his family moved to Cannes. Jacques always felt himself more Provençal than Parisian.

The Monod family originated from a Swiss pastor who came from Geneva to France in 1808 and whose descendants now number several hundreds. Professors, civil servants, pastors, and doctors have been the dominant products of this Huguenot family. Monod's paternal grandfather was a general practitioner and his wife belonged to a Protestant family from the Dauphiné. His father, Lucien Monod—born in 1867—was a painter, engraver, and art historian. Lucien Monod's watercolours, flowers, landscapes, and portraits reveal great sensitivity and talent. At the same time, he was a scholar with a lifelong and passionate interest in the work of the intellect. His admiration for Darwin was transmitted to his son; it was thus that Jacques became interested in biology. Moreover, Lucien Monod was a free thinker, imbued with a positivist faith in the joined progress of science and society. On the whole a remarkable exception in this puritan family.

Monod's maternal grandfather, Robert Todd MacGregor, the son of a Scottish minister, had emigrated to the States at the age of eighteen in 1852. His maternal grandmother was a New Englander, whose ancestor, Edward Elmore, had arrived in the colonies in 1632. Jacques Monod's mother, Sharlie Todd MacGregor, was born in Milwaukee in 1867. In

* Reprinted with the kind permission of the Royal Society from the *Biographical Memoirs of Fellows of the Royal Society* (vol. 23, p. 385–412, 1977). The reference numbers in parentheses refer to Monod's publications as listed beginning on p. 319.

1

Jacques Monod's writings one sometimes finds quotations attributed to MacGregor: they are from Jacques Monod himself.

Until 1928, Jacques Monod attended the College at Cannes—now a Lycée. One of his teachers, M. Dor de la Souchère, professor of Greek, was an excellent humanist. Jacques Monod freely acknowledged his debt to this highly cultured man, whom he admired and loved. M. Dor de la Souchère created the museum of Antibes, and is now—in 1977, at the age of 89—its curator. He loved Jacques and wrote me a moving letter after the death of his pupil.

In addition to his other gifts and interests, Jacques' father was a devotee of music. The musical as well as the intellectual life at the family home was intense, and Jacques himself learned to play the cello. Clos Saint Jacques, accordingly, provided an exceptionally favourable environment for the development of a sensitive and intellectually gifted child.

The Beginning

Jacques Monod passed the baccalaureat in the summer of 1928 and came to Paris in October to study biology. He studied zoology, geology, general biology, and general chemistry, and in 1931 became *licencié ès sciences*. Only later did he realize that the teaching in natural sciences in the Sorbonne was then twenty years or more behind the times. Only one professor, George Urbain, who taught thermodynamics, left a mark on him. Like many students in zoology, Jacques Monod came to the Station Biologique at Roscoff. There he met the four scientists to whom—as he said in an interview—he owed his true initiation in biology. "To Georges Teissier, the taste for quantitative descriptions; to André Lwoff, the initiation to the powers of microbiology; to Boris Ephrussi, the discovery of physiological genetics; to Louis Rapkine, the idea that only the chemical and molecular descriptions can give a complete interpretation of the functioning of living beings."

In October 1931, Jacques Monod received a fellowship to work with Edouard Chatton, professor of biology at the University of Strasbourg. Edouard Chatton was the great protistologist of his time. He had worked in the Institut Pasteur in Paris and in Tunis and was an accomplished microbiologist. Under his firm guidance, Jacques Monod became familiar with microbiological techniques and disciplines. Among other things, he learned to grow ciliates in bacteria-free cultures; the organisms were to provide the material for his first studies on growth. In Strasbourg, moreover, he was associated with the work on the stomatogenesis of ciliates:

hence publications (1), (2), and (3). It is strange that in his autobiographies and interviews, Jacques Monod never mentioned the name of Edouard Chatton.

In October 1932, Jacques Monod obtained another fellowship and returned to Paris, where he first spent two years in the Laboratoire d'Evolution des Etres organisés. He never mentioned the name of its director, Maurice Caullery, a good zoologist and a good teacher of modern biology—including genetics. Thereafter he became assistant in the Laboratoire de Zoologie.

From October 1932 he was thus free, that is, sentenced to discover for himself, painfully, the problem which would satisfy his exacting mind. The search lasted three years. Its course is marked by a few papers on axial gradients in ciliates, on galvanotropism, and on the role of symbiotic chlorellas (4–7).

In the summer of 1934, Jacques Monod had embarked on the *Pourquoi pas?*, visited Greenland, and published a preliminary account of his observations dealing with natural history (1935:8). This account remained preliminary.

Incursion into Genetics

In the spring of 1936, Monod was preparing to take part for the second time in an expedition to Greenland. Boris Ephrussi was going to spend a year with T. H. Morgan's group; he convinced Monod that genetics was interesting and important, and helped him to obtain a Rockefeller Fellowship; they went together to Pasadena. This very year, the *Pourquoi pas?* was lost with all hands on the coast of Greenland. Genetics had saved the life of Jacques Monod, a debt that he would later repay.

At the California Institute of Technology, with Morgan's group, Monod not only learned genetics but discovered a scientific world very different from the old Sorbonne: easy personal relations with scientists of all ages, free exchange of ideas, lively critical discussions, friendly cooperation.

Back in Paris, Monod spent a few months in Boris Ephrussi's laboratory at the Institut de Biologie Physico-chimique. There he attacked some problems of physiological genetics, implanting imaginal disc in drosophilas. However, this did not correspond at all to Monod's taste or tendencies, and he went back to the Laboratoire de Zoologie de la Sorbonne, which he had entered in October 1934 as an assistant and where, under the influence of Georges Teissier, he started to work on growth.

Growth

The first paper on growth was published in 1935 (9). The growth rate of *Glaucoma* (later *Tetrahymena*) *piriformis* was measured as a function of the concentration of nutrient. I was then working on the nutrition of *Tetrahymena*—very little was known at that time—and Jacques came to discuss his work. I told him that ciliates were the worst material to attack the problems of growth, and advised him to use a bacterium able to grow in a synthetic medium, for example *Escherichia coli.* "Is it pathogenic?" asked Jacques. The answer being satisfactory, Monod began, in 1937, to play with *E. coli,* and this was the origin of everything. For it is the systematic analysis of the various parameters of growth of *E. coli* which led to the study of induced enzyme synthesis—at the time enzymatic adaptation—a study which developed into the physiology of the gene and the laws of molecular biology.

Monod first showed that the growth yield as a function of the amount of the energy source provided is independent of growth rate. This means that the fraction of metabolic energy utilized for the maintenance of cellular structures is negligible compared to the fraction utilized for biosynthesis. The growth yield was measured with numerous sugars, as well as the growth rate as a function of the concentration of the limiting carbon source. The results suggested that, at low concentration, the growth rate is controlled by an enzymatic reaction. It turned out later that the controlling factor is, at least under certain conditions, specific permeation. The measurements of growth rate as a function of temperature permitted a determination of the activation energy of the limiting reaction(s).

After having considered growth in the presence of one sugar, it seemed of interest to study the interaction of two carbon sources. In some mixtures of two sugars one observes two distinct growth cycles, separated by a lag phase. This he called diauxy. Jacques Monod has told how, in December 1940, at the Institut Pasteur, he came and showed me the diauxic curve and asked, "What could that mean?" I said it could have something to do with enzymatic adaptation. The answer was, "Enzymatic adaptation, what is that?" I told Monod what was known—what I knew—and he objected that the diauxic curve showed an inhibition of growth rather than an adaptation. We know today that repression and induction are complementary, but I simply repeated that diauxy should be related to adaptation. Anyhow, I gave him Emile Duclaux's *Traité de microbiologie*, Marjory Stephenson's *Bacterial Metabolism*, and a few reprints I had secured, among them the precious Ph.D. thesis of Karstrom—which I never saw again.

It turned out that the glucose was inhibiting the synthesis of a few enzymes responsible for the metabolism of other sugars—catabolic repression—but the enzymes involved in diauxy were nevertheless adaptative. Induced enzyme synthesis was the key to diauxy.

In 1941, Monod was awarded his Ph.D. for his thesis, "Recherches sur la croissance des cultures bactériennes" (16). The importance and originality of this fundamental and now classical work were not perceived by the members of the jury. After the ceremony, the director of the laboratory where Monod was working told me, "What Monod is doing does not interest the Sorbonne." This was, alas, true.

The Transition

"From this very day of December 1940," wrote Jacques Monod, "all my scientific activity has been devoted to the study of enzymatic adaptation." Yet during the dark years, he had joined the Underground. He had even been arrested by the Gestapo, but cleverly managed to escape. However, he could no longer work in the Sorbonne and came clandestinely to the Institut Pasteur, where he could perform a few experiments. Among other things, he showed that the enzymatic adaptation coupled with biosynthesis was—and probably reflected—synthesis of the specific protein involved.

After the liberation of Paris, Monod joined the army and played a determining role in the integration of the resistance force. As a member of the staff of General de Lattre de Tassigny, he met American officers and had an opportunity to read some American scientific journals. In this way, he came across the Luria–Delbrück paper on the spontaneous character of some bacterial mutations and the epoch-making publication of Avery, McLeod, and McCarthy, which identified the transforming principles as deoxyribonucleic acid.

The war ended and Monod returned to Paris. At the Sorbonne, he worked in a small room which opened on a corridor lined with glass cabinets containing skeletons and stuffed animals. He was doing everything himself: washing glassware, preparing culture media, autoclaving. No one in the Laboratoire de Zoologie took any interest in enzymes and Monod was rather unhappy.

I invited him to join the Service de Physiologie Microbienne as Chef de Laboratoire. He was no longer obliged to wash his glassware and to autoclave. Moreover, the Centre National de la Recherche Scientifique provided him with a technician. He selected Madeleine Jolit, who, until 1971, participated efficiently in his researches. Moreover, a few microbiologists were active in the attic. They knew how to isolate bacterial strains and to select mutants. They were aware of the existence of lactose-positive (L^+) mutants in L^- strains of *E. coli*, and of the work of Massini and his followers.

Jacques Monod and Alice Audureau selected an L^+ mutant from an *L-mutabile* strain of *E. coli* isolated from my digestive tract (hence *E. coli m.l.*). They showed that the β-galactosidase is an adaptive enzyme. The hypothesis had been earlier proposed that the $L^- \rightarrow L^+$ mutation is an alteration of an enzyme precursor common to the different specific enzymes metabolizing sugars. A gene mutation was not—at the time—considered as the basis of the phenomenon.

Yet Monod, who had also studied the bacterial ability to synthesize methionine, concluded that it was controlled by a gene mutation (1946:29). The gene had entered the scene. However, it would take some time before the idea developed that the mutation controlling the utilization of lactose is a virtual genetic property revealed only in the presence of lactose, before the link between genetic and biochemical determinants was established, and, finally, before the problem of genetic regulation was posed.

Preliminary Games

The problem of the relations between gene and enzyme, and more generally of regulation, was not ripe for an attack. So for a few years, Monod played with various problems. The role of CO_2 in bacterial growth was analysed with André Lwoff (1947:31, 37, 43). It was shown that aspartic and glutamic acid could partially replace the requirement of CO_2 for the development of *E. coli*: a certain number of essential metabolites could be synthesized only by carboxylation. Then with Elie Wollman, Jacques Monod discovered (1947:32) that infection by a bacteriophage would prevent the adaptive synthesis of β-galactosidase, whereas the activity of the enzyme present at the time of infection is not affected.

This very same year, Monod was invited to discuss enzymatic adaptation at the Growth Symposium. This was the stimulus for reviewing the data concerning the induced synthesis of enzymes and their possible interpretations. In section IV, "Adaptive enzymes and genes," the fundamental problem of relation between gene and enzyme is posed. "The problem consists of evaluating the respective role of hereditary factors (i.e., genes or other self-duplicating units) and environmental factors (substrate) in the synthesis of an enzyme" (1947:36).

While writing this remarkable review, Monod realized that the phenomenon of enzyme induction was mysterious, that almost nothing was known about it. However, owing to its specificity, its regularity, and its dependence, on a mutation and on the environment, it necessarily involved an interaction between a genetic and a chemical determinant; and it appeared of such interest, of such profound significance, that Monod decided to go ahead. The respective role of the gene and of the inductive substrate in enzyme formation was posed. The enzyme, of course, was β-galactosidase.

By-Products

The goal had been defined, but the way toward the solution of the problem was far from being straight. It was necessary to learn as much as possible about the physiology of E. coli, and a few discoveries not directly relevant to the main problem emerged from these explorations.

The first was the discovery with Anne-Marie Torriani of a new enzyme, amylomaltase (1949:40, 41; 1950:46). It catalyses a reversible synthesis of amylose from maltose. The length of the amylose chain is controlled by glucose. Very long amylose chains are formed in the absence of glucose, for the degradation of the polysaccharide does not take place in the absence of the monosaccharide.

The second was an important experimental and theoretical contribution to the methodology of continuous bacterial growth, the bacteria being maintained indefinitely in a chemical and physiological stable state (1958:48). The method consists in creating conditions such that a limiting substrate is provided at a rate inferior to the total metabolic capacity of the organism. A stable state is thus automatically reached, characterized by a constancy of all the significant elements. The experimental potentialities of the method are wide. It provides a means of changing instantaneously the growth rate without modifying either the composition of the medium or the temperature. It also offers the possibility to select specific mutants—and this remarkable tool has been, and still is, widely used. It should be stated that a similar method had been devised independently and simultaneously by Aaron Novick and Leo Szilard.

The third was the effect of irradiation with ultraviolet light on enzyme synthesis. François Jacob had arrived in the attic in 1950. A year later, he began to be interested in β-galactosidase. J. Monod, A. M. Torriani, and F. Jacob showed (1951:52) that bacteria irradiated with a heavy dose of UV rays are unable to synthesize β-galactosidase, but can still produce bacteriophage after infection. The effect of the irradiation cannot be attributed to a general effect on metabolism; it is a specific alteration of the bacterial component responsible for the synthesis of the enzyme which accounts for the UV effect.

These discoveries were on side-roads, but nevertheless played a determining role in the solution of the problem.

The Very Nature of Enzyme Induction

Soon after the war, Alvin Pappenheimer visited the Institut Pasteur. He was deeply interested in the induced synthesis of enzymes and thought that immunological methods might be helpful for the analysis of the phenomenon. This led to the suggestion that one of his students should come and work with Jacques Monod. So, during the winter of 1948, Melvin Cohn arrived in Paris, a good immunologist and biochemist, a remarkable experimenter, hardworking, enthusiastic, lively, and friendly. He mastered not only the problem of induction but also the art of living on fellowships, and so managed to spend seven years in the attic. He played a major role in the characterization of the enzyme, in the study of substrate and inducer specifically, and, more generally, in the life of the laboratory.

The β-galactosidase extracted from *E. coli* (with A. M. Torriani and J. Gribetz [1948:39]) was purified and its properties studied with Melvin Cohn (1951:50). Monovalent ions are all activators, but differ in their effectiveness. The displacement of a strongly active ion by a less active one results in a decrease of enzyme activity. Moreover, the relative activating power of different ions is not fixed but varies with the substrate employed (with Germaine Cohen-Bazire [1951:49]).

Then Melvin Cohn and A. M. Torriani started the immunochemical study of β-galactosidase and of related proteins. The preparation of a specific antiserum made it possible to estimate the enzyme as an antigen and thus to attack the kinetics of enzyme formation. The experiments suggested that, in the presence of the inducer, a total biosynthesis from amino acids took place (with A. M. Pappenheimer and G. Cohen-Bazire [1952:57]). Later on, with David Hogness and Melvin Cohn (1955:65), it was proved that the enzyme is formed from amino acids synthesized after the addition of the inducer. Moreover, the molecule of β-galactosidase is fully stable in vivo, as are, under normal conditions of growth, the other protein molecules of the bacterium. The romantic dogma of "the dynamic state of living matter" was seriously shaken. Fierce counterattacks were launched, but they were unsuccessful and the dead god went down to the grave.

Thus the induced production of an enzyme was the total biosynthesis of a protein from newly formed amino acids. Therefore the increase of enzymatic activity during induction was a true measure of the synthesis of the specific protein.

Inducers and Inhibitors

Until 1952, only the substrates of enzymes were known to serve as inducers of those enzymes. Three theories had been proposed to account for the inducing activity of the substrates: (a) the synthesis of an enzyme is determined by its activity; (b) the synthesis of an enzyme is limited by a dynamic equilibrium controlled by the specific substrate–inducer complex (whatever this could mean); and (c) the substrate–inducer complex plays an organizing role by combining with the precursor of the enzyme. These were purely hypothetical and rather metaphysical notions.

A systematic study of many galactosides was undertaken by Jacques Monod, Germaine Cohen-Bazire, and Melvin Cohn (1951:51) and continued with Melvin Cohn (1952:55).

A number of unexpected—at the time strange—data emerged from these studies:

(a) Some substrates are inducers.
(b) Some substrates are devoid of inducing power.
(c) Some galactosides devoid of any affinity for the enzyme are powerful inducers.
(d) Some galactosides exhibiting a high affinity for the enzyme are not hydrolysed.
(e) Some compounds are substrates and anti-inducers.
(f) Some inducing compounds are not attacked by the enzyme and inhibit competitively its activity.

It turned out that only β-galactosides are substrates, whereas the inducing activity is associated with the presence of an intact galactoside residue either in α or in β linkage.

The interpretation of these data was complicated by the fact that the inducing activity of some substances could be modified by mutations of the bacterium, and also that permeability effects might be involved. Whatever the case, these observations led to the conclusion that the activity of the inducer could not be due to a reaction with the enzyme, but rather to a "catalytic" effect on the enzyme-forming system. The new data concerning induction were analysed in connection with those relative to repressors.

Negative Control of Anabolic Pathways: Repression
Vogel and Davis had shown that in *E. coli* strains which require arginine or acetylornithine, acetylornithinase was produced in the presence of acetylornithine but not in the presence of arginine. They had concluded that synthesis of the enzyme was induced by its substrate. Monod suggested that this could be interpreted as an inhibitory effect of arginine, rather than an inducing effect of acetylornithine. This was the origin of the concept of repression, and led to the discovery that the constitutive synthesis of β-galactosidase is inhibited by β-galactosides (with G. Cohen-Bazire [1953:58]). It also led to the study of the regulation of enzymes operative in two anabolic pathways: the tryptophan pathway with G. Cohen-Bazire (1953:59) and the methionine pathway with Melvin Cohn and Georges Cohen (1953:61).

It turned out that the synthesis of tryptophan synthetase is inhibited by tryptophan, and the hypothesis was proposed that specific inhibition could be a general property of enzyme-forming systems. In agreement with this hypothesis, synthesis of methionine synthetase proved to be inhibited by methionine.

Thus the synthesis of two enzymes operative in the biosynthesis of essential metabolites was shown to be under negative control. It was difficult to conceive that induction and repression were the expression of two different mechanisms. The inducer could be an antagonist of an endogenous repressor. A digression is necessary here.

Another By-Product: the Permease
Among the numerous mutants isolated from the original *m.l.* strain were the so-called cryptics. They are able to synthesize β-galactosidase but unable to metabolize β-galactosides. The mystery was solved by Monod, Rickenberg, Cohen, and Buttin (1956:68). Labelled thiogalactosides accumulate rapidly in induced wild-type bacteria, but not in uninduced ones. Nor do they accumulate either in cryptic mutants. The ability to accumulate galactosides is under inducible control. The various parameters of induction were analysed: kinetics, specificity, etc. The conclusion was clear: the factor responsible for thiogalactoside accumulation could only be a specific protein, controlled by a gene *y* distinct from the galactosidase gene *z*. The synthesis of this protein was induced by β-galactosides together with

that of β-galactosidase. The protein was christened galactoside permease. A novel category of enzymes which mediate the penetration of small molecules into the bacterium had appeared on the scene; a new chapter of cell physiology was open.

The existence of permeases was the unavoidable logical conclusion of a series of rigorous experiments. It was immediately objected that their existence rested on in vivo experiments. It was also objected that one should not give a name to a protein before it has been isolated. The same objection was made later to the "repressor." Many enzymes have been named before anything was known about their nature; also genes, and all viruses. A few years after the discovery of "permease," galactoside transacetylase was isolated by Zabin, Kepes, and Monod (1959:79; 1962:94). The permease, discovered in 1956, was isolated only in 1965 by Fox and Kennedy.

The study of the permease and the transacetylase had revealed an unexpected situation. A number of mutants constitutive for β-galactosidase synthesis had been isolated, and it turned out that the mutation was pleiotropic. Not only was the β-galactosidase constitutive, but so were permease and transacetylase. This was strange because each of the three enzymes was, of course, controlled by a distinct gene.

Once the physiological relation between β-galactoside and β-galactoside permease was understood, once it was known that they are controlled by two distinct genetic elements but nevertheless subject to the same determination of induction, the problem of the expression of these genes was posed.

Birth of the Repressor

François Jacob and Elie Wollman had discovered the mechanism of the sexual process in bacteria. Following conjugation the $+$ "male" bacterium injects its chromosome into the $-$ "female" partner. The process can be interrupted at will. Thus the kinetics of the entry of a given gene could be followed and the gene sequence determined as a function of the time of entry. A new method was available for the study of gene physiology. The problem of regulation could be attacked thanks to a new and powerful tool.

The work of Monod and his disciples had shown that in E. coli, the synthesis of β-galactosidase depends (a) on a gene z governing the capacity/incapacity to produce the enzyme and (b) on a genetic factor known to exist under the forms i^+, wild type, corresponding to inducibility, and i^-, mutant, corresponding to constitutivity. Genetic analysis revealed that the z and i genes are closely linked.

The synthesis of β-galactosidase was followed in zygotes resulting from the conjugation of male bacteria with females carrying respectively "opposite" z and i genes. The z^+i^+ and z^-i^- parents are mated in the absence of inducer. Both parents are unable to synthesize the enzyme, one because of the absence of inducer, the other because of the deficiency of gene z. It is necessary to add that the male is streptomycin sensitive whereas the female is resistant. This allows the male to be killed selectively at any time.

The results differ with the direction of the cross. With the system $\delta z^-i^- \times ♀z^+i^+$, no enzyme is synthesized. With the system $\delta z^+i^+ \times ♀z^-i^-$, enzyme synthesis starts 3 to 4 minutes after entry of the z^+ gene into the female. This means that the factors z and i, despite their linkage, belong to two functionally different units able to cooperate through the cytoplasm. The z^+ gene is immediately expressed in an i^- cytoplasm, whereas the constitutive gene i^- is not expressed in an i^+ cytoplasm. Contrary to expectation, the i^+ gene is dominant. This dominance is manifested in the zygote: the synthesis of enzyme stops after 2 hours because the zygote has become phenotypically inducible. Such is the epoch-making classical Pardee–Jacob–Monod (PaJaMa) experiment (1958:72; 1959:74), which led to the hypothesis that the i^+ gene produces a repressing substance which was called "repressor" and which blocks the expression of the z^+ gene.

One problem was solved; many more were posed. A systematic genetic and biochemical offensive was launched. About a thousand mutants differing in their capacity to produce β-galactosidase, galactoside permease, and galactoside transacetylase were isolated and used to construct a detailed genetic map. It turned out that the synthesis of the protein of the β-galactosidase system is controlled by one gene, distinct from the structural ones. This controlling gene, expressed in the cytoplasm, is responsible for the production of the specific repressor.

The problem of enzyme biosynthesis thus appeared in a new light: the two known regulatory effects, induction and repression, should be two aspects of the same fundamental mechanism. F. Jacob and G. Cohen studied various mutants in which the synthesis of tryptophan synthetase was altered: the sensitivity to the repressive action of tryptophan is controlled by one specific gene and the "repressed" allele is dominant over the "derepressed" one. Repression is determined by a gene producing a specific cytoplasmic repressor activated by a specific metabolite which is the end product of a biosynthetic chain of reactions: tryptophan.

Monod and Jacob discussed the problems of regulation in a series of critical reviews which are now classics (1961:96; 1963:97, 98, 99). The tendency was to consider induction as the result of the expression or neutralization of an anti-inducer. It turned out that it was an antirepressor effect. In fact, the hypothesis had been proposed by Leo Szilard during a seminar given at the Institut Pasteur. As will be seen later, Leo Szilard's intuition was correct.

The Messenger

During conjugation, structural genes of the male are introduced in the female: this allowed an attack on the problem of gene expression. It was generally believed at the time that genes produce stable structures which accumulate in the cytoplasm. Since ribosomal RNA was the only known RNA, it was presumed to act as a template for protein synthesis. A number of experimental findings were not in accord with this presumption and a new hypothesis was proposed: the structural gene produces a metabolically unstable RNA (1960:86). This RNA was christened *messenger* (1961:87). The messenger soon ceased to be an être de raison and became a molecule.

Operon and Operator

The concept of operon stemmed from the study of lysogeny. In a lysogenic bacterium the structural genes of the prophage are not expressed. Moreover, a lysogenic bacterium is "immune" against superinfecting homologous bacteriophages. The specificity of immunity is determined by a "C" region which obviously controls the activity of the rest of the prophage genome. Immunity is a dominant character and has a cytoplasmic expression. When a prophage is introduced by a male chromosome into the cytoplasm of a non-lysogenic female, its development is induced: all the structural genes are derepressed. This is the phenomenon of zygotic induction, discovered by F. Jacob and E. Wollman. The analogy with the Pardee–Jacob–Monod experiment is obvious. Moreover, the genes involved in the production of enzymes that mediate a particular biochemical sequence are frequently adjacent. The hypothesis that the phage DNA molecule is not only a unit of replication but also a unit of activity was proposed by F. Jacob in his September 1958 Harvey Lecture; it led to a series of experiments which established the operon as a respectable citizen (1960:80, 86; 1961:87, 90; 1962:93; 1963:97). Again, new mutants of *E. coli* were isolated and analysed. A new specific structure became necessary to account for the specificity of action of the repressor, a new structure subject to mutation. It was the operator gene or "operator." A single operator controls the expression of β-galactosidase, permease, and acetylase structural genes. The operator acts only on the adjacent gene located on the same chromosome. Certain mutations "inactivate" the operator, thus preventing the expression of the structural genes. A detailed analysis showed that the operator is the terminal part of the last structural genes—that of the galactosidase.

Thus, units of coordinate transcription exist in the chromosome: these units are the operons. An operon is composed of structural genes connected by an operator, subject to the action of a repressor produced by a regulator gene.

Nature of the Repressor

The coordinated regulation of enzyme synthesis is controlled by two genes: the regulator gene, responsible for the formation of the repressor, and the operator gene, responsible for the expression of the operon. The problem of the nature of the repressor was posed. The repressor has to recognize both the inducer and the operator. It is a privilege of proteins to form stereospecific complexes with small molecules. The repressor could only be a protein. The conclusion remained hypothetical until the repressor was isolated and proved to be a protein. In the meantime, the systematic study of numerous mutants of regulation led to the identification of the types predicted by the theory. The repressors produced by some mutants were either unable to recognize the inducer or unable to recognize the operator. And, necessarily, of course, some of the mutants of the operator gene were unable to recognize the wild-type repressor.

The regulator gene could only be a structural gene coding for the repressor. In the operon, only the operator is a pure receptor and transmitter of signals. A general scheme of the mechanism of regulation was proposed (1961:87, 88, 91).

Within ten years, the problems posed by the induced synthesis of β-galactosidase in E. coli had been solved. The new ideas were applied to a large number of catabolic and anabolic pathways, to viral development as well as to differentiation. Everything was clear. A coherent scheme accounted for the interplay of regulator gene, operator, structural gene, messenger, and repressor. It accounted also for the nature of various types of mutations affecting regulation. Out of the monotonous succession of nucleotides there emerged the concept of operon as a coordinated unit of integrated structures and functions.

These discoveries can be considered from another viewpoint. The nature of molecular communications had, for a long time, been a complete mystery. Nothing was known about the way messages coming from the outer world or emanating from metabolic systems could affect the genetic material. The problem was solved: the inducer, its reactions with the repressor, the reaction of the repressor with the operator gene, the effect of the operator gene on the structural genes of the operon, were clarified.

Allostery–Symmetry

In a biosynthetic pathway, the activity of the first enzyme is inhibited by the product of the last enzyme. The Novick–Szilard–Umbarger effect had not received an interpretation. J.-P. Changeux, a student of Jacques Monod, had shown that the activity of threonine deaminase is inhibited by L-isoleucine and that the enzyme can lose its sensitivity to the inhibitor while remaining active. The kinetics suggested a bimolecular reaction, and it seemed that threonine and isoleucine were bound to different sites.

The interaction between the inducer and the—at the time hypothetical—repressor was extremely rapid and entirely reversible. Probably, only a very small number of molecules were involved, which nevertheless triggered first the complex mechanism of the repeated transcription of the operon and, secondarily, the repeated translation of the messenger, that is, the repeated formation of thousands of peptide bonds. The inducer seemed to act as a chemical signal recognized by the repressor, but did not participate in any of the reactions for which it was responsible.

During the winter of 1961, David Perrin and Agnes Ullmann were working late one evening, in the laboratory, when Jacques Monod, pale and tired, entered the room and said: "Mes enfants, j'ai découvert le deuxième secret de la vie" ("I have discovered the second secret of life"). "Please sit down, rest and have a drink," said Agnes. Yet the secret had not vanished after the rest and Monod explained what it was. The activity of enzymes depended on their conformation, which was controlled by the attachment—or detachment—of an effector. The observed actions of effectors were due to indirect interactions between distinct stereospecific receptors. The interpretation was applicable to the repressor which would have two binding sites, one for the inducer, the other for the operator gene. *Allostery* was born, the name as well as the concept. However, the theory was not at first accepted by the enzymologists of the laboratory, who were worried rather than excited by the "non-classical" aspect of the curves they observed.

In 1961, in Cold Spring Harbor, Monod and Jacob (91) gave the general conclusion of the symposium. The fact was stressed that the inhibition of an enzyme may be caused by substances which are not steric analogues of the substrate. The expression *allosteric inhibition* was coined to describe the phenomenon. The discussion led to the conclusion that two distinct, albeit interacting, binding sites exist on *allosteric enzymes*. The effector acts by altering the conformation of an enzyme; the alteration is stabilized by the formation of a complex. It is interesting to recall that in the discussion which had followed the presentation of Changeux's paper, during the 1961 Cold Spring Harbor Symposium, B. Davis called attention to the effect of oxygen, which modifies the affinity of haemoglobin for oxygen.

The concept of allostery was further discussed by Jacob and Monod (1962:98) and applied to the "induced-fit" theory of Koshland, the mutual effect of substrate and enzyme on molecular configuration. The following year (1963:100), Monod, Changeux, and Jacob extended the concept and insisted on the fact that allosteric effects are entirely due to reversible conformational alterations induced in the protein when it binds to the specific effector.

From the symmetry of the curve of saturation of haemoglobin by oxygen, Jeffries Wyman had been led to suggest that a structural symmetry of the molecule was involved. This was the origin of the classical Monod–Wyman–Changeux paper (1965:106). The general properties of allosteric systems are stated. (1) Most allosteric proteins are oligomers made of several identical units. (2) Allosteric interactions are correlated with alteration of the quaternary structure of the proteins. (3) Heterotropic effects, involving interactions of different ligands, may be either positive or negative; homotropic effects, involving identical ligands, are always cooperative. (4) The individual isolated units, the monomers, when associated to form the allosteric protein, are called protomers. (5) Protomers are linked in such a way that they all occupy equivalent positions. (6) Each allomeric molecule possesses at least one axis of symmetry. (7) One protomer possesses only one binding site for each ligand. (8) The conformation of each protomer is constrained by its association with the other protomers. (9) The association between monomers is specific—and most oligomeric proteins are stable—despite the fact that no covalent bonds are involved. (10) The protomers are probably linked by a multiplicity of noncovalent bonds. It is this multiplicity which confers stability on the association.

In an isologous association of monomers, when the domain of binding involves two identical binding sites, there is a twofold axis of rotational symmetry. The problem of symmetry was extensively discussed, as also was the fact that each protomer is somewhat "constrained" and should adopt the same quaternary conformation. Lastly, the finality of size and structure of proteins is evoked. The authors conclude their classic paper by stating that they "have tried to develop and justify the concept that a general and initially simple relationship between symmetry and function may explain the emergence, evolution and properties of oligomeric proteins as 'molecular amplifiers,' of both random structural accidents and of highly specific, organized, metabolic interactions." The pinnacle of the theory of allostery was Monod's discussion on symmetry and functions in biological systems (1968:122). Allostery made it possible to interpret and to integrate a great number of isolated observations into a coherent unifying concept. In almost all papers attention had been called to the danger of inconvenience of a concept endowed with such explanatory power that it did not exclude anything. It is why Boris Magasanik called it "the most decadent theory in biology." A decadence which must have triggered in Jacques Monod a secret feeling of deep satisfaction.

Regulatory Systems and Evolution

To understand the regulation of cellular functions at the molecular level, mutants were widely used, and proved to be a most powerful tool. The genetic control of regulatory mechanisms was also essential for the understanding of the mechanism of evolution. It is easy to demonstrate that efficient regulatory systems confer selective advantages. In a medium devoid of β-galactosides, the production by a constitutive bacterium of some 6,000 molecules of β-galactosidase represents a waste of amino acids and of energy. When constitutive and adaptive strains are placed together in a medium devoid of the substrate, the adaptive strain is selected: it multiplies more rapidly than the constitutive one.

This conclusion is also valid for anabolic systems. A dual regulation is at work in the machinery responsible for the biosynthesis of essential metabolites. The end product, through a repressor, controls the activity of the structural genes of the system. The end product also controls the activity of the enzymatic machinery. Here again, the regulatory mechanism confers a selective advantage.

Regulation is performed by small molecules; and a prerequisite for their action is the existence of a receptor site on the protein—whether enzyme or repressor. The properties of a protein—whether enzyme or repressor—are controlled by its tertiary and quaternary structure, in turn determined by the primary structure. Since the primary structure is determined by the genetic information, it follows that the evolution of regulatory systems is the consequence of mutations, necessarily random, of the genetic material: regulatory, operator, and structural genes.

The problem of regulation led to the problem of evolution. Jacques Monod's essay on the philosophy of modern biology, *Le hasard et la nécessité* (*Chance and Necessity*), is the by-product, or better the unavoidable consequence, of the work on regulation. It is in essence a modern version—accessible to the layman—of Darwin's ideas concerning evolution and selection. Francis Crick's comments on the book (*Nature [London]* 1976, **262**, 429–430) are the following: "Written with force and clarity, in an unmistakable personal style, it presented a view of the universe that to many lay readers appeared strange, sombre, arid, and austere. This is all the more surprising since the central vision of life that it projected is shared by the great majority of working scientists of any distinction." This very successful book was translated into many languages and provoked much discussion. Due to the limited space at my disposal, a critical examination of Monod's philosophical views is unfortunately not possible.

Head of a Department

The enzyme β-galactosidase, how beloved it has been, was only a tool for the understanding of the relation between genes and enzymes; how often have I heard Monod complaining that he was far away from the gene. When the work on the induced synthesis of enzymes was started in 1941, nothing was known except the phenomenon; the concepts developed essentially from 1948 on. In the first phase, biochemical, Melvin Cohn played a determining role. In the second, genetical and regulatory, François Jacob's intervention had been essential, and this was—Francis Crick *dixit*—the "grand collaboration." Between 1948 and 1963, the main problems posed by the induced synthesis of enzymes (that is, regulation) were solved, and molecular biology was created *ex nihilo*.

From 1945 on, Monod had worked in the Service de Physiologie Microbienne, in the attic laboratories. In 1953, he was made head of the Department of Cellular Biochemistry and moved into new quarters at the end of 1955. Both in the attic and in his new laboratory, Monod showed remarkable gifts as a leader. He received a very large number of students and postdoctoral workers, and oriented them in conformity with their tastes and aptitudes. As noted by Francis Crick, he "treated his students with affection and candour, as if they were members of his family." He proffered ideas generously and enjoyed discussions. The weekly seminars were exciting shows.

Director of the Institut Pasteur

In April 1971, Monod was appointed Director General of the Institut Pasteur. He was 61. It may seem strange that a dedicated passionate scientist could, in full activity, abandon the laboratory. Perhaps he sensed that with allostery and symmetry he had reached the peak of his scientific achievements. Possibly he did not fully realize that the directorship would destroy almost entirely his scientific activity—and most of his freedom. Monod liked to plan, to organize, to decide, to command. The directorship of the Institut Pasteur was an extraordinary challenge for a man of great energy, endowed with a clear vision of what the evolution of an institute of biochemical research should be. The most likely hypothesis is that his sense of duty played a determining role in his decision. Be that as it may, the Institut Pasteur deserved to be loved, and the encounter between Jacques Monod and the prestigious institute could not fail to be a great event.

All Pasteurians had suffered from the many errors in the organization and management of the institute, not to speak of the scientific planning— or absence of planning. The development of research, as well as the financial balance, was compromised, and Monod was well aware of the extent of the disorder. He abandoned the direction of the Service de Biochimie Cellulaire, where he was replaced by Georges Cohen, and he was soon obliged to abandon also his professorship at the Collège de France. He nevertheless continued to discuss the planning and results of research with his disciples, as evidenced by his last two publications of 1974 and 1976 (130, 131).

The Institut Pasteur is a research institute. But it has an industrial wing which provides about half the budget. The other half comes from various sources. These include such governmental agencies as the Centre National de la Recherche Scientifique, the Institut National de la Santé et de la Recherche Médicale, the Délégation Générale à la Recherche Scientifique et Technique, and private organizations like the Fondation pour la Recherche Médicale, and private donors. In April 1971, the financial situation was catastrophic.

Jacques Monod had first to learn how to run a business. Within a few months he had become an expert in management, to the point that he was asked by François Dalle, an industrialist, to write a foreword to his book *Quand l'entreprise s'éveille à la conscience totale*. He had analysed the situation of the institute, built and set in action the industry, and defined the main axes of the offensive of renovation. The industrial sector had to be restructured, priority given to new products, foreign markets sought. In addition, a rigorous administrative and financial structure had to be put in place. Impressed by the seriousness of these measures, the government decided to increase its financial aid.

This was only one part of the director's responsibility. Emile Duclaux had succeeded Louis Pasteur in 1895 as head of the Institute. Since his death in 1904, there had been a succession of directors and deputy directors, some of them eminent scientists, such as Emile Roux, Elie Metchnikoff, Albert Calmette, and Gaston Ramon. Yet the Institute had to be entirely rethought and reorganized as a consequence of the evolution of biomedical sciences. Moreoever, the golden age of tropical medicine, microbiology, and parasitology had passed, and, as a result of decolonization, the status of the numerous extraterritorial branches of the Institut Pasteur had changed. Jacques Monod was not an M.D., and yet he very rapidly dominated the problems posed by the evolution of medical microbiology, virology, immunology, and experimental pathology. A reorganization of research was necessary. Some departments had to be suppressed, others expanded or created. A number of difficult problems were solved with energy, sometimes in the face of fierce opposition.

Between April 1971 and June 1976, the scientific and industrial policy had been defined and put into effect. A most remarkable achievement, particularly in view of the fact that Jacques Monod was handicapped for six months, in 1972, by viral hepatitis, and after October 1975, by the disease which was responsible for his death. Yet illness never stopped him from assuming his responsibilities. He showed in his role as an administrator the rigour, eagerness, logic, and intelligence that had been evident in the direction of his laboratory.

Outside Science

Science was the dominant activity of Monod's life, the field where he expressed his creativity and originality. Yet science, inhuman science, could not by itself satisfy the aspirations of a man of rich diversity and tremendous energy, endowed with an intense curiosity, a great artistic sensitivity, and a deep humanity, conscious of his duties as an intellectual and as a citizen.

Monod was a lover of music and for years played the cello in a quartet. He had also created a Bach choir, "La Cantate," which he directed until 1948. With his friend François Morin, he had translated Sir James Jeans's book *Science and Music*.

He had been seriously tempted to make a career as a conductor. Certainly the direction—and domination—of an orchestra could have given him great satisfaction. In 1936, he was offered a position as a conductor in the United States and the temptation was great. Three events determined his choice. First, a conversation with Louis Rapkine, who, as he told me, convinced Monod that he lacked the basic musical knowledge necessary to conduct an orchestra; second, the fact that with the study of bacterial growth, Monod had found his scientific way; third, his marriage, in 1938, to Odette Bruhl, an archaeologist and orientalist, specialist in Tibetan painting, who became curator of the Musée Guimet. His wife, who died in 1972, was a person of great charm, sensitive and discreet, who brought to Jacques both stability and the enrichment of a complementary culture. They had twin sons. One is a physicist, the other a geologist.

So music was sacrificed on the altar of science. Science had won, but the love for music persisted throughout his life as a constant temptation and, perhaps, a regret. The radio station "France Musique" gives every week a two-hour *concert égoïste*. The programme is scheduled by laymen and lively discussions take place. Jacques Monod had been twice on the stage during the year 1975. These two concerts were revealing of Monod's taste; by the way, it was while listening to them that I learned that Monod's godmother had been the first wife of Claude Debussy.

The passion for music went hand in hand with intellectual interests. Monod was a great reader and possessed a good knowledge of classical as well as modern literature. He used to read books with the same critical rigour that he would put into the analysis of an experiment. In this connection, the interview he gave to *Lire* in 1975 is characteristic. Monod liked not only to read, but to write. His style is clear and incisive. The forewords to Ernst Mayr's *Populations, espèces et evolution*, Karl Popper's *Logique de la découverte scientifique*, and Medvedev's *La grandeur et la chute de Lyssenko* are revealing of the various aspects of Monod's intellectual gifts and talents. By the way, he was perfectly bilingual: his English—spoken or written—was as excellent as his French.

Monod was conscious of his responsibilities and duties as a citizen. During the war, he took an active part in the Underground. This action expressed his will to resist oppression and slavery. He had been very active in the "army of shadows" and exerted important responsibilities. In the position he occupied last, his three predecessors had disappeared. After the liberation of Paris, Monod played a determining role in the integration of the free French forces into the regular army, and was a member of the staff of General de Lattre de Tassigny.

The underground group to which Monod belonged was a communist one, and Monod felt he had to join the communist party. He left it soon after the war: he could not accept the rigid dogmatic attitude of the party, particularly the stand in the tragedy of Russian genetics. He could not accept an ideology which was a negation of truth, science, and rationalism, a negation also of human dignity (not speaking of the mass murders which have dishonoured so many communist states). He never stopped fighting for his conception of justice and for the respect of human values.

He soon found another battlefield. Contraception was unlawful in France. He supported the action of the "Mouvement français pour le planning familial" and became one of its honorary presidents. Later, he actively supported "Choisir," a movement which was fighting for the legalization of abortion.

A scientist who has been a professor at the Sorbonne was necessarily aware of the problems posed by the university. With Pierre Aigrain, Monod was responsible for the two "Caen Symposia" at which the university was reorganized, at least on paper.

Finally, Monod had participated in the creation of the "Centre de Royaumont pour une Science de l'Homme," which tried to develop a scientific and synthetic approach to the problems which face mankind.

His intense intellectual activity was balanced by physical activity. Despite a handicap which was the sequel of poliomyelitis, Jacques Monod became a good rock climber, practising on Sundays at Fontainebleau. During the summer he performed difficult ascents in the Alps. Later, he abandoned the mountains for the sea and became an accomplished yachtsman. Those who cruised with him on the *Tara* have told me that he was not only an excellent skipper, but also a kind one, which seems to be relatively rare.

Monod had decided not to ask for the renewal of his six-year term as director. He wanted to live his own life and to write *L'Homme et le temps*, a book which will never see the light.

In October 1975, an inexorable disease was diagnosed. He knew the prognosis but continued to assume his directorship. From time to time, he went to Cannes for a "rest." The last rest, at the end of May 1976, was active as usual: walking and sailing. Some of the pictures illustrating this biography were taken on 29 May by Jean Hardy. A day later, he realized that the end was approaching. He died quietly. His last words were "Je cherche à comprendre." All his life, he had tried to understand.

The Man and the Monument

"Good looking, though small of stature, he commanded attention by his intelligence, his clarity, his incisiveness, and by the obvious breadth and depth of his interests. Never lacking in courage, he combined a debonair manner and an impish sense of humour with a deep moral commitment to any issue he regarded as fundamental." This is the portrait sketched by Francis Crick. Many of those who have known Jacques Monod will agree with this picture. But there is something more to be said and it would be dishonest to mask the shadows. Martin Pollock, in an obituary, wrote the following: "I have often wondered how many scientists there were from all over the world who struggled to get accepted as visiting workers under his stimulating guidance at the Pasteur and now carry with them the fruits of a contact, however brief, with a real master of enlightenment. Perhaps the light that emanated was too dazzling sometimes. It was just this tendency to dazzle, and to exercise—indeed to demand— intellectual predominance over his fellow scientists which one might legitimately criticize. It was often very difficult to think independently in his presence when others were around. In open conference, he could be a tough and uncompromising opponent rather too ready to condemn without proper consideration. At times he could be exasperating and many found him arrogant, elitist or condescending. But it was quite a different matter in private discussions: there one was listened to attentively, with courtesy, if not always with respect. The polemics were no longer necessary."

The two facets of Jacques Monod's dual personality are well illustrated by these comments. On the one hand, a man of extreme courteousness and charm showing great warmth to his friends. On the other, a man who could not accept public opposition to his views, who liked to impose his ideas and decisions, to dominate, conscious of his intelligence and his gifts and eager to manifest his authority. One should bear in mind that the construction of a scientific monument through forty years of uninterrupted effort implied a considerable amount of experimental work and a constant intellectual tension; it could be achieved only thanks to a certain hardness, a corollary of rigour and exactness. This may in part account for the negative traits which, by the way, were already apparent when Monod was a student.

Yet the defects were minor when compared to Monod's work, which, from the growth of populations to molecular language, is marked by an uninterrupted series of discoveries. The impressive monument, crowned by the most elegant spire of allostery, bears witness to a great talent. The success was due to a conjunction of eminent gifts and to a preestablished harmony between the nature of the gifts and the nature of the task to which Monod devoted his activity. He was an excellent experimentalist. Rigour and precision were served by an implacable deductive logic. Critical sense never hindered imagination nor audacity.

The development of the work from diauxy to allostery is wholly admirable. It started with the interpretation of growth curves, and ended with the solution of the problem of regulation at the molecular level. The molecular language was deciphered: how molecules receive and transmit messages, obey, and command. New classes of structures and phenomena were brought to light; new concepts were built. Each step generated new questions until the central problem was solved. A scientist has to give birth to his own problems, step by step, in pain—and enlightment.

In addition to the gifts and talents of Jacques Monod, a number of factors played in his success. The right problem was posed at the right time in the right environment. The right bacterium was selected and the right enzyme. For an outsider, the success appears to be the highly improbable combination of improbable events. Yet one should not forget the numerous trials and errors and the fact that selection intervenes constantly, at each step of the phylogeny of a scientific construction. The unfit is eliminated. Finally, one should also bear in mind that preadaptation plays a role in the choice of the working place; that cooperative effects exist, here as elsewhere; and that the whole process becomes autocatalytic. That the name of Jacques Monod is so intimately associated with the birth, development, and triumph of molecular biology is not a matter of chance but of necessity. The necessity was Jacques Monod.

The work of a creative genius sometimes outsteps the man. The scientist—or the artist—may dread this transcendence which would destroy the persona he tries to shape; neither scientist nor artist is aware of the secret of his genius. However, any powerful construction of the mind or spirit engenders through its resonances the image of its creator, at one its reflection and its symbol. Such will perhaps be the fate—or privilege—of Jacques Monod, architect of molecular biology.

André Lwoff (1902–1994): Remembrances*

Agnes Ullmann

I first met André Lwoff in January 1958, in rather special circumstances. I had just arrived in Paris from Hungary to spend some time in Jacques Monod's laboratory. It was a Friday and I was immediately taken to the monthly "Club de Physiologie Cellulaire," at the Institut de Biologie Physico-Chimique. At the end of the lecture, I was studying the Paris Metro map, when a distinguished gentleman asked me where I wanted to go. It turned out that my destination was on his way and he offered to take me by car. I looked at it with stupefaction; it was a Citroën 2CV ("deux chevaux"), which I had never seen before: it was nonexistent in Hungary. When I asked him whether he had made it himself he burst out laughing and explained to me that it was a genuine, factory-made car. Next Monday, everybody in the laboratory had heard the story and I learned that the gentleman with the 2CV was André Lwoff. I had never read a word of his and I was yet to discover the richness and the generosity of his personality.

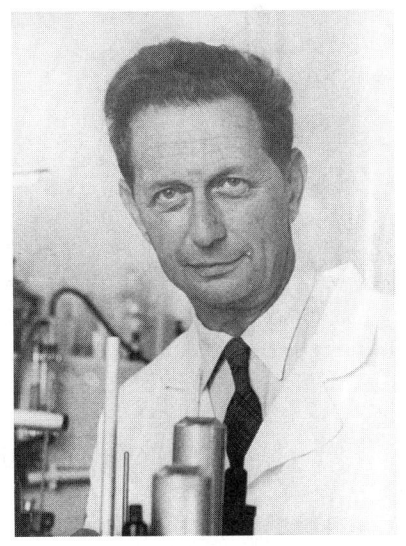

One of the fathers of molecular biology, André Lwoff began his scientific life at the age of 19, but had already seen his first microbe under a

*Reprinted with the kind permission of the *EMBO Journal* [**14**(14): 3289–3291, 1995].

microscope at the age of 13, shown to him by Elie Metchnikoff, a friend of his father. His father, born in Russia, was a chief physician in a psychiatric hospital. As a child, Lwoff was taken often to the ward. He wrote: "...I am under the impression that the contacts, perhaps somewhat premature, with the manifestations of mental disorder considerably reinforced my inclination for scientific disciplines" (Lwoff, 1971).

His first (and perhaps only) master was Edouard Chatton, whom Lwoff considered as the greatest protozoologist of all time. His passion for ciliates led him, in collaboration with Chatton and with his wife Marguerite Lwoff, to work for almost 15 years on their morphology and morphogenesis at several marine biological laboratories during the summer months.

At the age of 20, Lwoff entered the Institut Pasteur in the laboratory of Felix Mesnil (who had been the secretary of Louis Pasteur), where he devoted his work to the nutrition of ciliates. This seemed to be an enormous task at that time, as no pure cultures in synthetic media had previously been obtained. He finally succeeded. He went on to study the nutritional requirements of some protists and realized that growth factor requirements could be interpreted as a loss of some biosynthetic functions, as an evolutionary consequence of adaptation to parasitism (Lwoff, 1932). His theory of regressive physiological evolution conflicted at that time with the dominant ideology, implying that progressive evolution is linked to enrichment of functions. In his introduction to *L'Evolution physiologique* (Lwoff, 1944), Lwoff relates in his typical delightful style a discussion that he had with one of his colleagues who accused him of being ignorant of evolution, and arguing that he (the colleague) was able to do things that a *Chlamydomonas* would be unable to do. Lwoff concluded by wondering how this eminent colleague would have survived, exposed to sunlight in a potassium nitrate solution?

During the pre-war period, André and Marguerite Lwoff played a leading part in determining the vitamin requirements of microorganisms and were the first to demonstrate that a vitamin may function as a coenzyme, which led to the recognition of their role in cellular metabolism. These discoveries permitted an important conclusion, namely that vitamins constitute a class of essential growth factors for organisms that are unable to synthesize them.

During World War II, Lwoff's laboratory, the "Service de Physiologie Microbienne," the famous attic, became an active centre of the Underground. Jacques Monod, before definitively joining his laboratory in 1945, spent some time there clandestinely, to perform a few experiments. With the end of the war started a "Golden Period" in the laboratory. Lwoff had brought together a remarkable staff, which included Jacques Monod,

François Jacob, Elie Wollman and Pierre Schaeffer, and a great number of foreign scientists, including Mel Cohn, Annamaria Torriani, Martin Pollock, Seymour Cohen, Alvin Pappenheimer, Mike Doudoroff, Dale Kayser, Dave Hogness and many others. All of them considered that it was André who generated the exceptional enthusiastic, exciting and friendly atmosphere: the attic became the stimulating centre of the emerging new field of molecular biology.

Around 1949, André Lwoff started to work on lysogeny. He had fixed his choice on a lysogenic *Bacillus megaterium*, a particularly large bacterium, because he wanted to study single bacteria. With a microscope and a micromanipulator he inoculated them into individual microdrops and let them grow. Why this methodological choice? Because he "disliked mathematics, and wanted to avoid formulas, statistical analysis and, more generally, calculations as much as possible" (Lwoff, 1966). By following the kinetics of phage production, he realized that only a relatively small fraction of the bacteria lysed and produced bacteriophage. Why did some cultures lyse, while others did not? With Louis Siminovitch and Niels Kjeldgaard he tried, day after day, to obtain lysis of the totality of the bacterial population. After several unsuccessful months, he decided to irradiate the bacteria with ultraviolet light—a UV lamp, used by Jacques Monod to mutagenize *Escherichia coli*, was available at the other end of the corridor. The bacteria were irradiated for a few seconds, and after one hour . . . the culture was entirely lysed. He wrote: "As far as I can remember, this was the greatest thrill of my scientific career—for the first time in my life, I had the feeling of having discovered something" (Lwoff, 1966).

This was a far-ranging discovery. First of all a clear picture of lysogeny emerged: the lysogenic bacterium perpetuates the genetic material of the bacteriophage, which he dubbed the prophage; induction of the prophage would lead to the vegetative multiplication of the phage. The subsequent discovery of zygotic induction by François Jacob and Elie Wollman, and the uncovering of the mechanism of genetic repression by Jacob and Monod, were direct consequences of the induction of phage production by the cells. The mechanisms by which inducing agents and environmental factors act in altering the chromosome-prophage equilibrium to lead to the production of virus particles are still of great topicality in the field of retroviruses and cancer.

Around 1954, Lwoff decided "to intrude into the jungle of animal virology" (Lwoff, 1971), abandoning the field of lysogeny. He then spent a sabbatical year in Renato Dulbecco's laboratory to learn about tissue culture and animal viruses. ". . . I suppose there are a number of recipes for the rejuvenation of ageing scientists. One of them is to become a student" (Lwoff, 1971). Back at his laboratory at the Institut Pasteur, André and Marguerite Lwoff immersed themselves in the study of poliovirus reproduction, which led them to a number of important accomplishments such as the characterization of the thermosensitive step and identification of "reproduction temperature" (rt) mutants, important for live polio vaccines.

In 1965, Lwoff, together with Monod and Jacob, was awarded the Nobel Prize in Physiology or Medicine for discoveries concerning the genetic control of enzyme and virus synthesis. Three years later, he abandoned the Institut Pasteur to take the directorship of the Cancer Institute in Villejuif to help a new generation of scientists develop research on cancer. One may wonder whether in accepting this position he recalled his feelings toward a former director of the Institut Pasteur... "I had decided long ago and once and forever that the scientists transcended the director and the board of trustees..." (Lwoff, 1971).

André Lwoff was a master of language. He loved to consult dictionaries and to use classical quotations. He attached great value to accurate definitions and terminology, defending even paradoxical definitions such as "viruses are viruses" (Lwoff, 1957). Publicly and in private he corrected wrong terms or language mistakes. Most people disliked it. I greatly appreciated his habit of correcting my French by a literary quotation—after many decades I still remember them.

His concern for intellectual clarity revealed a constant effort to relate scientific ideas to the whole of a cultural tradition. He was a man of immense culture; interested in literature, philosophy, fine arts, and last but not least, in the future of our culture and civilization. I remember, around 1959, having had lunch at his home, I was admiring his library. He then started to question me on my knowledge of French literature, pointing to several of the books. Seemingly satisfied, he added with his mocking smile (that many people interpreted as haughty) that he did not realize that French literature was commonly read in Hungary. As a kind of reward André then offered me a book, *La trahison des clercs* by Julien Benda, written in 1927, and unobtainable in libraries; he had two copies of it. After a while, I was subjected to a thorough oral examination to check whether I had really read the book. André considered Benda a very great thinker and the book a very great book. Many years later, he prefaced a new edition of the book—a masterpiece of literary analysis (Lwoff, 1975).

André Lwoff started to paint at the age of 57. He was encouraged by a friend of his, Marcelle Wahl, who was a physician and painter. He realized his first still life in her studio. From then on, he never stopped painting.

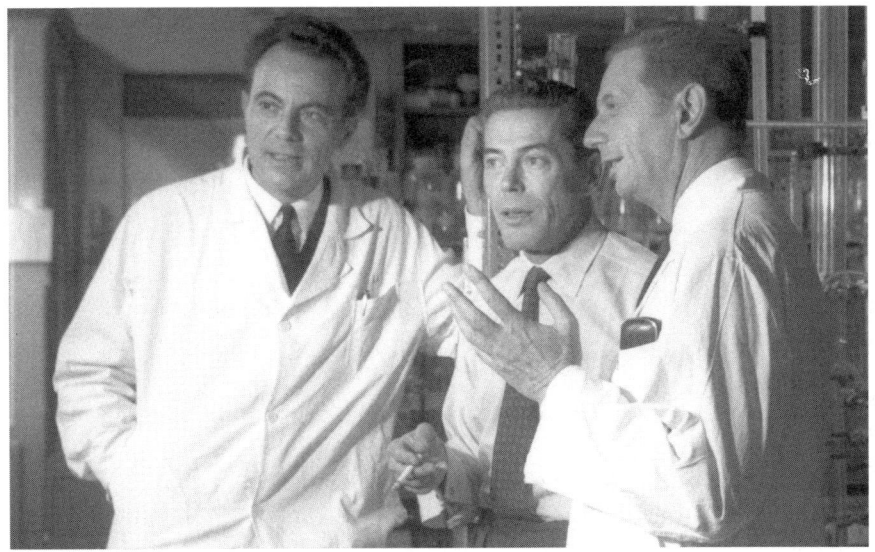

F. Jacob, J. Monod, and A. Lwoff, 1965.

While we were working in 1977 to edit two books to honour the memory of Jacques Monod (Lwoff and Ullmann, 1978, 1979), André told me about his beginnings as a painter, and offered me one of Marcelle Wahl's etchings, saying that he always considered her a great artist.

André Lwoff had a profound sense of justice: he committed himself in all fights related to the tragedy of our times: he struggled against nuclear armament, against racism and antisemitism, and in all circumstances and in all the scenes, he struggled for human rights.

His warmth and generosity were most of the time hidden behind a proud reserve. When Jacques Monod organized my escape from Hungary, Jacques needed some financial help. When I arrived safely in Paris, Jacques told me that André Lwoff had made a substantial contribution, but when I wanted to express my thanks to him, he answered in a haughty tone that he did not know what I was talking about. So, I left his office... Many years later, I took my "revenge" on him. He was exhibiting his paintings at a gallery. The income of the sale was in aid of Soviet dissidents and of the Pasteur-Weizmann Council. When I showed him the gouache I wanted to buy, he warned me that it would be too expensive for me. I answered that this was the only way I could repay my debt to him. He remembered, he laughed and I bought the gouache.

André Lwoff, leader of one of the great schools of biology, was, above all, an artist. As Jacques Monod noted (Monod, 1971), he revealed the same talents in every domain: elegance, sensibility, precision, finesse, freedom; in a word, style.

References

1. **Lwoff, A.** (1932) Recherches Biochimiques sur la Nutrition des Protozoaires. Masson, Paris.
2. **Lwoff, A.** (1944) L'Evolution Physiologique: Etude des Pertes de Fonctions chez les Microorganismes. Hermann, Paris.
3. **Lwoff, A.** (1966) In Cairns, J., Stent, G. S., and Watson, J. D. (eds), Phage and the Origins of Molecular Biology. Cold Spring Harbor Laboratory Press, Cold Spring Harbor, N.Y., pp. 80–99.
4. **Lwoff, A.** (1957) J. Gen. Microbiol., **17:** 239–253.
5. **Lwoff, A.** (1975) La Trahison des Clercs. Grasset, Paris.
6. **Lwoff, A.** (1971) Annu. Rev. Microbiol., **25:** 1–26.
7. **Lwoff, A. and Ullmann, A.** (1978) Selected Papers in Molecular Biology by Jacques Monod. Academic Press, New York.
8. **Lwoff, A. and Ullmann, A.** (1979) Origins of Molecular Biology. A Tribute to Jacques Monod. Academic Press, New York.
9. **Monod, J.** (1971) In Monod, J. and Borek, E. (eds), Of Microbes and Life. Columbia University Press, New York, pp. 1–9.

The Outer and The Inner Man*

Roger Y. Stanier

The prize of the general is not a bigger tent, but command.
OLIVER WENDELL HOLMES, JR.

I n April 1971 Monod became Director of the Pasteur Institute. A previous attempt to put him in this post had been blocked at the highest level by President Pompidou, who had not forgotten some disobliging public criticisms offered by Monod a few years earlier. It is rumoured that Pompidou, who had modest literary pretensions and even more modest literary talents, relented as a result of the success of *Chance and Necessity*. Why did Monod accept this onerous and unrewarding task, which consumed the last years of his life? A profound sense of indebtedness was probably the key factor. Since 1945, the Pasteur Institute had given him the freedom to pursue his own work, without making other demands on his time. Now its very existence was menaced; and Monod believed that nobody else had the drive, ability, and prestige to save it. As we shall see, he accomplished a great deal in five years; whether these reforms were sufficient to preserve the historical structure of the Institute still remains to be decided.

Monod's years as Director were shadowed by bereavement, sacrifice, and illness. His wife, the former Odette Bruhl, an orientalist and the curator of the Musée Guimet, was gravely ill in 1971, and died early the following year. The projected sequel to *Chance and Necessity* had to be put aside, and work on it was never resumed. Experience soon convinced him that even the light duties of a professor at the Collège de France were incompatible with the directorship, and he resigned from his chair. He suffered a severe attack of viral hepatitis in 1972, and became ill again in October 1975. This time, the disease was terminal. He had developed an aplastic anaemia, but

*Reprinted with permission of the editors of the journal from an obituary published by *Journal of General Microbiology*.

continued working to the end, kept alive (as he casually remarked one day) by the blood of others.

A brief historical introduction is necessary to explain the multiple problems which confronted Monod in 1971. The Pasteur Institute had been created by subscription, the result of a wave of public enthusiasm, following Pasteur's development of a rabies vaccine; the original building is still called "le Batiment de la Rage." A special law gave it the status of a private foundation recognized to be of public utility, making it one of the very rare scientific institutes in France largely independent of governmental control. Despite this, the Institute has always played a national role of fundamental importance: it is the central microbiological reference laboratory of France, the repository of nearly all national culture collections, and a major teaching centre for microbiology and immunology. Since its foundation, the Institute has engaged in the production of serums, vaccines, and other biological products, on the sale of which its financial independence was originally based. Only after 1945 did this source of revenue begin to be supplemented to a significant degree by state support, which thereafter took an increasingly important place. The sources of state support were multiple and, in part, indirect; for example, many of the scientists and technicians who work at the Institute are employed by governmental organizations, such as the Centre National de la Recherche Scientifique.

The infusion of funds from external sources for some time prevented recognition that the Institute's industrial operations had been going downhill for many decades, as a result of antiquated production methods, failure to develop new products and improve old ones, and the absence of cost accounting; in short, a complete lack of sound business practice. Competition from private industry had made serious inroads into the traditional markets, both domestic and foreign. The gravity of the situation was further concealed by the absence of any overall control of the Institute's finances: nobody seemed to know where the money came from or where it went, and deficits were met by the dissipation of endowment funds. By 1971, however, it had become clear that bankruptcy could not be far away. The state would at that point intervene, to maintain those activities which it deemed essential; but the independence of the Institute would be lost. Monod assumed the directorship to try and avoid this outcome.

He had no previous administrative or business experience, and it took him some months to grasp the parameters of the problem. He then acted with energy and firmness. Since it was evident that the Director could not simultaneously head a research institute and run a business, a subsidiary company—Institut Pasteur Production—was created, and its key officers were recruited from private industry. Monod's success in staffing the new company with men who were able, devoted, and loyal says much for his powers of persuasion; objectively, he had very little to offer them. The slow and painful task of revitalizing the industrial activities of the Institute was thus begun. However, financial projections revealed that even under the most optimistic assumptions, the Institute could not hope to attain financial equilibrium from this source; increased state support was indispensable. Monod therefore appealed to the government. A special commission was created to study the situation, and shortly before his death, the government agreed to help out.

Needless to say, the building up of the scientific activities of the Institute was also on Monod's agenda when he became Director. But when the financial preconditions were on the way to fulfilment, his time had run out. By an irony of fate, Monod will be remembered among the directors of the Pasteur Institute for his financial and administrative reforms, not for his role in determining its research programme.

The Outer and The Inner Man

The attempt to sketch a likeness of Jacques Monod is perhaps foolhardy; it would tax the powers of an observer as subtle and perceptive as Henry James. For this, there are several reasons. First, his many-sidedness: scientist, teacher, musician, writer, sportsman, administrator, and public figure. Second, his deeply ingrained sense of privacy; he kept his associations well compartmentalized, and few (if any) of his friends had more than a partial knowledge of the various facets of his life. The greatest difficulty of all stems from the complexity of his character, with its seeming contradictions: he could be considerate or brutal; courteous or rude; contemptuous of authority or eager to exercise it.

In the personality of Monod, it is possible to perceive three superimposed and often incompatible elements. The most obvious personage was the liberal cosmopolitan, cultivated, urbane, and ironical, equally at ease in two languages and three cultures. His familiarity with Anglo-American speech and folkways trapped many English-speaking acquaintances and friends into imagining that he was, at heart, one of them. Nothing could be further from the truth. Under the cosmopolitan facade, Monod was quintessentially French: a patriot and a conscious representative of a very distinctive subculture, the Protestant grande bourgeoisie. Failure to recognize this led to misunderstandings, made easier by Monod's overt behaviour, notably his oft-expressed admiration for some Anglo-Saxon attitudes and his trenchant criticisms of the defects, as he saw them, of French society. This behaviour did not reflect a sense of estrangement from his own country and culture, but a passionate desire, almost Gaullian in its intensity and arrogance, to improve that country and culture.

Underlying the cosmopolitan and the Frenchman was a third persona, recognition of which is essential to an understanding of the man. Although separated by at least one generation from Protestant religion, Monod was imbued with the sternest and most pitiless version of the Protestant ethos: at heart, he remained a Calvinist. The novels of André Gide, another distinguished and wayward product of French Calvinism, provide a useful literary guide to this aspect of his character. Since France is still largely permeated by the Catholic ethos, Monod's Protestant conscience, important in determining both his own destiny and his relations with others, was not clearly recognized by many of his compatriots, who remained perplexed by its manifestations. This component of Monod's character was without question a fertile source of conflicts, internal and external, sometimes painfully overcome, sometimes unresolved.

An anecdote permits a glimpse of one internal conflict, which was overcome. Shortly after his arrival in Lwoff's attic, Monod had collaborated with Elie Wollman in a study of the effect of phage infection on β-galactosidase synthesis. Some time later, when the work on β-galactosidase had run into temporary difficulties, he confessed to one of his coworkers that he had been strongly tempted to work on phage, but had resisted the temptation "parce que j'en avais trop envie" ("I wanted too much to do it"). It would be difficult to find a better illustration of Samuel Butler's aphorism about the Protestant ethos, that virtue is the condition in which the pain precedes the pleasure!

Pitiless with himself, Monod could also act harshly towards others, particularly during his years as Director. For example, having decided to retrench by abolishing some small research units, mostly of mediocre quality and peripheral to the main interests of the Institute, he arbitrarily forced their heads into premature retirement, when half an hour of sympathetic discussion would in many cases have elicited a voluntary retirement. One of the individuals so affected had accompanied Monod to Greenland on the *Pourquoi pas?* in 1934 and had watched Monod's subsequent rise to fame with admiration and affection. He had served the Institute well in the past, but he was permitted to leave the scene of his life's work without a world of thanks from the Director.

Although Monod could be authoritarian and inflexible, he could also act with great generosity and kindness. The scientists, technicians, and secretaries who worked with him could always count on his sympathy, advice, and support; in return, he received an unreserved loyalty and devotion from nearly all of them. His time, his prestige, and his personal fortune were freely used to aid humanitarian causes and individuals. Although the fact never became public knowledge, several victims of both fascist and communist tyranny owe their escape and opportunity to make a new life in the free world to him. As a leader of the movement (finally successful) to legalize abortion in France, Monod often received anguished private letters appealing for his help. He always responded in the only way possible before the law had been changed: by sending money.

One common French quality which Monod did not share was the Latin brand of sentimentality. It differs from the English one in being lavished not on animals, but on national shibboleths and (in lesser measure) on fellow humans. Desperately searching as Director for solutions to the problems of the Institute, he hit on the notion of rebuilding it on the suburban campus at Garches, financing the operation by the sale of the very valuable land which it occupies in Paris. One day while he was expatiating on the virtues of this solution (never, of course, put into effect), the author of this memoir asked him what he proposed to do about the tombs of Pasteur and Roux, both on the Paris campus. "No problem," he replied. "We'll dig them up and take them along with us." Fortunately, no French colleagues were present to hear him utter this blasphemy!

If Jacques Monod's faults have been evoked here, that is because one cannot write other than honestly about such a man. At the risk of eliciting sniggers from readers for whom the very notion of the hero is a medieval anachronism, it must be said that he was of heroic stature. I hope this is evident from the preceding account of his life.

Roger Yate Stanier

 Roger Stanier obtained his Ph.D. under C. B. van Niel's direction at the University of California-Los Angeles in 1942. In 1947 he joined the faculty of the University of California-Berkeley Department of Bacteriology. His 24 years in Berkeley encompassed the period of his greatest scientific productivity. He made innumerable fundamental contributions to our understanding of microbes, especially with regard to taxonomy, metabolism, physiology, and structure of bacteria. He was a revered teacher and research director of students. His book *The Microbial World*, published in 1957 in collaboration with two Berkeley colleagues, had a worldwide renown.

In 1971, at the age of 55, he retired from the University of California and moved to the Institut Pasteur of Paris, where as a professor he worked as head of the Unité de Physiologie Microbienne for the last 10 years of his life.

Roger Stanier was a Fellow of the Royal Society of London, a member of the National Academy of Sciences (U.S.), and a Foreign Member of the French Academy of Sciences. He was awarded the Leeuwenhoek Medal and was named Chevalier of the Légion d'Honneur. He passed away in 1982 at the age of 66.

A Bit of Luck

Madeleine Jolit

In June 1945, I graduated in chemistry from the E. N. P., a vocational school in Bourges, and went back home to the Jura. A friend of my family, who lived in Paris, had just heard, from Parisian friends of hers, that a scientist, who was then teaching at the Sorbonne, was looking for a technician. In fact, a visit to Paris just after the war was the main attraction for me when I made the journey to the Institut Pasteur to present myself to Monsieur Monod.

I waited sitting on a stool in a tiny laboratory located at the top of the building under a mansard roof, when a very young man, to whom I first did not pay much attention, came in whistling. He was wearing canvas trousers and a sports shirt with an open collar. Still whistling he put aside a vial that he was holding in his hand and gathered a few papers that were scattered on the desk. Then he noticed me and introduced himself. Where had I gotten the idea that scientists were strict and severe-looking old gentlemen?

I only knew a little of mineral chemistry and had some vague notions of organic chemistry; my school year had ended by a period of training in a foundry. What was I doing in this service of "Microbial Physiology," whose obscure name did not mean anything to me? He looked at my notebooks, asked me a few questions, and then said, "At any rate, I prefer that you know nothing, because no school could teach you what we are going to need: I am in search of the secret of life." Astounded, I accepted the position . . . just for a try.

"It is not he, it is not he." That's all I could think of and kept repeating to myself during that endless morning when members of his family and a few friends were waiting in the garden of his home in Cannes to accompany him to the Jas (Cannes cemetery). No, it could not be

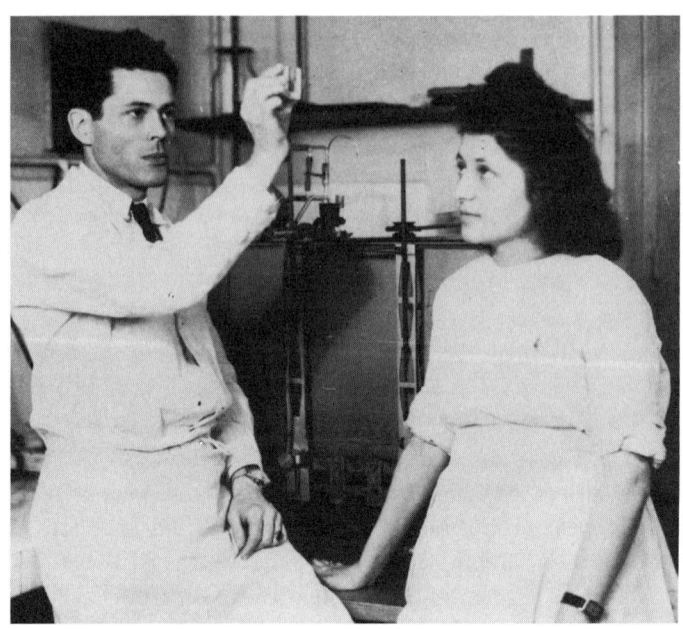

J. Monod and M. Jolit, 1946.

Monsieur Monod whom I had just seen lying motionless, absent. It was not possible. I could not realize it altogether, nor did I want to understand it, and yet I was already feeling deep in myself that it was all over—that never again would I find this extraordinary presence.

It is his silhouette in the lab that comes first to my mind, in the days when we were working in the "attic," when he used to stride into the lab, a smile on his lips, whistling, happy, just as he would do later into his "service"; whether leaning over the bench explaining the work to me, or sitting and dictating the numbers he was reading on the Warburg manometer . . . the first numbers only, for, as soon as he had understood the result, he would leave the rest of the experiment to me. We were working on the first assays of amylomaltase.

I can also remember his laughs mixing with those of Melvin Cohn as they happened to meet in the small cluttered corridor on which the labs opened, and the doors were rarely closed. Mel Cohn was always carrying out several experiments at the same time, so that, whether working beside the centrifuge, the incubator, or in the cold room, he kept moving about a lot. And he always had a minute to tell a story, start a conversation, or speak of his "last finding." Ideas were exchanged, but scientific matters were not the only subjects that caused the joyful explosions that made the lab resound with roars of laughter.

Lunches in the lab were also great moments, although savoring the food did not matter much to Monsieur Monod. It was a time when discussions started again, ideas were continuously arising, matters of everyday life were not excluded. Simone de Beauvoir and her book *The Second Sex*, which had just been published, provided, among other things, the theme of joyous polemics. By contrast, he had a respectful and admiring friendship for Louis Rapkine, who used to visit him quite often. They would enter into serious and calm conversations exchanged in a low voice.

I remember his arrival at the lab one Monday morning: "Madeleine, what is necessary is catalase! Yes, it is catalase which partly reactivates the bacteria sterilized by UV. I thought of that yesterday while climbing rocks at Fontainebleau." It was indeed the answer to all the strange results we had recently obtained and to that series of experiments which he had made me carry out again and again. Everything seemed to yield to his will; that is why I told him once, "With you there is no fun. We always find what you have announced." And I can see again the grave and amused smile which showed on his face at hearing this.

He always gave me several experimental projects at the same time. "Like that," he used to say, "there is always something that comes out right. It keeps up our spirit and in that way we get several results every day." At that time we were isolating mutants and carrying out the first assays of amylomaltase and galactosidase.

And then he invented the "bactogène," an apparatus designed— following the description he himself wrote on the patent request—to maintain "a continuous process for the cultivation of microorganisms, involving continuous and simultaneous addition of nutrient medium into, and removal of culture liquid from a fermenter. A method characterized by the fact that the culture is kept homogeneous." How enthusiastic he was when he talked of his toy! Every preparation was a festival, and yet it was not a small affair. René Mazé, the laboratory assistant, was in charge of this impressive machine, which occupied a place of honor in a large laboratory of a neighboring "service," in the midst of a maze of pipes. In a corner, a camp-bed was installed, where Monsieur Monod would spend the night to keep watch on the growth of cultures, because the three of us had to take turns during the two days which were necessary for the process.

And the same man who was busying himself, for two days in a row, around the bactogène was also capable of remaining completely silent for a long afternoon, laboriously isolating bacteria. I mostly remember him absorbed in his thoughts, absentminded. His absence of mind was famous and made all of us smile. I often would knock at the door of his office, enter, wait, then go out after a few minutes, unnoticed; I would have to come back later. Sometimes I would again knock at the door, and he would immediately grasp the telephone, say "Hello," and then, realizing his mistake, put down the receiver.

One day he came striding up the big stairs, as he did every morning, but laughing even more than the other mornings. Madame Odette Monod, who used to drive to work, had taken him in her car to the Institut Pasteur. When getting out he had held out some money to his wife . . . for the price of the fare! I do not know if she had accepted it, but I do hope she had, just for the fun of it.

Speaking of cars, there was a day when he could not find his own car. He had to declare to the police that it had been stolen; they soon found it exactly where he had parked it.

And yet, I remember what an attentive and passionate look he had when one day he entered the tiny lab where I was working holding the two bacterial strains which he had just received from Lederberg; one was a male strain, the other was a female. He began telling me about the sexuality of bacteria and got lost in dreams . . . he had already become a geneticist.

We were not very many in the group working in the attic; we were like a family. When I got married we had a celebration. For the Friday tea, Monsieur Monod asked Agnès Thébaut to cook two splendid strawberry pies in the sterilizing oven, and tea was replaced by Banyuls wine provided by Monsieur Lwoff.

Around that time, a professor of the Institut Pasteur, who was retiring, paid him a visit and took the opportunity to give him some advice about retirement pensions. It was with a fantastic laugh that he greeted this anxious kindness. "But I shall never retire, I shall never finish what I am doing!" "My young friend, you will see, time goes fast." To me it was perfectly obvious that Monsieur Monod could never be an old gentleman and give up laboratory work.

Later, the atmosphere changed when he created the Service de Biochimie Cellulaire, located on the ground floor, whereas André Lwoff and François Jacob remained in the attic. We were then a larger group; the laboratories were clear and spacious, with high ceilings and modern equipment. He got his own laboratory and an office next to it, new impressive apparatuses, and students who came to crowd the "service."

For a while, when measuring the optical densities of our cultures, we kept referring to the "blue of the Meunier from upstairs" (the old photometer which we were no longer using); some nostalgia was still lingering. But, after some time, we spoke solely of galactosidase. How many assays we made! Later I heard someone say that the success of galactosidase was due to the fact that its assay was yellow; "otherwise Monod would have given up," since he pretended to be color-blind and maintained that he could only recognize one color, yellow. Many times he gave us opportunities to make fun of his mistakes in this respect, but I suspect him to have greatly taken advantage of this defect, for he could very well differentiate every single shade, from white to purple, going through all the spectrum of pinks, of the bacterial strains on the eosin-methylene blue medium which was our daily bread. He nevertheless asked Madame Monod, whose good taste he trusted and admired, to come specially to choose the fabric of a curtain which was to hide some old cupboards in the library.

Monsieur Monod was busy setting up his new service, welcoming and taking care of his students both in the lab and at the university. Time was rapidly becoming short. He would work only occasionally at the bench now, participating in some permease assays, having a look at the petri dishes. And yet he was still fascinated by the laboratory, which attracted him from his office. I can still see him coming in with a cigarette in his hand, making a little round, reading the figures over my shoulder when I was installed at the Zeiss spectrophotometer or at the Geiger counter.

And then, as Monsieur Jacob says, came the time of the "i^-, o^c, i^s, y^- and z^+, and all that junk." Monsieur Jacob, anxious and ill at ease in his narrow lab, used to come down every morning with an "Everything all right?" Then both he and Monsieur Monod, after glancing at the petri dishes and reading the last results in my notebook, would resume their discussion. Monsieur Jacob would emphasize his remarks by making such funny faces that Monsieur Monod would burst out with laughter—at times, a break was necessary—but all of a sudden everything was beginning again. The same scene would repeat itself several times a day. In his office, comfortably seated in an armchair, I was taking notes for the pursuit of the work and Monsieur Jacob was pacing up and down the room. Monsieur Monod was writing on the blackboard. Next, it was Monsieur Jacob's turn to take the piece of chalk; after further information, everything was obvious and clear. Then Monsieur Monod would get up, erase the blackboard, and say, "No, François, we are going to demonstrate it differently, it will be more elegant: we shall start by showing that . . . " Completely lost, I was waiting.

Since that time, how often I have heard his colleagues, friends, and foes pronounce the word "elegance" when speaking of his works! And each time, the vision of Monsieur Monod standing in front of the blackboard comes back to my mind.

I also remember him with his students, discussing, criticizing, giving advice, either standing next to them at their benches or meeting them in the corridor, or again at lunchtime during the meals which were taken in common in the lab. At five o'clock he liked to have a cup of tea with us in the lab. Thinking that some work was keeping him busy at his desk, we would bring him tea to his office. He then would come out, his cup in his hand, asking us whether we were punishing him; for it was time for a break and all would take this opportunity to talk of their own problems. I heard two of his disciples say, "What is wonderful with Monod is that he puts himself immediately in the mood; we can corner him at any moment." On the contrary, one of his most brilliant students reproached him for abandoning him in quandaries and leaving him alone entangled with problems. But much later that same student admitted that it had been much better for him and that he had benefited a lot more from this attitude. But there were others whom Monsieur Monod would lead more attentively, encouraging them frequently.

In May 1968, he told me, "It is always the same story. When I was their age we were told that we were impossible. What would our professors say now? At any rate these young people are not at all that much convinced. I keep telling them, 'Why don't you get rid of François [Jacob] and myself? You should do it.' But they are only making noise."

After this period conflicts often arose. The traditional paternalistic attitude had to be given up. One day when several young research workers were vehemently arguing, I made the remark that in Monsieur Monod's time they would not have had such ideas. One of them, probably the most obstinate, replied, "Monod is out of the question, because *he* had an indisputable authority."

This authority was felt by all the personnel. We had a washing room where eight women were busy sterilizing the laboratory glassware and preparing culture media. He had placed me in charge of the organization of this "kitchen" and never used his authority to interfere in the quarrels that might arise. Tremendous work was actually done in the kitchen, which was nonetheless carried out with songs and laughter. He would smile at everyone and welcome this cheerful agitation. I saw several of these women cry when we heard the news of his death, and one of them told me, "Why did we like him so much, Monsieur Monod? He did not spare us, how hard he used to work us!"

The news of the Nobel Prize had been a great festival for all of us. The lab looked like a huge work yard where everyone was looking for everyone. All the photographers wanted to take the snapshot of the day; seated near the Zeiss counter we were waiting for the picture of the laureate "hard at work" to be taken. I seized that opportunity to ask him, "Are you happy to have this award?" He did not answer me immediately, but after a while said, "Yes, for all the little things that flatter a man's vanity, I am pleased. But it will be heavy to bear; think of it, Madeleine, from now on I shall have to think very carefully and to pay great attention to what I shall say. It is a new responsibility; it will no longer be my personal opinion but that of a Nobel Prize laureate." And he fell into deep thought. And then, just a little later; "Madeleine, I shall take you to Stockholm with me to receive the prize. It will be to celebrate our 20th anniversary in the lab." Of course I thought that he was out of his mind with excitement. But he did take me to Stockholm, together with the other Madeleine (Brunerie), his secretary, and we participated for ten days in all the ceremonies and feasts which usually take place on such occasions. He also gave me a part of his prize and wrote on the check, "This is your share of the N. P." After the royal ceremony it was snowing a lot and to return to our cars we had to go through a narrow passage and a small back staircase. At a curve of the passage Monsieur Monod was there holding Madame

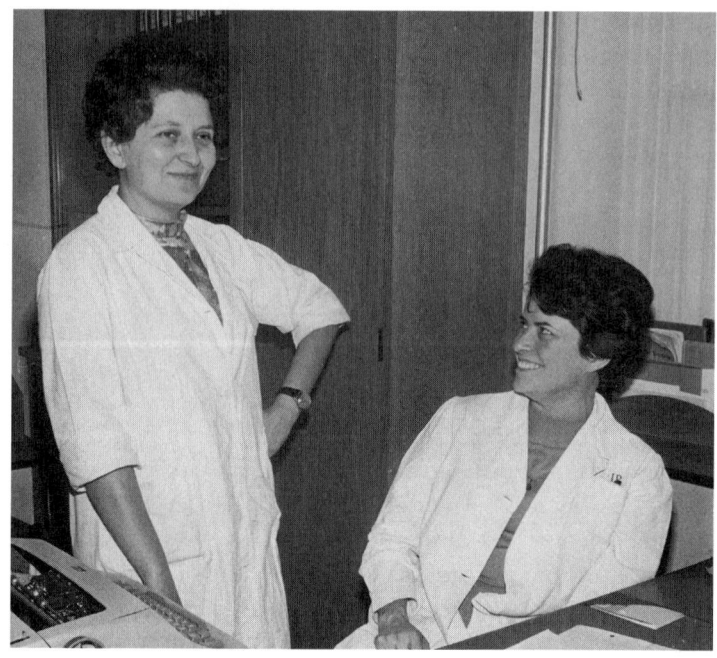
M. Jolit and M. Brunerie, ca. 1965.

Monod's hand and exchanging with her the most happy and radiant look.

I have to make an effort if I want to remember him as the director of the Institut Pasteur. I paid him several visits in his huge, sad, empty office, so different from the happy pandemonium of a lab. He always greeted me with warm and joyful friendship. After asking me news of my family, of Maxime Schwartz, his disciple with whom I am working now, of our work, he would soon go back to the "good old time," that is to say the time of the attic. Then memories would come back with the same images, the same anecdotes which did not age for us who had lived them, and each time they made us laugh just as much. And he would say, "You know, Madeleine, it has been an extraordinary time and we were all so happy. It is thanks to this unique atmosphere which we all had created. Such a group is very rare. I have known many labs in France and abroad, and I have never seen that. We have been very lucky."

Madeleine Jolit

 After the awarding of his Nobel Prize in 1965, I continued to collaborate with Jacques Monod, but as he became more and more busy, he asked Maxime Schwartz, one of his students, to supervise my experimental work. In 1971, when Jacques Monod became Director of the Institut Pasteur, for a while I was working with both Maxime Schwartz and Maurice Hofnung. I still remember their great qualities and I have excellent memories from this period.

Later on, I collaborated exclusively with M. Schwartz, who headed the unit of Molecular Genetics. He abandoned me because he also became, in 1983, Director of the Institut Pasteur. I continued to work in the same unit with a new postdoctoral student (C. Desaymard) until my retirement in 1985 after 40 years of work and satisfaction.

Portrait by Lucien Monod, 1940.　　　Monod in the military, 1944.

Aboard the *Pourquoi pas?*, 1934.

Once Upon a Time . . .

Madeleine Brunerie

Once upon a time, there was a teenage girl whose most beautiful dream was to meet a great scientist. I was that girl. And in late October 1946, when necessity made me look for a job, I wonder if it was only chance which led me to work at the Institut Pasteur, which I had once visited when I was studying in a high school in Paris.

Today, I shudder remembering how much I had wanted to leave the Institut Pasteur after the death of my first patron, Professor Michel Macheboeuf, in 1953. I owe it to his close family that I stayed on, and I anxiously waited almost a year for his successor.

On June 29, 1954, in the reserve room for laboratory material of the old Service de Chimie Biologique, Jacques Monod appeared forever in my life as a charming and very courteous man. I first believed I would not be as impressed by him as I had been by Professor Macheboeuf. It was true: he was much smaller in stature.

When I told him how much I wished to work as a technician, Jacques Monod answered quite frankly: "You can choose your job; I need a secretary as well as a technician." I then thought he had given me the chance of my life. A week later, letting him know I had chosen to work with him as a technician, he promptly argued that above all he needed a secretary. The die was thrown . . .

I cannot remember for how long I had a grudge against him, if indeed I ever had. A few days later, on vacation, I started learning English shorthand in order to help him more efficiently as a secretary.

From then on, day after day, he impressed me more and more. I was discovering the scientist as well as the human being. I cannot say how many times I had the opportunity to appreciate him as a great-hearted man. Competent people will think of him as a scientist. But I would like to emphasize that even if I could not understand very well the fundamental researches he was pursuing (which was, in fact, molecular biology pioneering), it soon (1955) intuitively appeared to me, from the reports he dictated, that Jacques Monod and his team were doing very important scientific work. From that moment, he kept me enthusiastic about the researches he undertook, and I waited faithfully for the results he expected.

Sometime in late October 1959 I understood that my secret hopes might come true. A close friend of Jacques Monod wrote me confidentially, asking for a curriculum vitae and reprints to support a proposal for a scientific prize to be shared with André Lwoff (in fact the Nobel Prize). I felt it was a confirmation of my high opinion of him, and afterward, each fall, I waited hopefully for Nobel news.

From his correspondence, I knew that Jacques Monod was an eager supporter of human rights, and I even participated in one of his plans: the "extraction" of Agnes Ullmann and her husband from Hungary. When Agnes left, after her second stay in the lab, in 1959, Jacques Monod entrusted me with the organizing of secret communication with her; for instance, news from Budapest reached us, hidden in a Béla Bartók record sleeve. As soon as the plan was ready, he left for Hungary to try and make sure the operation would succeed. In fact, nothing was certain until that morning in August 1960, when I stood waiting for Agnes and Tom at the Gare de l'Est, in Paris.

Work continued for everyone on the cellular biochemistry team. I loved the life in our very active laboratory, whose soul was Jacques Monod. Experimental work as well as new ideas or concepts were discussed at length not only in his office, the labs, and the library, but more often just about anywhere else. I mean in the dining room, during lunch or afternoon tea, and in the corridor which was called the "pensotron" ("thinkotron"). "Modesty befits the scientist, but not the ideas that inhabit him," Jacques Monod used to say. Discussions were usually very animated, and sometimes funny. They often were punctuated by jokes, and I used to enjoy hearing, for instance, Mel Cohn and Jacques Monod bursting out with laughter!

I was supposed to protect him from too many "undesirable" people from outside, but he always wished to be approachable by any of his coworkers, collaborators, students, or scientists (even from the outside) who needed his advice or wanted to discuss scientific results or personal problems. Unfortunately, year after year, with his increasing responsibilities, I know it was one of his deepest regrets not to be as available for this kind of contact as he would have wished to be.

M. Brunerie and J. Monod, 1965.

To be his secretary was very exalting. It has been for me a source of both enrichment and modesty. Because of the nature of this job, I grew to know more about him and his close collaborators and friends, sometimes more than I would have liked to. But at the same time, it created a confident symbiosis between us and this was most precious to me.

As a secretary, I very much appreciated the fact that Jacques Monod was a fine writer in French as well as in English (his mother was born in Milwaukee of Scottish and New England parents). Publishing either a scientific work or a book (*Chance and Necessity*) usually took him a fairly long time, as he would not hesitate to rewrite several times parts of the manuscript to make it as perfect as possible. He was as rigorous for himself as he was for others.

But I should also say that he often tried to spare me the retyping of some letter or report on which he had added some minor corrections: "Please, Madeleine, don't bother to retype it: it really does not matter."

While he wrote his publications and the chapters of his book in longhand, he preferred—in order to save time—dictating not only current letters, but also reports, grant applications, and sometimes lectures, either in French or in English. Endowed with a great readiness of mind, he had at the same time the capacity of isolating himself when necessary. I remember some entertaining moments when he paced my small office, dictating at length. He did not seem aware of anything happening around him, following only his thoughts. Several times, at tea time, not only did he finish his cup of tea and cakes, but also mine. If somebody knocked at the door, he frightened me by shouting, "Come in!" and went on with his dictation, either not seeing or forgetting the visitor, who, after a while, just tiptoed out.

I will never forget another great moment when I was taking down in shorthand an application for a grant, when suddenly Jacques Monod stopped dictating and began writing some mysterious formulas and drawings on the blackboard. I was fascinated by this interlude. Suddenly, coming back to earth, he then exclaimed, "Look at this! I have it!" My pencil at rest, I felt fired with enthusiasm, not daring to move a finger for fear of dispelling his train of thought. In fact, he soon went back to his lab to discuss the new idea or concept just born in his mind.

It was a pleasure to work with Jacques Monod. He was a very patient patron, and he always remained courteous and thankful. He knew I was doing my best for him, so when he saw me overwhelmed with urgent work, he told me: "Don't worry! When one has done everything one can do, one has done everything one was meant to do."

In August 1963, he was sailing around Spain for his vacation and could not reach Marseille in time to attend a symposium there. I knew the cable he sent from Gibraltar to excuse his absence was received with some irony by the participants, and it hurt me deeply. A few days after he returned to Paris, he accepted halfheartedly an invitation to participate in some summer school program the following year. I told him I disapproved of his reply, guessing he would cancel it later on, as he had done for the Marseille symposium. He did not like my remark at all. But finally, he declined the invitation. For some time after, in similar cases, he used to call me "Ma conscience."

On Tuesday, October 12, 1965, owing to some leakages from Swedish sources, Jacques Monod revealed to me that the "Prize" had some chance to be given to André Lwoff, François Jacob, and himself. In the lab, only a few people shared the secret. Therefore, when the cable from Stockholm arrived the following Thursday, the surprise was great. Everybody, both in the lab and in the whole Institute, welcomed the news. For several days, the team of close collaborators lived in a kind of whirl. I thought I had imagined the perturbation such an event could raise, but I discovered how far I was from reality. Jacques Monod himself told me, four days later: "It's funny. Since the Nobel was announced, my little inner dialogue has been cut off, and I felt rather sorry. Happily it came back this morning!"

This is where my fairy tale really starts.

A week or so after the official announcement and those terribly exciting days, Jacques Monod and I were filing the Nobel Foundation forms for the ceremonies in Stockholm, in December. He gave the name of each member

M. Jolit, J. Monod, and M. Brunerie, ca. 1965.

of his close family and, looking at me affectionately, he said: "Please, add Madeleine Brunerie: I will need my secretary in Stockholm," leaving me no chance to believe he only did it to please me. That was one of the happiest days of my life. It sounded like a dream. I thought I knew him well, but I did not measure his generosity until that moment.

He, of course, invited Madeleine Jolit, his lifelong technician, to join us. What a happy and exciting period we both experienced, "the other" Madeleine and myself!

During ten days in Stockholm, we both shared every honor rendered to Jacques Monod, who not only paid all the expenses of this unforgettable event, but gave each of us a large check on which he wrote, "Your part of the Nobel."

Jacques Monod was generous with youth, to whom he devoted most of his efforts and his time, and he never refused to receive young students interested in biology. To the best ones who wanted to become scientists, he first gave a very black picture of the career, especially to the girls. "If the future prospects of such a job do not frighten them, then they can try their luck," he used to conclude. How many owe him the choice of a scientific career?

He was also generous with people he did not even know. He never answered Christmas cards, but he never failed to reply to letters from troubled or ill people who asked him for moral help or medical advice. Not a physician himself, he always tried to give either the adequate information or the opinion from a specialist.

In the same way, this great man, sometimes terribly impressive, had the gift of being accessible to modest people, who had a profound and sincere affection for him. When he sized someone up, he did not dissociate intellectual qualities from human ones. He told me several times that one finds exactly the same proportion of able and respectable people as well as stupid and worthless ones in every class of society, whether among academicians or among street sweepers.

Jacques Monod was not only a good chief, he was also a warm friend to his close collaborators and fellow workers. I had the opportunity to realize this many times in rather sad circumstances, for instance, when one of our American postdoctoral fellows had a dramatic accident in which his young boy was killed; and in my own case, when I suddenly lost my father in 1963, and then during the long illness of my mother, as well as when she died in 1967. Everybody knew they could rely on him, and how precious his discreet and efficient help could be.

Very fond of literature, Jacques Monod also liked the theater, and often went to see plays even given by amateur companies (he came several times to the performances of the one I belonged to and encouraged us). In 1964 or so, he himself wrote a play entitled *Le puits de Syène*, a kind of antique allegory dealing with the relationship between science and society, and more specifically debating the question of the scientist's responsibilities and his connection with political power. After the Nobel was announced, he told me, half jokingly, he was happy to be a laureate, adding that he would have been happier still if his play had been produced!

I shared the best; I shared the worst too. I was near him in the office when he learned on the phone from Professor Jean Bernard in February 1971 the diagnosis of his wife's inexorable disease. He at once told me about it, and I admired him even more, if it was possible, for his courage. Eleven months later, Odette Monod passed away.

On April 15, 1971, Jacques Monod became General Director of the Institut Pasteur. Accepting this leadership, he undertook it with the same passion he exerted in whatever he did. I kept his personal and scientific secretaryship, hoping I could go on working for him when he retired, after six years of directorship, in order to write his second book, *L'Homme et le temps*. He only had time to put down a number of ideas for it on some scattered sheets of paper, found on his desk at home.

He often thought of his retirement, and discussed it several times with me. And because of this idea (or because he thought his illness was inexorable), Jacques Monod wanted me to become the assistant of Joël de Rosnay, a man he first met at the Institut Pasteur, and whom he had chosen long ago to help him give a new life to the prestigious Institute. How can I be thankful enough, now, for making it possible for me to be associated with the great enterprise he started for the salvation of the Institut Pasteur!

I keep in my heart two precious images of Jacques Monod when, in 1976, I went to his home, during his illness, to work with him.

The first one occurred sometime in January. Among the mail I brought to him was a letter from a very young boy called Bruno, who wanted to know what was the most important thread of his life. Jacques Monod dictated to me: "It is courage and love of truth, or rather, the hate of lies."

The second occurred on Thursday, May 10. We were working on his ultimate publication (Catabolite modulator factor: a possible mediator of catabolite repression in bacteria, A. Ullmann F. Tillier, and J. Monod, 1976, *Proc. Natl. Acad. Sci. USA*, **73**, 3476–3479) with Agnes Ullmann. He was seated at his desk, discussing with Agnes and then dictating to me some corrections, just as during our best moments at the Institut Pasteur. He looked much better. But did I inwardly feel he was near the end? I could not help staring at him, fascinated by the great man I knew. The manuscript being completed, the three of us went out for a walk through the Champ de Mars with Vicky, his young and impetuous dog. Then, we shared a delicious paella prepared for us by Rosario. During the meal, he told us he could not bear the idea of losing, because of illness, either the use of his brain or of his limbs....

A few days later, on Monday, May 31, 1976, Jacques Monod died in Cannes, where he had gone for a rest. It was a relaxed and happy holiday before the last attack of the inexorable disease, which, in a few hours, carried him off, leaving without an answer his very last question, "Je cherche à comprendre," and leaving behind his family, friends, and close collaborators facing an irreparable loss.

Can the thread which led me to the Pasteur Institute, and then to Jacques Monod, be called chance? Let us say it can. In fact, I encountered a double chance, not only by working with one of the world's most famous biologists, but also by living for almost a quarter of a century in close contact with a great man who enlightened, and still enlightens, my life.

Madeleine Brunerie

Thanks to Jacques Monod, on April 20, 1976, I became the assistant of Joël de Rosnay (later named Director of Research Applications). Joël taught me how to use a computer, for which I'm grateful to him every day! After he left the Pasteur Institute I started to classify Jacques Monod's archives. Historians being interested in molecular biology, the Direction of the Institute decided to create (April 1987) the Service des Archives, headed by a curator of the Archives nationales. When I retired from the Institute (1990), I was particularly authorized to work full-time and freely on Jacques Monod's archives. The "fonds Monod" was expanding and since 2001 has had a website: http://www.pasteur.fr/infosci/archives.

This collection is quite exhaustive with around 100 boxes containing Monod's biographical details; laboratory notebooks; manuscripts (some of them unpublished) and texts of lectures; an important collection of correspondence, especially scientific (around 3,000 letters); notes from the Service de Biochimie Cellulaire which Jacques Monod headed; and papers on university reforms in France and personal opinions on political affairs.

I am now writing my own memoirs (from personal notes since the last war) to pay a tribute to my modest family to whom I owe so much, and also to pay a warm tribute to my illustrious "patron" and to the scientific community who gave shape to my craziest teenage dreams. I think I would be very unhappy if life did not let me end this ambitious enterprise...

Le Labo de Jacques

Annamaria Torriani

A mansard roof covered the top floor of the main brick and stone building at the Pasteur Institute, which comprised a wide wooden bench in
the center, another long bench under the slanting ceiling, and a small desk
on the other side—my desk. Jacques, young and living as if on a tightrope,
was exploiting his energies, his intelligence, his charm, to the maximum.
He would move quickly along the corridor in short jumps (a way of dis-
guising a small limp from polio in one leg), always whistling the same part
of a Bach cantata. Arriving from Italy (February 1948) and leaving behind
a profound desperate love and a most warmhearted, serious companion,
I found the charm and excitement of Jacques a bit obnoxious. Here is my
first written account of him in a letter to Luigi.* "He was teaching me how
to use the slide rule and took advantage of a good occasion to gently hold
my hand in his (a beautiful, nervous, expressive hand with a golden ring
and a green dark stone). It was a wrong approach, but it was gently done.
For two hours he explained to me his work and our project. I have the impression of an intelligent man, with a vast, profound, and personal view of
biology, but a bit fast in his mathematical deductions. As an enzymologist
and chemist I don't think he is a genius. But . . . I may be totally wrong."
For a while I was his only assistant and it was a real pleasure to see his
mind at work. He had recently joined the Institut Pasteur as a member of
André Lwoff's department. His thesis became my bible and was in fact the
point of departure for our early research.

* When I went to Paris, Luigi Gorini, who was to become my husband, was in Milano. After
the antifascist underground, he had to start anew his interrupted scientific career, which
would then bring him to Paris (Laboratoire de Chimie Biologique, Sorbonne) and to Boston
(Department of Microbiology, Harvard Medical School). His achievements in science and
in the advancement of human rights are well known to his friends and to the scientific
community.

55

A very young woman (Madeleine Vuillet, now Mme Jolit) who had just finished her two years of school as a chemical technician was helping us. She knew all about "le labo" and was a great help. She had a very even, steady character and was an anchor for me in a new and stormy sea. Jacques had a small office just big enough for a desk, a bench, and a table, around which we usually had lunch. Food was still rationed and everyone cooked or warmed a small meal. Lunchtime was the most exciting time. With us were usually Pierre Schaeffer, tall, blond, and gentle, with a short trumpeting laugh; Elie Wollman, short, peppery, and always well informed; Ileane Jonesco, friendly and lonely. Bussard frequently joined us, as did Louis Rapkine for too short a time (since his short life was at the end). The talks were mainly about highly controversial subjects like politics, DeGaulle's past and present roles, and the daily editorials published by *Combat*. But music, art, America, "the atomic scientist," and Lysenko were also part of Jacques' interest and part of our arguments. Never in these spirited lunchtime conversations did I hear anyone talking about families or everyday difficulties, personal problems or . . . solutions. Lunch was for intellectual pleasure—but sometimes the language or local jokes constituted a frustrating barrier for me. One day at lunch, shortly after my arrival, Jacques asked me if there were many atheists in Italy. I told him that the question was difficult to answer, since I thought that man is incapable of real atheism. To which Mme Lwoff (Marguerite) smilingly added, "That makes sense: Jacques has the cult of himself!"

It was in the laboratory, between one series of Warburgs and another, while thinking of the results, that Jacques would say to me (or to himself), "Mon petit, j'ai raté ma carrière, j'aurai du être un grand couturier [he loved elegant women's dresses] ou bien chef d'orchestre [he was proud of having been one during his time in the USA]." Of course he would say that after an experiment in which his brilliant serendipity had been successfully challenged.

The laboratory was blessed by the efficient, industrious, and intelligent help of a technical factotum: Mazé, always serious in his work, yet smiling, friendly, and pleasant. He could do everything—bleed a rabbit or fix countless Warburg flasks, side-arms, hooks, and all in a magistral fashion. The smooth flow of research was supported by a group of hard-working and competent women from Bretagne: Mme le Naour (the boss—I never knew her first name), her charming sister Agnès, and friendly Mimi. Day after day they boiled, rinsed, and sterilized heavy buckets of glass petri dishes. They also cut glass tubing and pulled it through the flame to make hundreds of Pasteur pipettes. Who was Mr. Petri? No one ever asked, but Monsieur Pasteur! No one would have dreamed of writing Pasteur pipette without a capital "P," as I see it written now in the American catalogues!

In the laboratory population there was, every year, a substantial dose of English-speaking scientists to whom André Lwoff would only speak French. He required them to answer the same, constantly (but not always patiently) correcting them. Jacques found it more comfortable to exchange opinions in English (which was his mother's tongue), never answering in French to their painfully constructed, often incomprehensible phrases. This made me furious: I had come to a French-speaking laboratory to understand science, not to learn English. But at the end, through all that hard listening, some English rubbed off and was of great use to me. Our guests were all pleasant, very friendly, and a bit lost. They frequently complained of diarrhea at first and then of the lack of friendliness toward foreigners in Paris. Seymour Cohen was the first American I met there when I arrived. He looked sort of lonesome, was very sweet, and danced very well. Then came Martin Pollock, preceded by long discussions and hypotheses on his very special system: the mechanism of control of penicillinase. Jacques was excited by Martin's results, and liked his high-class British accent and manners. With Martin we were constantly measuring penicillinase activity in Warburgs. Jacques would come in behind us and put a friendly hand on our shoulders, see the partial results in a glance, and rush to the corridor to discuss them with "the others" (there were always two or three people ready to talk in the corridor). Martin, thus deprived of the best pleasure in science, was getting upset: "C'est le lab oratoir!" Later came Mike Doudoroff. It made us all very happy since he was one of our scientific gods. Tall and bent with a smiling face and swollen hands which appeared clumsy, but were not, he was frequently inebriated, but always intelligent and warmhearted. We were purifying the amylomaltase and the "precipitate" had precipitated on the floor! Mike was accustomed to misfortunes of that kind and quickly sucked it up by vacuum and used it! He also would open a Warburg flask every so often to add another substance he had just thought of, which could modify the reaction. At the end I had no idea what was going on, but he did, and Jacques, who was a very neat and precise worker, nevertheless enjoyed the messy, bubbly way of Mike.

Mel Cohn, Seymour Benzer, Francis Ryan, Lane Barksdale, Alvin Pappenheimer, Dave Hogness, Aaron Novick, Cy Levinthal, and Howard Rickenberg, each brought a new personal input of experience, intelligence, and knowledge to the development of research. Each created lifelong friendships. Jacques would tell me that "they [the Americans] are the experts,

Luigi Gorini, Jacques Monod, Annamaria Torriani, and Mike Doudoroff, 1948 or 1949.

I am a self-made man." In a way it was true. But his originality and spark was of continuous excitement and provided profitable discussions.

The core of the question in those days (1948–1950) was to understand why there was an increased rate of enzyme formation upon addition of the substrate (adaptation) or a diauxic inhibition. The working hypotheses in the lab were that "many different enzymes may stem from a common precursor or pool of precursor molecules" (*Growth*, 1947) and that the "master pattern configuration determining the specificity was not the enzyme itself but a pre-existing self-duplicating unit (the gene)" (*Annu. Rev. Microbiol.*). One day Monod suggested an experiment to demonstrate that enzyme and gene were two independent entities. The reasoning was the following: A mutation provoked by UV irradiation means a modification in one or more particles (or molecules of enzyme). Bacteria adapted to maltose, for instance, have a very large number of enzyme molecules, while the unadapted have very few. If a mutation required the inactivation of most of the enzyme molecules, it should be much easier to obtain such mutation in unadapted than in adapted cells. If, on the contrary, the gene hypothesis was correct, then the probability of mutation should be independent from the adaptation. But Wollman objected with the alternative hypothesis that the unadapted cells may contain a large number of inactive precursor molecules which would have to be eliminated to detect the mutation. Thus an equally low frequency of mutants would be expected in both adapted and unadapted cells. The experiment was never performed, like hundreds of others eliminated by active intellectual exchange: it was like a continuous game of checkers. Jacques dominated the scientific discussions, particularly at seminars organized at Rue Pierre Curie (now Rue Pierre *et Marie* Curie) and jokingly called by friends "le Club des Monod-theists." The seminars were followed by dinners at the Brasserie Alsacienne, where Nissman, Szulmajster, and Latarjet rivaled with Jacques, Cohen, and Bussard in the exchange of "contrepetries," i.e., scrambled words with hidden, mostly sexual, double meanings.

In the laboratory the exciting moments were many! With Jacques I learned that in science one gets excited every day: either by a new hypothesis or by the results supporting it, or by those which one day later will shake the hypothesis and require a new one. It was an exciting moment when, one day, at the end of an experiment, the addition of iodine to a Warburg flask unexpectedly produced a deep blue color (Monod and Torriani, 1948). We were studying the activity of the adaptive enzyme maltase on maltose (Gl α1, 4–D-glucose); no starch was expected to form, since we strongly believed in the recently described "Cori ester" (Gl–1–PO$_4$) as a requirement for starch synthesis (Cori, 1945, *Fed. Proc.*, **4**, 232). Why did we add iodine then? To measure the hydrolysis of lactose and of maltose, Jacques had the idea of using a glucose oxidase (or Notatine from *Penicillium notatum.*) and measuring in Warburg the O$_2$ liberated by the following chain of reactions:

$$\text{maltose} \xrightarrow{\text{maltase}} \text{glucose} \xrightarrow[\text{oxidase}]{\text{glucose}} \text{gluconic acid} + H_2O_2 \xrightarrow{\text{catalase}} O_2$$

(Monod, Torriani, and Gribetz, *Comptes Rendus*). It worked beautifully from lactose but from maltose it gave half of the expected glucose. Could it possibly be that one glucose from each maltose hydrolyzed by the enzyme was utilized to form a polysaccharide? It was, but we could not have seen it in the absence of Notatine because the accumulation of glucose inhibited the formation of starch-like material. We were very happy and Jacques wrote a nice note on the use of Notatine, only to find out, opening the *Biochemistry Journal* that day (1948, **42,** 230), that Keilin and Hartree had just published the very same method! Disappointing, but "quoi qu'il en soit" (as Jacques frequently wrote in his papers); at any rate, the interesting fact was the synthesis of polysaccharides. To clearly convince ourselves that the starch-like substance was produced by a single enzyme and that it was produced in the absence of any phosphate took us quite a few months, until we purified the enzyme in total absence of phosphates by using Veronal or citrate buffer (Monod and Torriani, 1950, *Ann. Inst. Pasteur*, **78**, 65; Torriani and Monod, 1949, *Comptes Rendus Acad. Sci.*, **228**, 718). We had discovered a new enzyme in *E. coli* and Jacques, who was terrific at inventing new all-explanatory words, baptized it "amylo-maltase." This was probably the only time I felt stronger than Jacques! To do the balance of the reaction and to analyze the products we needed some chemistry; if mine was not great, Jacques' was worse. He did not pretend otherwise! For instance, to eliminate the iodine used for precipitating the polysaccharides we used alcohol. Jacques was boiling the mixture over an open flame to eliminate the alcohol; I suggested a reflux column, which helped a lot! The biochemistry I knew; I learned it from Luigi.

Why were we studying amylomaltase and lactase? The working hypothesis was, as mentioned earlier, the existence of a preenzyme which was a precursor to both lactase and maltase and possibly to other enzymes. We studied also a trehalase which was never worked out further. The approach was to study a number of inducible and constitutive enzymes. For each enzyme, mutants had to be isolated, and Madeleine was doing this work directly with Jacques. The enzymes had to be extracted and purified and their molecular and catalytic properties analyzed. Immunological analysis of wild type versus mutants would suggest if a precursor existed. The rate of synthesis would be compared for each enzyme when induced singularly or together. Jacques outlined this whole program for me the second day I was with him. And all of it was done by his group in the eight years I spent with him, and so much more!

Our experiments required a "system artificially set up by establishing a constant limited supply of an essential metabolite, while all other nutrients would be in excess" (*Annu. Rev. Microbiol.*, 1949, **III**, 371, see pp. 378 and 386). Furthermore, H. Virtanen had just published a paper in which he suggested that in conditions of limiting N source, "the proteolytic enzyme of *E. coli* retains its activity [while] the dispensable enzymes, saccharase and lactase, decrease very sharply . . . The adaptive enzymes which are necessary only in definite nutritional conditions seem, as a rule, to decrease or disappear with the lowering of the nitrogen content of the cells" (Virtanen and Winkler, 1949, *Acta Chemica Scand.*, **3**, 272–278). Jacques thought that to keep a culture in continuous exponential phase would help to understand the kinetics of enzyme adaptation. Was it autocatalytic or not? And was

Virtanen correct? On April 11, 1949, we spent the day at 37°C to give birth to the "bactogène"! Once the culture was in exponential phase on glycerol as C source, we started diluting ("rejuvenating") by half with fresh medium and lactose every 30 minutes. The results were encouraging and Jacques wrote in my notebook the theory which was the basis of the bactogène (Monod, 1950, *Ann. Inst. Pasteur*, **79**, 390–410). A few days later we were repeating the experiment using a rotating flask with a simple method of feeding and collecting. This method, or a very similar one, was also devised by Novick and Szilard (*Science*, 1950, **112**, 715–716) almost at the same time and gave us and them and many others long years of pleasant work.

Life around Jacques was exciting, challenging, and not always easy. Those whom he did not like or did not consider intelligent enough Jacques looked straight "through" as if they were transparent and did not exist. If they were rude enough to ask a direct question repeatedly, he would wake up from a sort of dream and snap a short phrase at them. He had a particular disdain for "les sorbonnards" (the Sorbonne faculty): "Ils sont bêtes comme mes pieds!" So they, the Fac Profs, chose not to have his arrogance among their ignorance!

Sometimes he would arrive at the lab by taxi, borrow from me some francs for the fare, and never repay me, not knowing that I was in a sort of voluntary exile and very poor indeed. If I asked him, he would of course rush to give the money back, but he would forget about it between the time he put his hand in the pocket, which was frequently empty, and the time he took out the coins, since something more interesting, such as a result or a telephone call, would certainly intervene.

On Sundays Jacques would frequently go to the forest of Fontainebleau: sneakers, rope, a little rug, and some pof-pof (a rosin powder) to climb rocks that rose only a few feet high from the sand in the forest. It was all a ritual: the rug and the pof-pof were necessary to clean the sneakers from the sand, which made the climbing very difficult. The same sand, properly washed, made the extraction of enzymes so easy! I remember going there with him and Jean Weigle, both well-trained climbers. The next day my arms were stiff since it had been many years since I had climbed the Dolomites, and even the Warburg flasks seemed very heavy!

Thirty years passed. Rapkine, Weigle, Ryan, Doudoroff, Mimi, are all gone. Luigi left us shortly after Jacques. Luigi was my whole life and Jacques was champagne to our intellectual pursuit.

Annamaria Torriani

 I was born in Milano on 19 December 1918 during the huge Spanish Flu pandemic, the daughter of Carlo Torriani (of Milanese nobility dating to the XIIIth century) and Ada Forti (of an ancient Jewish family). I was 10 years old when my father died of cancer. During the Second World War, my mother, sister, and I were in Milano during the day, but at night we were "sfollati" a few miles away to an ex-convent of medieval origin. It was of solid stone, big and cold (we were guests of the family of a friend, Merlini). There we could admire the umbrella of flares that the U.S. airplanes were parachuting over Milano before the bombing of the city. The earth was shaking and the town was 30% destroyed.

From 1941 to 1948 with Luigi Gorini (a biochemist and socialist, born in Milano in 1903) we studied the synthesis of penicillinase by *Escherichia coli*. From 1945 to 1948, Luigi and I were also busy organizing and feeding children who had survived the Nazi concentration camps, in a former Fascist summer house for children in the pre-Alps (Selvino, Bergamo). In the course of three years, about 800 children (in groups of 200 to 300) recovered physically and psychologically. By the end of 1948 they were all gone, by boat, to Israel, the "Promised Land."

In 1948 I went to Paris, to the laboratory of Jacques Monod. In 1956 we went by boat to New York, to work at NYU Medical School. It was a trip with no return. Luigi and I got married in 1959 at Reno, Nevada, and in 1960 Daniel was born. I was a research associate in biology at MIT; Luigi was professor of biochemistry at Harvard Medical School. I had a good group of collaborators studying the regulation of enzyme biosynthesis in microorganisms. In 1971 I became associate professor; in 1976, full professor; and professor emerita in 1989.

When Luigi died of leukemia in 1976 and Dan was in college at the Rhode Island School of Design, I was alone and my passion for mountain and rock climbing (in the Dolomites in 1928–1939) reemerged. I met Arlene Blum and, inspired by her book about the tragic ascent of Annapurna, I started trekking in the Himalayas of Nepal. I made one trek (of four or five people) every three years between 1980 and 1997. In 1999 and 2000 I traveled to Italy and Greece, including Crete, in summertime with my grandchildren (Marco and Nika). Now, in 2003, Dan is married to a lovely young Indian wife. I traveled a lot in India, and many of my collaborators (in particular, Narayana N. Rao) have been from there.

Remembrance of Things Past

Germaine Stanier (Cohen-Bazire)

'L'art du chercheur, c'est d'abord de se trouver un bon patron.'
A. M. LWOFF, CITED BY F. JACOB IN *Of Microbes and Life*, 1971

During my university career at Toulouse, I had the exceptional luck to come under the influence of Monsieur Vandel, a man who offered a five-year course, continuously renewed, in general biology. It was then that I decided to become a research biologist, with the naive ambition to understand how a cell became two and how cells differentiate.

At that time, and indeed for several more years, no training in biochemistry (apart from the medical course) was offered in Toulouse to students in science. M. Vandel suggested that it would be desirable to learn this subject if I wanted to pursue a meaningful career in biological research. He accordingly wrote to Professor Javillier, who held the chair of biochemistry in Paris, and asked him to take me as a graduate student in his laboratory. The episode lasted, all told, for eighteen months. Javillier, close to retirement, had delegated younger assistants to help out with research. The lectures, exclusively concerned with structural biochemistry, were held in overcrowded lecture halls. I was supposed to study the hypothetical relationships between lycopene and vitamin A_2 and was provided with a barrel of spoiled tomato juice from which to extract lycopene.

I shall always owe a profound debt of thanks to Professor Edgar Lederer, my godfather at the CNRS, who listened to this sad tale and remarked, "You have a hopeless problem." He did not, however, suggest another one! Moreover, I was a long way from learning biochemistry.

When, in 1947, the late Professor Fromageot succeeded the chair of biochemistry in Paris, a general cleaning-up of Javillier's laboratory ensued. By that time, however, I had already renewed an acquaintance with an old friend, Georges Cohen, first encountered as a fellow student after the debacle in Toulouse. Following the completion of his doctorate in Monsieur Macheboeuf's laboratory at the Pasteur Institute, Georges was starting a new field of research: the analysis of fermentation products of clostridia, in Monsieur Prévot's laboratory at Garches. He generously offered to take me as an assistant, and in so doing, started my real education as a biochemist. There, we worked together in almost complete isolation, apart from rare visits to the Pasteur Institute in Paris and its library. The red letter days were the monthly meetings of the "Club de Physiologie cellulaire," which we faithfully attended. The reunions of the club were a unique educational experience. We had the opportunity to listen to and meet, in an informal and congenial atmosphere, some of the leading contemporary biochemists and experimental biologists. The club was Jacques Monod's creation and reflected his generosity and devotion to the education of younger scientists. It was at one of these reunions that we first met.

After obtaining my doctorate in the spring of 1950, I plucked up the courage to approach Monsieur Lwoff to find out if it would be possible to work for a time in his laboratory. I explained that, although I was very happy to continue collaborating with Georges Cohen, a change of subject, even a temporary one, would be salutary.

I had attended the meeting "Les unités douées de continuité génétique" at which André Lwoff and Jacques Monod had participated, and read with great fascination the recent review published in the *Growth Symposium* by Jacques Monod. By training and inclination a biologist, I desired to become better acquainted with some aspects of research being pursued in his laboratory. André Lwoff very kindly listened and said, "Why don't you ask Boris Ephrussi?" My answer was "No, thank you." His amused smile did not ask for an explanation. Then quite seriously he said that there was no space in his laboratory, and he would be absent from Paris for several months, but I could work under Jacques Monod's supervision, if he agreed, for a month, starting the first of September.

On that first beautiful September day, Jacques Monod showed me how to measure the activity of β-galactosidase with orthonitrophenyl β-D-galactoside (ONPG) as substrate. He had just received from Melvin Cohn a sample of this new chromogenic substrate, discovered by Joshua Lederberg. I was then charged with the determination of the molar extinction coefficient of o-nitrophenol at neutral and alkaline pH in order to determine units of activity of the enzyme.

A few days later, Jacques Monod asked me to determine the Michaelis constant of β-galactosidase for ONPG. "Michaelis constant, what's that?" I said. "You find out," he answered laughing, while handing me his copy of

Baldwin's *Dynamic Aspects of Biochemistry*. I spent the evening studying that particular chapter in Baldwin's book and, at the end of the next day, brought him the answer. I must have passed the test! The following week Jacques Monod asked if I should like to remain in his laboratory. I was absolutely thrilled by a prospect which was beyond all my expectations. I thanked him profusely, said that I would like nothing more, but that it would mean leaving Georges without any collaborators. After a few sleepless nights, I opened my heart to Georges. With great generosity, he replied that I shouldn't think of him, but only of my scientific future. My conscience was somewhat salved by Jacques Monod's assurance that Georges would join his group as soon as space became available. This eventually happened in 1954; but by that time I had already left for the United States. Little did I know, at the time, that Georges had refused, on my behalf, a similar invitation a year previously because we could not both be accommodated in the crowded attic. Many years later, I also learned that before leaving Paris on his travels, André Lwoff had suggested to Jacques Monod that I might join his group, where, he thought, I could be a "valuable addition." (I leave the responsibility of these last two words to André Lwoff himself!)

If I have evoked at such length the roadblocks to the pursuit of a career in developmental biology, it is because—incredible as it now seems—it was the way it was in 1945–1950. I was indeed exceptionally fortunate to meet André Lwoff and Jacques Monod, but the matter rested on pure chance. In respectable French academic circles their names and work were completely ignored.

Thus began the most fruitful and exhilarating three years of my life. Others have described better than I can the atmosphere of friendship, gaiety, intellectual excitement, in short of "gai savoir," which pervaded the part of the attic occupied by the Service de Physiologie Microbienne. In the fall of 1950, Annamaria Torriani, Mel Cohn, and Madeleine Jolit were the only members of the group working closely with Jacques Monod. Annamaria was busily running Warburgs and cooking large quantities of rice for our lunches. Madeleine, cooped up in her tiny room, was searching for mutants and kept our cultures going. Mel Cohn was in England for a time, collecting and synthesizing all sorts of galactosides which we would try as inducers, substrates, or complexants of β-galactosidase.

At that time the challenging mystery was the role played by the inducer in the synthesis of the enzyme. Every day a theory was built and another buried. There was great excitement when we found that thiophenyl-β-D-galactoside, a very potent inhibitor of the enzyme, was not an inducer and that melibiose, which was neither a substrate nor a competitive inhibitor and not even metabolized in any detectable way by the cells (*E. coli* strain ML), was a very potent inducer: all theories involving the enzyme per se in its own induction were invalidated.

The study of the effect of cations on the activity of β-galactosidase and the affinity of its substrates was also part of my occupations. Jacques Monod thought that the first results on the competition between sodium and hydrogen ions on the activation of β-galactosidase should be published as a note to the Academy of Sciences. I must have rewritten that little paper at least five times before he was satisfied and finally agreed to present it to Monsieur Tréfouël for publication in the *Comptes Rendus*. On that occasion, Jacques Monod introduced me to M. Tréfouël, then director of the Pasteur Institute, who, after a desultory look at the manuscript, entertained us with an account of the qualities of his secretary's new electric typewriter.

It is not without nostalgia that I remember those times of close everyday collaboration with Jacques Monod. Looking back through my old notebooks I find protocols of experiments written either by myself or by him. Each day, the purpose of the experiment was noted and subsequently what we had learned from it. Sometimes, he could not refrain from writing his reflections on the day's results or simply what crossed his mind at the time. On the 5.II.51, I find: "O surprise! O beauté simple et naïve et intelligible, l'activité est proportionnelle à la concentration de l'inducteur" ("O surprise! O simple, naive and intelligible beauty. The activity is proportional to the concentration of inducer") (at low concentrations and in certain conditions). Another day (9.II.51): "Triste jour, j'entre dans ma 42ème année. Ce n'est pas que cet age me déplaise mais il me rapproche d'un age qui me plait moins. Pour célébrer ce jour..." ("Sad day, I enter my 42nd year. It is not this age that displeases me, but it brings me closer to an age I like less. To celebrate this day..."); then comes the purpose of the experiment. He was always full of wonder about the beauty of the simple mathematical laws which biological systems obeyed: the beauty of a growth curve, for example.

In the fall of 1951, A. M. Pappenheimer, Jr.—"Pap"—appeared, on sabbatical leave, and shared my very small room in the laboratory. He wanted to improve his French and work on the induction of β-galactosidase; no doubt, that is the reason why Jacques Monod had put us together, since my spoken English was nonexistent.

Pap's imagination was challenged by Annamaria and Mel's recent discovery of the protein Pz present in uninduced cells, which cross-reacted immunologically with β-galactosidase (Gz). Moreover, the intracellular level of Pz decreased when Gz was synthesized. After a lengthy discussion, Pap and I decided to do a little simple experiment (secretly withheld from Jacques Monod) to test a possible relationship between Pz and Gz. We proposed to induce a culture in which growth had stopped from nitrogen limitation and to follow the kinetics of β-galactosidase synthesis after the addition of a nitrogen source.

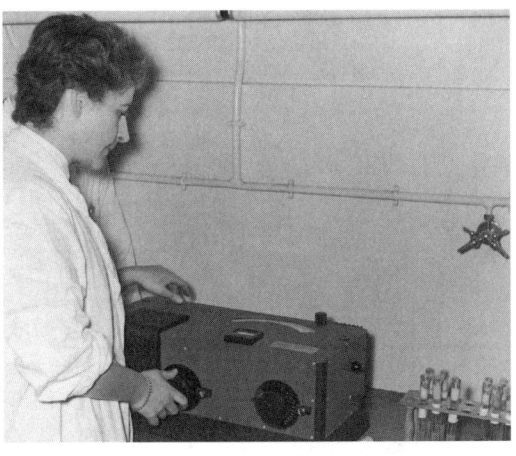

Germaine Cohen-Bazire, 1954.

Jacques Monod, who was always in and out of the lab, naturally wanted to know what we were doing. We did not want to tell him for the very good reason that we did not really know what to expect. If we found something interesting it would be a good surprise. He was a little annoyed with us, I could feel, but it did not take him long to guess. He was very sharp with me when I confessed that the reason for my secrecy was the lack of a precise hypothesis. "Always try to formulate one," he exclaimed. He then told us, "The way to do the experiment is to use amino acid auxotrophs." It became evident that the amount of enzyme synthesized was rigorously proportional to the amount of growth of the culture. From these experiments, Jacques Monod produced a conceptual masterpiece, a fine example of his extraordinary hypothetico-deductive acumen. The hypothesis of a dynamic equilibrium between Pz and Gz—or, for that matter, between any proteins—was given a serious blow. Thus was defined the differential rate of enzyme synthesis or what became known later (in the United States) as Monod's plot. The masterpiece was handed to me to be transmitted to Professor Fromageot, then editor of *Biochimica Biophysica Acta*. A few days later, Fromageot called and told Jacques Monod that he thought the paper far too theoretical. He answered, "If you publish this as it stands, I promise that my next paper will contain only facts."

One of the most agreeable aspects of life in the attic was the ease with which one acquired the elements of adult education. Conversations at lunchtime often dealt with politics or the most recent movies, but also with art, music, and literature. A typical product of the French pedagogical system and of life in a provincial town, my ignorance, particularly of literature, was profound. When I confessed one day to Jacques Monod that

I had never read Camus, Stendhal, or even Anatole France, he laughed: "I envy you. You do not know how lucky you are to have all these pleasures in front of you!" Later on, when I was hospitalized for surgery, he brought me an armful of books.

After almost two years spent in Jacques Monod's laboratory, he began to cast about for a postdoctoral position in the United States. This eventually materialized in 1953, in the Department of Bacteriology of the University of California at Berkeley; and I sailed to apply my newly acquired biological insights to the problems of regulation of pigment synthesis of photosynthetic bacteria, little knowing that I should remain away from France, and my real home, the Pasteur Institute, until 1971.

By then, Jacques Monod had become the director of "la Maison," as old Pasteurians used to call it. Immersed in his worries and responsibilities, I didn't see him very often, sometimes in the hallways, where he would give me a kiss on both cheeks and ask if all was well. All was well as long as I knew he was there, somewhere.

Germaine Stanier (Cohen-Bazire)

Throughout her scientific career, Germaine Stanier (Cohen-Bazire) kept her fascination for biological phenomena, showing the same enthusiasm whether studying organisms, their structure, or molecules.

Her discovery with Jacques Monod, in 1953, of the specific inhibition of the synthesis of tryptophan synthase by tryptophan led to the concept of end-product repression of biosynthetic pathways. In 1956 Germaine married Roger Stanier and became his collaborator in Berkeley, California. There she began research on the physiological, biochemical, and structural properties of the principal groups of photosynthetic prokaryotes, the purple and green anoxyphotobacteria and the cyanobacteria, and became one of the world's specialists on these microorganisms.

After her return to Paris in 1971, she continued her successful research on cyanobacteria in the laboratory headed by Roger Stanier. After his death in 1982, she succeeded him as head until her retirement in 1988. In 1985 she was nominated professor at the Pasteur Institute, and she also acted for many years as treasurer of the International Cell Research Organization, promoting training courses in developing countries. She passed away in 2001.

Whatever Happened to Pz?

A. M. Pappenheimer, Jr.

I first met Jacques Monod at the 1946 Cold Spring Harbor Symposium. It was this symposium, I think, that finally convinced biologists everywhere that bacteria and even viruses, crystallizable as nucleoproteins, really did have mutable genes and chromosomes that could undergo sexual recombination and replication just as did the genes of animals and plants. At the meeting, André Lwoff gave a paper on "Biochemical Mutations in Bacteria" in which he reviewed Jacques' work on the new phenomenon of "diauxie" and its relation to adaptive enzyme formation. Lwoff discussed at some length a working hypothesis of Monod's which he summarized as follows: "Enzymes allowing attack of carbohydrates by bacteria are all derived from a *common precursor*. This precursor (preenzyme) has a slight general affinity for carbohydrates. Transformation of the precursor into an adapted, specific enzyme occurs as a result of the substrate–preenzyme combination (which would account for the specificity of enzymatic adaptation)" (1).

After the symposium, Jacques and André spent the weekend with us in Scotland, Connecticut, where, in addition to helping me select stones from an abandoned mill to build a terrace and stone wall, they agreed to save a place for me in André's Service de Physiologie Microbienne at the Pasteur Institute during my sabbatical leave from New York University in 1951–1952. It was fortunate indeed for me that I applied so early. By the time I arrived with my family in Paris five years later, the "grenier" of André and Jacques was bursting at the seams and had already become the exciting and stimulating center of the emerging new field of molecular biology. Space was at a premium, and the atmosphere in the laboratory was incredibly exciting. At lunch, speculations and hypotheses on the mechanism of enzyme and temperate virus induction, how to test the theories and how to tear them apart, were feverishly debated in French and English simultaneously, in every kind of foreign accent and with everyone talking at once. The conversation was aided by a constant stream of visitors including my friend and former student, Mel Cohn, who had already been visiting for two years before my arrival. To the laboratory that year came Seymour Benzer, Max Delbruck, Luigi Gorini, Martin Pollock, Francis Ryan, Roger Stanier, Gunther Stent, Louis Siminovitch, and a great many others. It was an idyllic year both for me and my family. Weekdays, our three children went to the Ecole Alsacienne. On Sundays we would join Jacques, Odette, and their two boys to "faire les rochers" and picnic at Fontainebleau or one of the other "forêts près de Paris." One evening a week we would play chamber music at the Monods' apartment, 24 rue Monsieur le Prince. As I wrote for my chapter in André Lwoff's jubilee volume (2), I think the year our family spent in Paris was probably the happiest and most rewarding of my life.

On arriving at the Pasteur Institute, I told Jacques that I had come there to learn about adaptive enzymes and that I wished him to suggest a problem for me to work on. The idea that Lwoff had discussed at the 1946 symposium, namely that adaptive bacterial enzymes arose by transformation of a preexisting precursor, was still the prevalent view five years later, but apparently Jacques was beginning to have some doubts about the hypothesis. At any rate, the question which he proposed I try to answer was the following: When an inducer such as a β-galactoside is added to a culture of E. coli (growing in the absence of glucose), does induction of β-galactosidase production involve modification of one or more preexisting precursor polypeptides, or is the enzyme a totally new protein synthesized from scratch? As a matter of fact, a strong candidate for a "preenzyme" had already been found in noninduced E. coli by Mel Cohn and Annamaria Torriani (3). Cohn and Torriani had shown that there was present in extracts from noninduced cells an enzymatically inactive protein which they called Pz. This protein cross-reacted with antibodies raised in rabbits by immunization with purified β-galactosidase (Gz). The Pz protein was found in all strains of Enterobacteriaceae capable of fermenting lactose (4). Cohn and Torriani found that 80% or more of all the antienzyme in anti-Gz sera could be precipitated by Pz and that only 10 to 20% was truly specific for the enzyme itself. However, when mixtures of Pz and Gz were added together to their antisera, the enzyme was preferentially precipitated. Did the addition of an inducer to an uninduced culture of E. coli somehow bring about the modification of Pz so as to yield active Gz, analogous perhaps to the conversion of trypsinogen to trypsin?

The experiment that Jacques proposed was indeed very simple. He had recently received a series of twenty auxotrophic mutants of *E. coli* from Bernard Davis, each requiring a different amino acid or accessory factor for growth. Would conversion of precursor to enzyme take place in the absence of an amino acid essential for growth? Jacques suggested that we test each mutant strain in turn by adding inducer (β-methylgalactoside) just after growth ceased due to having used up the required amino acid. I was assigned to a laboratory with Germaine Cohen-Bazire, who taught me how to measure Gz by following the rate of hydrolysis of niphégal (*p*-nitrophenyl-β-galactoside) at 420 mμ.

I must confess that I did not really understand how all this was going to answer the question. I think what bothered me at the time about the suggested experiments with auxotrophic mutants was that I could not see how an enzyme could be synthesized without cell growth, although my reasons for this (essentially erroneous) conviction were far from clear. I remember asking Jacques just what he was really driving at. He said that what he hoped to accomplish eventually was the synthesis of β-galactosidase in a test tube using a cell-free system. I recall saying, "Jacques, if you succeed in doing that, you will be doing what God did. It will be equivalent to creating life." I little dreamed that within less than twenty years this would be a fait accompli, not only for β-galactosidase and other proteins (5), but that Jack Murphy in my own laboratory would have produced diphtheria toxin in vitro using a system extracted from *E. coli* and DNA isolated from corynephage β (6).

Nevertheless, Germaine and I tested the auxotrophic strains. We soon found that no enzyme was produced if inducer was added to a nongrowing culture, no matter which amino acid had become limiting. Enzyme synthesis began almost immediately upon adding back the required amino acid or growth factor. At this time, the kinetics of enzyme production were being followed as a function of time. It seemed to me that if growth were indeed required, we should be measuring enzyme production with respect to the increase in bacterial mass. In fact, when Germaine and I plotted our results in this way we found that enzyme synthesis, after adding inducer, was directly proportional to the new growth from the earliest moment that an increase in bacterial mass could be detected. At first, I cannot say that I found our results particularly surprising or novel, but when Germaine and I showed them to Jacques he became quite excited. Our results with the amino acid-requiring mutants had, of course, suggested that if Pz were indeed a precursor, its activation required the synthesis of another polypeptide containing all the amino acids. Somehow, the inducer would have to trigger the synthesis of this polypeptide de novo. Jacques was struck by the fact that our method of plotting enzyme production suggested that *from the very outset* enzyme was synthesized as a constant proportion of the total bacterial protein synthesis. He called this the "differential rate of protein synthesis." The fact that this rate was linear as soon as it could be measured after addition of inducer would not have been predicted by the precursor theory. The maximum differential rate of β-galactosidase synthesis was found in the constitutive mutants (requiring no inducer) that Germaine had just succeeded in isolating, and amounted to a surprising 5% of the total protein synthesis! Obviously, under normal circumstances, organisms must have a way of regulating the synthesis of their own enzymes so that the synthesis of any given enzyme is only turned on in time of need (7).

But the question still remained: Why did uninduced *E. coli* contain the protein Pz, which appeared to be so closely related to β-galactosidase both structurally and serologically? After a good deal of discussion, it was decided that the only way to eliminate the possibility that Pz might form part of the active enzyme would be by the use of radioactive isotopes. After discussions with Mel Cohn, the following experiment was proposed: *E. coli* would be grown in the absence of inducer on a medium containing $S^{35}O_4$ as the sole source of sulfur until growth ceased. Inducer and unlabeled sulfate would then be added and samples would be withdrawn at short intervals, extracted, and precipitated by a Pz-absorbed anti-β-galactosidase (anti-Z) serum. Since Pz is a sulfur-containing protein, if it formed part of the induced enzyme molecule, we would expect to find radioactivity in the specific precipitate from the Pz-absorbed serum. If it did not, then the newly synthesized enzyme should contain no labeled sulfur. Before leaving Paris, I performed the first such experiment. Although inconclusive, we did find that the radioactivity in the specific precipitate was a good deal less than would be expected if Pz were incorporated directly into the first enzyme formed. Our controls suggested that what radioactivity was precipitated with Gz was almost certainly due to contamination of the precipitate with other nonspecific sulfur compounds present in the crude bacterial extracts. It was obviously going to be necessary to purify the newly formed enzyme before its specific precipitation by anti-Z, to reduce the nonspecific precipitation of label, in order to obtain convincing results. This was accomplished in elegant fashion by Hogness, Cohn, and Monod (8). Thus if any part of Pz were present in the induced enzyme, it must be lacking in both its methionine and its cysteine.

The final and complete elimination of the Pz protein as a precursor having anything to do with induction of β-galactosidase synthesis came a few years later when Cohn, Lennox, and Spiegelman (9) transduced the *lacZ* gene into an i^+z^- *Shigella dysenteriae* containing no Pz, and obtained *lac*$^+$ recombinant *Shigella*. Their finding demonstrated that the Pz locus was not linked to the *lac* operon and that synthesis of the gene product occurred de novo and was totally independent of Pz. That was in 1960. I have only found one paper referring to Pz since that time. In 1972 Erickson and Steers (10) reported on a cross-reacting antigen present in a number of *Enterobacteriaceae* species which, like Pz, reacted with anti-β-galactosidase antisera. Of course, by 1961 there was no longer any reason to even think about precursors anymore, because the *lac* operator had been defined and it was recognized that the *i* gene product was a repressor. Induction of enzyme synthesis resulted from derepression of the operator. Moreover, mRNA and DNA-dependent RNA polymerase had been discovered and the role of ribosomes in translation of mRNA was beginning to be appreciated. In *The Lactose Operon* (5), which appeared in 1970, the last entry in the index under "P" is "PyJaMa experiments." Pz is not mentioned.

So we return to where we started in 1951. What is Pz and whatever happened to it? I find it very difficult to believe that its close serological and structural similarity to the β-galactosidase protein was the result of some purely fortuitous accident. Could Pz have evolved from a relocated stretch of DNA that arose by duplication of an ancestral Z gene? If this were so, then according to Zipkas and Riley (11) we might expect to find the Pz gene located approximately 90° or 180° away from the *lac* operon on the *E. coli* chromosome. Several laboratories (12–14) have recently described mutant strains of *E. coli* which carry a β-galactosidase gene that lies outside the *lac* operon. For instance, Campbell, Lengyel, and Langridge (13) demonstrated such a β-galactosidase in strains derived by successive mutations from an *E. coli* K-12 strain that already contained a large deletion in its Z gene. In this mutant strain, LC110-Ebg$^+$ (where Ebg stands for *evolved β-galactosidase activity*), the *ebg* gene lies at 59 minutes on the *E. coli* map, almost exactly opposite the *lac* operon. One would like to know if the *ebg* gene product bears any relation to Pz. According to Campbell et al., extracts from the *ebg*$^+$ strain do not form a specific precipitate when mixed with anti-Z serum. However, it is not clear whether the anti-Z serum which they used would have given a precipitate with known Pz-containing extracts from uninduced bacteria. Perhaps the possibility of a relationship between Pz and the products of the *ebg* and Z genes should be explored further. For the present, however, the nature of the Pz protein and its function remain a mystery.

References

1. **Lwoff, A.** 1946. *Cold Spring Harbor Symp. Quant. Biol.* **11,** 139.
2. **Pappenheimer, A. M., Jr.** 1971. *In* J. Monod and E. Borek (ed.), *Of Microbes and Life.* Columbia University Press, New York, N.Y.
3. **Cohn, M., and Torriani, A.** 1952. *J. Immunol.,* **69,** 471.
4. **Cohn, M.** 1957. *Bact. Rev.,* **31,** 140.
5. **Zubay, C., Chambers, D. A., and Cheong, L. C.** 1970. *In* J. Beckwith and D. Zipser (ed.), *The Lactose Operon.* Cold Spring Harbor Laboratory Press, Cold Spring Harbor, N.Y.
6. **Murphy, J. R., Pappenheimer, A. M., Jr., and Tayard de Borms, S.** 1974. *Proc. Natl. Acad. Sci. USA,* **71,** 11.
7. **Monod, J., Pappenheimer, A. M., Jr., and Cohen-Bazire, G.** 1952. *Biochim. Biophys. Acta,* **9,** 648.
8. **Hogness, D. S., Cohn, M., and Monod, J.** 1955. *Biochim. Biophys. Acta,* **16,** 99.
9. **Cohn, M., Lennox, E., and Spiegelman, S.** 1960. *Biochim. Biophys. Acta,* **39,** 255.
10. **Erickson, R. P., and Steers, E., Jr.** 1972. *Immunochemistry,* **9,** 29.
11. **Zipkas, D., and Riley, H.** 1975. *Proc. Natl. Acad. Sci. USA,* **72,** 1354.
12. **Warren, R. A.** 1972. *Can. J. Microbiol,* **18,** 1439.
13. **Campbell, J. H., Lengyel, J. A., and Langridge, J.** 1973. *Proc. Natl. Acad. Sci. USA,* **70,** 1841.
14. **Hartl, D. L., and Hall, B. G.** 1974. *Nature,* **248,** 152.

A. M. Pappenheimer, Jr.

Alwin Max Pappenheimer, Jr., called Pap by his friends and students, was born in 1908 and passed away in 1995. During his highly productive career he contributed to our understanding of the mode of action of the microbial toxins at the molecular level. He received his Ph.D. in organic chemistry in 1932. In the late 1930s Pap isolated, crystallized, and characterized diphtheria toxin, for which he received the Eli Lilly award in 1942. From that time on, Pap devoted a major portion of his career to understanding the mode of action of diphtheria toxin, and for more than three decades—from 1945 at New York University and then from 1957 on at Harvard—he inspired a great number of students and colleagues to perform research in this area.

In the early 1960s, Pappenheimer and R. John Collier elucidated the fundamental biochemical reaction underlying diphtheria toxin action: inhibition of the eukaryotic protein synthesis machinery by inactivation through ADP-ribosylation of the elongation factor EF2. With T. Uchida and D. M. Gill in 1971, Pap provided the first evidence that the structural gene of a toxin resides within the genome of a bacteriophage (β phage of *Corynebacterium diphtheriae*). A member of the National Academy of Sciences since 1973, Pap also received with John Collier the Paul Ehrlich Prize and Gold Medal.

Cartoon by Francine Lavallé, 1957.

Origins of Molecular Biology: a Tribute to Jacques Monod

An Exciting but Exasperating Personality

Martin Pollock

L'univers n'est rempli que de bruits. L'homme, par choix, en compose à son image une musique dont il s'emerveille.*

MacGregor[†]

This book is not, I would suppose, intended to be a series of formal eulogies to Jacques Monod; nor is it simply a "tribute" to a very remarkable person. I imagine its aim, rather, is to provide a wide range of personal memoirs: the reactions, experiences, and subjective impressions of his friends and colleagues, compiled as a historical record of how Jacques appeared to those who knew him well. That, in itself, is sufficient tribute to his memory. There is no need here for facile expressions of adulation.

Anyway, this is how I look at it and I can do no more than describe as honestly as possible how I felt about him in the light of my own experiences. This must mean what some may take to be too deep an intrusion of my personality, but that cannot be helped. It also means that I have not attempted here to view the grand sweep of his superb achievements in molecular biology; I have already done this elsewhere (1965 Nobel Prize for Medicine, *Nature* [*London*] 1965, **208,** 1250; Obituary notice, *Trends Biochem. Sci.* Sept. N 208 [1976])—and anyway most people are already familiar with the story.

A picture of a personality drawn by a single individual is inevitably highly subjective and therefore to some extent tendentious. But to be real it must be honest and complete within that context. Moreover, in order to understand the deep structure that lies beneath the edifice of

---

* "The universe contains nothing but noises. Man chooses to create from them a music, in his own image, which he then proceeds to marvel at."

[†] A pseudonym that Monod liked to use when he wished to quote himself.

personal achievement, it is often the apparent trivia of everyday life, the small things that go wrong, the reactions to difficulties and personal relationships, that provide important clues. And these do not emerge in the published research reports or grand lecture reviews that claim public attention.

There are a few who, through ignorance or envy, have regarded Jacques mainly as a conceited and arrogant egoist. There are others, dazzled by his brilliance and charm, who could see nothing but genius and virtue. But most of us, I suspect, feel that he was a complex character who combined exceptional talents with great ambitions. Looking back now over the years, it still seems to me that his most outstanding characteristic—the key to understanding a number of otherwise puzzling and paradoxical features in his behavior towards others—was a supreme self-confidence in his own ability. I have never met anyone who had one-half such a high opinion of himself as had Jacques. And I could almost argue that it was justified in every sense except in the measure of the humility that is just as important as self-confidence. From this high sense of superiority came an egoism which was not, at first sight, any more marked than that from which we all suffer, but which penetrated more deeply into the structure of his behavior and could sometimes lead to the hurtful expression of an unfairly low opinion of others.

I began to sense his charm almost as soon as I first met him in André Lwoff's office in January 1947. But I only got to know him properly during my two visits to work in his laboratory: for three months in the spring and summer of 1948 and for a year or more over 1952–1953, when I really fell under his spell.

I soon had reason to be extremely grateful to him for his help and encouragement. I was already thirty-two years old when we first met, but my career as a research worker (into which I had, so to speak, gate-crashed, through having a medical degree that facilitated acceptance into the British Medical Council's employ without the traditional Ph.D. training) was considerably delayed by my medical education and the war. I was really very inexperienced and unsophisticated and Jacques took on the role of a sort of Ph.D. supervisor-substitute (without the thesis or the degree) who guided and inspired me, and I revered his opinions with almost filial devotion. Not only was he working on *my* subject of "enzyme adaptation" and did he think it was as crucially important to biology as I did, but he shared my political orientation and general practical philosophical approach (insofar as I had one) and we had so many personal tastes in common.

It was superb to have found someone to admire and feel so much affection for in this way, especially in the exciting atmosphere of postwar France. I believed I had found a friend with whom I could identify myself and attempt to emulate.

It was Jacques who organized for me an invitation to my first International Congress (of Microbiology in Copenhagen in 1947) and it was Jacques more than anyone (with the possible exception of Marjorie Stephenson) who encouraged me in the work I was then doing on tetrathionate and nitrate reductase adaptation in coliform bacteria. He and André arranged for me to give lectures and seminars under the auspices of the Pasteur Institute and provided space in their laboratories where I could work in that wonderfully stimulating atmosphere of discussion and banter and challenge that was both nerve-racking and exhilarating.

Jacques would criticize my draft papers and discuss my sometimes rather naive ideas paternalistically (but not too obviously patronizingly) and on many occasions gave me opportunities to present my work under the aura of his blessing. He was sufficiently older and more obviously successful than me for it to be possible to accept his rather authoritarian manner—expressed with a sort of modest insistence that only someone who was supremely self-confident could afford to adopt—without irritation. But he was *almost* my contemporary and we were, for a time at least, in principle, rivals in the field of enzyme adaptation, and I could not suppress a certain degree of envy of his superior intellect and the ease with which he could command attention in debate. It nearly always seemed I was trailing behind as, indeed, I was.

He was equally helpful in ordinary day-to-day affairs such as finding living accommodations for our family and advising about schools for the children and would put himself to some trouble in order to smooth over difficulties.

When I was living in Paris, Jacques, Odette, and their two sons were still occupying the rue M. le Prince apartment, a somewhat dilapidated place with a delightful atmosphere of the early nineteenth century. I still remember my deep regret when he finally left for a rather larger home near the Eiffel Tower. Nothing ever seemed quite the same after he had moved to this new ("horribly bourgeois") abode on the avenue de la Bourdonnais, which may have symbolized his worldly success but seemed to rob him of the romantic charm that emanated from M. le Prince.

"It's terrible, Martin, when you have to stop being a 'bright and promising young scientist.' It happened quite a long time ago for me; but you yourself are just on the brink."

We often discussed our individual problems together at that time, mainly in a scientific context. We were both ambitious; that made us very egocentric (though I think Jacques concealed it better than I did). I had to admit that I found my own character very difficult to deal with. Nothing ever seemed easy: there were so many temptations. Yet there were some "innocents" who so rarely seemed to suffer from them.

"And you, Martin, have *all* the temptations. . . ." And so it seemed.

"And you, Jacques?"

"I'm *certainly* no innocent!" He spoke vehemently, without the slightest hesitation.

When I pressed him further about his ambitions and (rather wistfully) admitted that I envied him his successes, he emphatically pointed out that they were due to the tremendous effort he had put into his work; they had been achieved only at the expense of "great sacrifices" and immense struggles.

He would understand only too well how I suffered by feeling I was not successful or clever enough and he would rather condescendingly tell how in his childhood home persons were never judged by cleverness or success, but by their charm and character. I only once visited his home in Cannes and met his courteous and cultured parents; but even from so brief a contact I could well appreciate how his early background of gentle good taste could have played an important part in the expression of his artistic and somewhat indirect approach to scientific problems and the originality he introduced into the development of molecular biology. But it had not apparently curbed the appetite of his ambitions! One such ambition was certainly his desire to exert influence. . . . When, very much later on, he was discussing the possibility of accepting the position of Director of the Pasteur Institute, one of his main arguments in favor was his ability, so he explained to me, in committees, to persuade others toward his own point of view. He may have been attracted by the vision of power such an appointment would provide; but it was not perhaps so much just power for its own sake—it was more important for him to feel that he would have an opportunity to put his own ideas into practice because he believed so firmly that they were right.

Jacques was an *aristocrat*, in the original, literal meaning of the word. He had extremely high standards in many respects and on the whole lived up to them. I remember challenging him once, on the spur of the moment, when we were walking up some great broad stone steps on an island (I think Isola Bella) in Lago Maggiore during the Pallanza symposium of 1952.

"Do you feel, Jacques, that you are alone in the world? That the world consists in a way of *you* (on the one hand) and all the rest of humanity on the other?"

I meant to imply that basically he felt himself superior to, or at least better qualified than, most others. Looking (or pretending to be) rather self-consciously embarrassed, he agreed at once, with almost shattering candor. I was fascinated and even a trifle shocked.

I am sure he adored the idea of being a "Father to his People." Indeed, I believe this gave him a sense of responsibility which he took very seriously.

"Alors, mes enfants, comment ça va?" He would stroll in through the door of the laboratory where Annamaria Torriani and I were trying to interpret some recent results, beaming with a benevolent conviction that he would be able to make sense of the mess he half-liked to suppose we were creating. And, indeed, he would sometimes succeed. One of his favorite occupations was, in fact, the interpretation or *re*-interpretation of other people's research in a more significant fashion than he supposed they could do, or had done, themselves. And it must be admitted that he was rather good at it, although it was not an offering that was always happily accepted by those concerned! I used to challenge him about this, too. He would accept a little gentle tease, but one had to be terribly careful. Although he certainly did not lack a sense of humor, I never attempted a real "leg-pull" because I did not believe he would have taken it. The cartoon I finally plucked up courage to pin up on his door at the laboratory (see "In Memoriam," next chapter) was an indication of what I was pretending to feel about him and his work at that time. He took no offense and, according to Mel Cohn, was actually quite pleased.

The immediate world for Jacques was Jacques and his "équipe": his circle of "pupils" and admirers. It applied mainly, of course, to science, but it extended in many directions: to groups he would take for climbing expeditions at Fontainebleau or to the crew on his yacht.

Warmly generous in so many ways to his friends and colleagues, he was often harsh in his assessment of their scientific abilities. So many of them were "charming," "delightful," or "attractive" but (alas!) "rather *stupid!*" But I soon discovered that his judgment was too frequently influenced by the extent to which their work and opinions conformed with, or supported, his own ideas. There were times when I began to think that the only person who constantly and consistently enjoyed Jacques' favor was Max Delbruck, in whom I am sure he recognized someone with talents *almost* as great as his own! By extrapolation from some of the remarks he made to me about some pretty able individuals, I shuddered to think what he thought of me—at least as a scientist. One very upsetting incident will, indeed, indicate how he must have felt. It concerns the time when I was fortunate enough to be elected into the Royal Society and have the added pleasure of receiving the traditional polite letters of congratulation from friends and colleagues. I was under no illusion, even during the initial euphoria, that this could have been anything other than a pretty borderline case (if not quite unjustified). A word from Jacques would have been appreciated probably more than from anyone else. When nothing came I could pretend to myself for a while that there was no reason why he should have heard about it. But the truth was far worse. Later on in the same year (1962), during a small meeting of "Pasteuriens" I attended in Paris, someone asked Jacques if he knew about the election of his old friend and "pupil," to which Jacques answered quite openly to the group, "Oh, yes, it was a mistake." I rapidly and emphatically endorsed his opinion before any possible embarrassment could permeate and everyone cooperated by changing the subject. Somehow or other I did not feel at the time so upset as I have been, on occasions, looking back on that rather shattering experience. It was shattering, however, not so much because of what Jacques may have thought about my FRS. I really would not have been much put out if I had heard privately about his real opinion; I was becoming inured to his severe judgments and anyway I truly supposed that it *was* a mistake—at least in the sense that there were plenty of better people at that time who merited election more surely than myself. But expressing himself so openly in my presence seemed to indicate that he did not care for me enough to consider my feelings in the slightest degree. It was—and still is—a most puzzling affair because I cannot understand what conceivable purpose it was intended to serve. Did he truly wish to be so wounding? That is *not*, perhaps, entirely impossible.

There were other times when he could be atrociously insensitive to people's needs or feelings. There were several stories, often quoted. I myself remember noticing a curious habit he had of choosing French or English for conversation (he was completely bilingual) frequently according to what was the *least* convenient for the person he was talking to, even in a straight tête-à-tête discussion. Was this simply a blind lack of consideration or refusal to recognize what language he was using, or the other person's needs? Or was it some curious streak of perversity in his nature?

Somewhat analogous was the story (possibly apocryphal: I cannot relate it from firsthand experience) of how Jacques, at the end of his day's work in the laboratory, would shut up the place, turn off the lights and apparatus that did not run through the night, and go home, regardless of whether anyone else was still working there! For *him* the day had ended and that meant the end of the day.

Yet all through there was his gentle courtesy, his desire to enlighten and set people on the "right" path, and his genuine concern for the welfare of those around him. He would take great pains to expound his ideas simply and clearly. Perhaps the most impressive of his qualities was his ability to *illuminate:* to interpret a puzzling situation so that everything seemed suddenly to fall into place and become comprehensible. This was due almost as much to the force of his personality as to his penetrating intellect. The same words from someone else would have had much less impact. Small wonder he inspired respect, admiration, and even devotion as well as exasperation.

I have dwelt previously on his intellectual domination and the way that his need to demonstrate this could weaken the self-confidence of those around him who could not easily face open confrontation. It can well be argued that this partial suffocation of lesser minds was a small price to pay for the great gifts that Jacques bestowed on the development of molecular biology and the clarifying vision he could offer to biologists generally. But why should it be necessary to pay such a price? It was certainly a personal problem for many of us who worked under the glare of his exceptional mind. A genuine sense of the privilege of being associated with him was sometimes tarnished by a feeling of being helplessly unappreciated—even in areas where there was some reason to believe a real contribution had been made to a problem. And I suspect that this may have been partly because he could seriously underestimate the quality of work done outside his "entourage." Like all scientists, however great, he must have often drawn heavily, if unconsciously, from the work and ideas of others, but he was not always perhaps fair in the manner or context in which he attributed his sources of inspiration.

I was myself so fearful of being intellectually swamped that I tended sometimes to become pettily unreasonable in my opposition to his ideas and stupidly reluctant to benefit from his advice.

I remember, for instance, arguing fiercely against a "Letter to *Nature*" on the "Terminology of enzyme formation" that he drafted on behalf of the five main workers on what was then called "enzyme adaptation." Jacques' and Mel Cohn's work on β-galactosidase formation had shown that inducers need not be enzyme substrates, and the teleological implications of the word "adaptation" so irked Jacques that he could no longer bear the idea of it being used in this context. I argued against the dogmatic presentation of the opinions of a self-appointed clique; I lamented the loss of the element of specificity (which seemed critical) in the word "adaptation" and I cited the realm of immunology, where a teleologically useless or dangerous immune reaction did not seem necessarily to have implied that the phenomenon in general should not be regarded as essentially adaptive. There is here, incidentally, an interesting hint of Jacques' philosophical distress engendered by the element of purpose supposedly contained in the concept of "adaptation," which emerged so powerfully in his *Chance and Necessity*; but surely the idea of biological adaptation, in its evolutionary context, could be quite satisfactorily sustained without necessarily implying any sort of grand design or purpose.

When I refused to sign, pointing out that the "edict" would be just as influential without my name to it, as long as it contained Jacques', Jacques resorted to (subtly flattering) blackmail by threatening that in that case it would not be sent for publication at all. I was thereupon told by another member of the "cabal" that I was being "purely obstructive" and I weakly submitted. Jacques knew exactly how to get his own way!

It is rather ironic to note that (a) only five years later Jacques was putting quotation marks around the word "inducibility" (the introduction of which had been the essence of our letter) in the title of his famous paper, with Pardee and Jacob, on β-galactosidase repression in inducible strains of *E. coli*, though I maintain that this was quite unnecessary even from his point of view; and (b) the suggestion, in our letter, that what Stanier had previously referred to as "simultaneous" induction/adaptation was, after all, in some cases and with respect to the action of the inducer itself, truly simultaneous!

Another illustration of Jacques' need to dominate the whole field of enzyme induction (and my need to feel at least *slightly* independent!) concerned the first review I wrote on the subject (1958). I asked Jacques for, and was generously provided with, details of his recent unpublished findings and was also sent a copy of his (unpublished) Jessup lectures, which I quoted where appropriate. But I did not send him a draft of my article until after it had been sent to press; i.e., too late for the major modifications which I feared he would insist upon and which it would have been difficult to refuse without offending him, or so I thought. However, I did not succeed in escaping his annoyance, for he wrote back protesting, perhaps understandably, in no uncertain terms, and I had to explain frankly why I had not brought him in at an earlier stage. I was convinced that he needed to be involved more than just to check the few points where I had quoted his unpublished results; there would anyway have been no difficulty in correcting details of fact had I unwittingly misquoted any of his own findings.

It was mainly during 1952–1953, in close association with Jacques at the Pasteur Institute, that I was able to look behind the scenes and become involved in some of the scientific problems of enzyme induction as they were being studied in Jacques' group in the early days before the Great Enlightenment. It was the time of the Pz red herring and the protein turnover controversy. (Pz was a protein in *E. coli* devoid of known enzymatic activity that was closely related to β-galactosidase [referred to then as "Gz"] immunologically and [apparently] metabolically.) I got into immediate trouble by not giving enough prominence to the Pz/Gz relationship in my paper at a symposium organized by Jacques at the International Biochemistry Congress in Paris, 1952. Then again, Jacques had been mainly responsible for showing (in conflict with Schönheimer's principle of the dynamic flux of protein breakdown and resynthesis in living tissues) that in exponentially growing *E. coli* there was negligible protein breakdown. But he seemed to be far too rigid in assuming this must be true for all bacterial species, to the extent of almost discarding the work done by those who found otherwise, e.g., in *Bacillus cereus*, where protein breakdown does occur to a small but significant extent even in logarithmically growing cultures.

Analogous again to this was his reluctance to appreciate the significance of the so-called "basal enzyme" (the level of specific enzyme activity found in uninduced inducible strains of microorganisms) as being a vital clue to the induction mechanism. Jacques had worked almost exclusively with *E. coli* β-galactosidase, where basal enzyme was so low in the commonly used wild type that it *could*, in principle, have been explained (as he was always emphasizing) by the few constitutive mutants that were known to be present. My own experience was founded on tetrathionate and nitrate reductase and penicillin β-lactamase, where basal enzyme was far too high to be explained in this way. It was thus reasonable to suppose at the time that basal enzyme should be an important clue, whereas Jacques, somewhat fixed on β-galactosidase, remained relatively uninterested.

In 1953 there were two formal possibilities for explaining basal enzyme: the presence in small amounts of an endogenous inducer or the existence, before induction, of the necessary genetic information for producing the enzyme *without* an inducer. It seemed more difficult, especially in the penicillinase system, to postulate the formation of a penicillin-like substance spontaneously in *B. cereus* than to suppose that *E. coli* might be producing small quantities of a lactose-like molecule. I was thus naturally forced to consider seriously the hypothesis of preinduction information, which directly implied that induction was due to the relief of some endogenous inhibition mechanism. I discussed the idea of looking for an inhibitor of enzyme formation that might be present in inducible strains. Jacques agreed at once that this was a formal possibility, but was obviously not very impressed with it. When I returned home I did a number of experiments using the *B. cereus* penicillinase system. These mainly consisted of preparing crushed cell extracts of the uninduced inducible strain 569 to see if they caused any inhibition of induction by penicillin in this strain, or of penicillinase formation in constitutive mutants. The results, of course, were totally negative and I soon lost heart. With hindsight it is easy to see why they failed. But of course their failure did not mean the hypothesis of relief from inhibition was incorrect, and I was finally led to conclude that it was the most probable explanation by more theoretical arguments. That, however, was not until 1957, when, ironically enough and unknown to me, Jacques and colleagues were hard at work with definitive experimental evidence.

So, in a way, I was right, but little credit is due because the idea at the time was only one of many possible hypotheses and I never pursued it with any determination. It is relatively easy to have a good idea; the real genius is to know how and when to follow it up with sufficient persistence to make something of it. Jacques and his group drove through with proper evidence to a final proof by using, with François Jacob, the genetics that he had long previously (in 1948) told me would be the way in for solving the problem of "enzyme adaptation." But he never referred to our earlier discussions when he acknowledged Leo Szilard's suggestions on the repressor hypothesis later on; perhaps he had forgotten.

There is an obvious enough moral in that story which applies to most research workers!

Jacques had plenty of wrong ideas, like the rest of us, along with good ones. His talents lay particularly in his ability to feel his way through the intricate maze of encouraging/discouraging and indicative/misleading evidence to the right solution.

But there was powerful logic there too, as well as intuitive inspiration. For instance, it was during those 1952–1953 days of the "Belle Époque" at the Pasteur Institute which I am mainly writing about that Jacques conceived the absolute necessity—and feasibility (using Mel

Cohn's immunological expertise)—of showing whether or not induced enzyme formation really corresponded to de novo biosynthesis of the enzyme protein itself. He argued, with inflexible logic and persistence, that all our molecular hypotheses about enzyme adaptation would continue to be a waste of time until we knew what molecular event we were measuring. This was strictly analogous to the need, ten years earlier, to know whether "enzyme adaptation" was a populational or cellular phenomenon. With hindsight, this seems obvious enough; but it was not so clear at the time. So, with Dave Hogness's and Mel Cohn's brilliant technology and excited enthusiasm, backed by the "Master's" encouragement and blessing, the famous piece of work on β-galactosidase was carried out with complete success and an unequivocal answer.

Against this background (with all its ups and downs, including a long frustrating period when it was not even possible to assay the enzyme properly because chromic acid from the cleaning fluid was unsuspectedly contaminating the pipettes), Annamaria Torriani and I were attempting to isolate the B. cereus-induced penicillinase by "straight" biochemical methods. Being largely extracellular, this enzyme was "ahead" of β-galactosidase as a candidate for purification, and we had Jacques' enthusiastic support, no doubt partly because he understandably enjoyed the idea of the work at least being *started* in his laboratory. When, finally, this work was completed to give a firm figure for the molecular activity of the enzyme, it was Jacques who argued that its main importance was to show that the inducer must act *catalytically*, since it could be calculated that there were at least 40 molecules of the enzyme formed on the average per cell for each molecule of penicillin fixed by the cells under conditions where no external inducer was present in the medium. This, he pointed out, formally disposed of the hypothesis that the inducer acted stoichiometrically by becoming itself part of the enzyme molecule. It was, I think, the only credit he ever bestowed on the B. cereus penicillinase system as an exclusive contribution to the understanding of enzyme induction. Ironically, pleased though I was with the "Master's" favor, I did not consider this to be of such critical importance because the "stoichiometric" hypothesis had never struck me as at all likely. But if Jacques was not directly responsible for the first purification of an inducible enzyme, he could at least be responsible for having interpreted its significance in the study of induction! But, alas, I do not think it *was* very important in that connection because it was off the mainstream of evidence that finally led to a solution. The significance was really rather limited and I think also Jacques privately considered it so. The "bouquet" he bestowed was for a piece of work not important enough to worry him unduly. But perhaps it seems a little cynical to suggest that had it been of really crucial importance, he would have found it much more difficult to acknowledge its value.

We had Jacques' encouragement for that work but, to be fair, we really owed more—at least at the practical level—to Mel Cohn. It was Mel who finally spurred me on to undertake what I felt would be a very difficult task (there were no simple well-worked-out techniques for protein purification available in those days) by actually starting to do some work on it himself, explaining that it was about time "to make that enzyme *talk.*" That was enough to get me going, with the technical knowledge and infectious vigor of Mel behind me and with his ability to plunge into a practical problem in biochemistry and throw it about until it showed some response.

Other features of our meanderings during those early days—four or five years before the great clarifying discoveries of 1958 et seq.—were not so clear-cut as the demonstration that enzyme induction was truly a de novo synthesis of specific protein from amino acids. The mechanisms by which this biosynthesis was switched on and off were as obscure to Jacques as to the rest of us. We talked about the "métabolisme des inducteurs" which was embodied in the hypothesis of an "organizer," formed as a result of inducer metabolism, that catalyzed enzyme production. Induction kinetics were not as simple as they seemed, even after Jacques' clarifying application of an "allometric plot" ($\Delta Z/\Delta B$) expressing increase in enzymatic activity against increase in cell mass instead of against time. This "trick" disclosed, in the case of β-galactosidase, a constant differential rate of enzyme production from the moment inducer was added, provided that induction conditions were "gratuitous" (i.e., not limited by metabolic dependence on the inducer itself). The apparent autocatalytic induction kinetics that supported the former "plasmagene" (self-replicating) hypothesis were thus shown to be totally misleading.

But *B. cereus* penicillinase induction kinetics were quite different from those of *E. coli* β-galactosidase. Their main feature was a prolonged lag phase of about 14 minutes after addition of penicillin before a steady rate of enzyme formation was established. And I fondly hoped that this lag would provide an important clue as to what was happening to the inducer (how the so-called "organizer" was formed or activated) in the cells before maximal induction was attained. To Jacques this lag was at variance with the $\Delta Z/\Delta B$ dogma and was considered rather "bizarre." This convenient word seemed to carry a slight implication of being more than just "odd": instead of the hope that it might be of positive value, it was rather something that had to be explained away. I looked at all those straight lines of the β-galactosidase $\Delta Z / \Delta B$ plot going through the origin so beautifully with mixed feelings. Then one day it was pointed out to me that they were just a *bit* selected. There were almost as many experiments giving plots showing a distinct lag corresponding to 2 or 3 minutes: small, but possibly significant. However, not much attention was paid to them; they did not conform. I was encouraged to think that perhaps the two systems did not differ so much after all; the lag might be universal and highly significant. But there was a double irony here because when, many years later, the mechanism of induction was more or less completely elucidated, the lag came into its own again as reflecting the short interval between initiation of transcription (corresponding to the onset of derepression) and termination of translation for the first molecule of β-galactosidase and its release from ribosomes. That allometric plot did not quite go through the origin after all and those curves showing a slight lag developed an enhanced prestige. But the penicillinase induction lag is still unsolved and it seems unlikely that it is due to the transcription–translation interval that apparently operates for β-galactosidase. It is perhaps a bizarre irrelevancy after all.

There is a third ironic curiosity arising from this problem: namely, the little-known phase in Jacques' pilgrimage toward the true explanation of enzyme induction when, for a short period, he seemed to have abandoned the lesson of the allometric plot (at least in its simplest form) and began to believe that some sort of autocatalytic process contributed to induction after all. I do not believe these ideas of his were ever published and unfortunately I cannot remember the supposed evidence or his arguments; they were abandoned anyway soon afterward. But they had an unfortunate aftermath, of no general significance, but personally a little upsetting. Annamaria and I had set our hearts on exploiting the long lag phenomenon in penicillinase induction. We felt that it could provide some useful information and we started studying the differential effect of irradiation with ultraviolet light applied before, during, and after the lag period. Differences in sensitivity (as measured by rates of enzyme formation) were indeed found and we were hopeful of trying to interpret them in terms of nucleic acid metabolism. Perhaps rather lazily, I left my colleague with the task of finalizing some of the experiments and of preparing the first draft (which of course had to be in French) after I had left Paris. When the draft arrived, it worried me by the extent to which Annamaria had, in my opinion, been overinfluenced by Jacques' "autocatalytic" ideas, to which I was very much opposed. So I decided, rather churlishly perhaps, that I could not add my name to hers on the paper; it would be better for her to go ahead and publish the work in her own way without me. However, when the final version eventually appeared it was very different. Jacques' "autocatalytic" phase was over; the paper had been considerably modified and became something altogether more acceptable. There were times when Jacques' opinions were almost whims, but they were always very influential!

Following the 1958 breakthrough I began to see less and less of Jacques. I think I must have become (for him) a bit of an irrelevancy, at least from the scientific point of view. He must have thought that the B. cereus induction system was no longer of any interest (if it ever had been) because the organism had no studiable genetics. It was difficult to persuade him to spend much time discussing these problems, let alone to come over to London (worse still, Edinburgh, where I had moved in 1965) to give a seminar; it would have been a waste of his time. But he was obviously delighted when Bill Hayes and I asked him to come over to inaugurate our new department of molecular biology in 1968. He loved the sort of "official importance" of such an occasion and the opportunity it gave him of bestowing his blessing on us all.

The last time I saw him was when he gave me a very private lunch at his home in Paris toward the end of 1974. I felt it truly an honor. He was as gently charming as ever and I longed to discuss everything under the sun with him as I always wanted to do—science, politics, philosophy, personal problems. But conversation did not run so smoothly; he was very preoccupied with his responsibilities as director of the Pasteur Institute, which was in great difficulties. I had the sensation that what I said and how I felt were not of much interest to him. Indeed, why should they be?

Like so many others, I feel immensely privileged to have known Jacques Monod as I did, especially over a revolutionary period of absolutely fundamental developments in biology, to which he himself contributed so much. It could, at times, be painful, frequently personally disappointing, and occasionally it was quite exasperating. But I would not have missed it for anything.

Martin Rivers Pollock

Martin Pollock was born in 1914 and studied medicine in Cambridge, England. He held some hospital appointments before he entered the National Institute of Medical Research (NIMR), Mill Hill, London. In 1949 he became head of the Bacterial Physiology Division, where he worked mainly on the *Bacillus cereus* penicillinase system. He then studied the various forms of penicillinases in different bacilli. The finding of the instability of a bacterium's capacity to produce penicillinase showed later that the gene for penicillinase (now called β-lactamase) is carried on a plasmid. Later he studied the mechanism by which β-lactamases are involved in the development of bacterial resistance to antibiotics.

In 1962 Pollock was elected Fellow of the Royal Society. In 1965 he was appointed Professor of Biology in Edinburgh, where he established with William Hayes a molecular biology teaching department. Martin Pollock took early retirement in 1976 and died in December 1999.

1960.

1962.

In Memoriam*

Melvin Cohn

Did I take on that awesome gift when death parted my limp form from his protective clasp?

MECHKONIN

The organizers of this symposium have asked me to trace in a personal way the contributions of Jacques Monod to the origins of our present concept of induced enzyme synthesis. I have chosen to deal with the Monod of the preoperon era of induced enzymes because it is a largely unknown chapter which is particularly illustrative of his creativity. This is appropriate because, in the last years of his life, Monod was intensely preoccupied with the creative process. He set the study of it as one of the goals of the Salk Institute, which he helped found. In Jacques Monod, this process was characterized by taste, elegance, and parsimony.

Monod in writing his own rather personalized curriculum vitae begins by saying:

> I was born in 1910 in Paris but in 1917 my parents moved to the south of France where I spent my youth. Consequently I consider myself more of a southerner than a Parisian. My father was a painter, a vocation rare in a Hugenot family dominated by doctors, pastors, civil servants, and teachers. My mother was American, of Scotch descent, born in Milwaukee; another anomaly when one considers the mores of the French bourgeoisie at the end of the last century. I came to Paris in 1928 to begin my studies in the *Faculté des Sciences.*

Monod then recalls his debt to his teachers, André Lwoff, Boris Ephrussi, and Louis Rapkine. He tells us that in 1934 he was a Fellow of the Rockefeller Foundation at Caltech working with Thomas Hunt Morgan. In 1936 he returned to France, soon to be faced with the Second World

---

* The first part of this article is reprinted from the 1976 Cold Spring Harbor proceedings on "Lactose."

War—terrible years which he never mentions, leaving it as a blank in his curriculum vitae—during which time he was in the French Underground. After the liberation, in 1945, Monod joined André Lwoff's laboratory at the Pasteur Institute.

I met Monod in 1947 at a Cold Spring Harbor symposium. He presented a paper entitled "The phenomenon of enzymatic adaptation and its bearings on cellular differentiation." He made the explicit point in his talk that we would have to understand enzymatic adaptation before we could understand differentiation, in particular antibody synthesis. This allusion plus the enthusiastic support of my teacher, Alvin Pappenheimer, Jr., is what sent me packing for Paris.

In the winter of 1948 I began my postdoctoral work at the Pasteur Institute in Paris. We were housed in an attic; at one end was André Lwoff's closed laboratory, on the door of which was a cartoon showing the Duke of Wellington addressing his officers after the battle of Waterloo, under which was the caption "Tea cleared my head and left me with no misapprehensions." At the other end of the attic was the laboratory which Jacques Monod, Annamaria Torriani, and I occupied. That year the Paris winter without heat was merciless. The glacial acetic acid remained frozen on the shelf until noon, at which time I had the distinct feeling that it was the heated discussion at the lunch table that thawed it out. Jacques was a choirmaster and during a good deal of that winter spent afternoons rehearsing the Bach Requiem he was to conduct that Christmas. Sundays we practiced rock climbing at Fontainebleau. There were many things to decide about the direction of the work but we simply could not settle down to any problem.

The most important preoccupation was that Monod, who symbolized reactionary Mendel–Morgan genetics, came specifically under vitriolic attack by French Marxist biologists who looked upon the very existence of adaptive enzymes as proof that the substrate induced a directed mutation or a permanent hereditary modification in the cell. This position had a certain respectability since Sir Cyril Hinshelwood was defending the same point. Even J. B. S. Haldane felt constrained to write only apologetic essays in defense of genetics. We spent one Thursday evening of every month at the meeting of the Michurin–Lysenko Society, at the Sorbonne, superficially debating the facts of genetics, but in reality what concerned us was the meaning of the scientific method. For Jacques Monod, who was *engagé* in the Sartre sense, the debates were ugly and degrading and they stomped on his sense of elegance and parsimony. He was moved to make his life's goal a crusade against antiscientific, religious metaphysics, whether it be from Church or State. The last time we strolled together on the beach at Torrey Pines, in 1974, he was bitter. "The battle against such ignorance will never be won," he said. "All that one can do is die without calling a priest to the bedside."

In the spring of 1949, we settled down to work. I remember that I felt like Alice in Wonderland when Monod identified three key characteristics of adaptive enzymes for study.

1. The response to a given substrate was specific for that substrate, i.e., the phenomenon was adaptive. The consequences of the existence of systems which paradoxically seem to have a purpose yet arise blindly by variation and selection was a constant theme in his thinking, culminating in his book *Chance and Necessity*.

2. The ability to metabolize a new substrate appeared as an autocatalytic function of time. This had led to the "plasmagene" hypothesis of Spiegelman, in which a gene produced a cytoplasmic self-replicating unit which in turn synthesized the adaptive enzyme.

3. Substrates competed for each other in the induction of given enzymes. This was the striking "diauxy" phenomenon, where an organism faced with two growth substrates metabolized one or the other preferentially. Today we call this *catabolite repression*. There was competition between substrates for the attention of the cell. For Monod, this implied competition for precursor subunit molecules.

Given what we now know, it seems remarkable that these three facts could have provided a solid basis for us to begin because they were so misleading. Yet Monod singled them out as he brought exquisite taste to bear on complexity. Today we know that of all the misleading truths at the time, only these three could have led to the creation of the modern field of regulatory biology.

The Monod concept to explain these three facts was the following: A group of genes coding for a pool of precursor subunits could be complemented in various combinations to make different enzymes. It was the directive influence of the substrate which caused an aggregation of some of the subunits to make the corresponding enzyme. Once seeded, the crystallization process was autocatalytic. If two substrates were involved there was competition for subunits. In other words, a large number of induced enzymes could be constructed from combinations of a small number of subunits which preexisted the appearance of the substrate in the milieu.

The way to test this hypothesis was to show that all substrates as well as competitive inhibitors were inducers. The hypothesis limited the choice of systems for study. *Escherichia coli* had ideal growth properties, as well as an emerging genetics analyzable by mating and viral transduction. It expressed an adaptive enzyme, β-galactosidase, which had a substrate, analogues of which were reasonably easy to synthesize. In 1950 I went to Bell's laboratory in Cambridge, England, and later to Helferich's laboratory in Bonn, Germany, to make the compounds which were sent back to Paris to test. By 1951, four findings changed our entire perspective.

1. Excellent substrates were *not* necessarily inducers, e.g., ortho-nitrophenyl-β-D-galactoside.
2. Excellent nonmetabolizable competitive inhibitors were *not* inducers, e.g., phenyl-β-D-thiogalactoside.
3. Poor nonmetabolizable competitive inhibitors could be excellent inducers, e.g., methyl- or isopropyl-β-D-thiogalactosides.
4. Noncompetitive inhibitors could be excellent inducers, e.g., the α-galactoside melibiose.

The realization that his hypothesis was false had already crossed Monod's mind, when on Oct. 14, 1950, he sent a telegram to me in England (Fig. 1) concerning phenyl-β-D-thiogalactoside, which I had last given him to test.

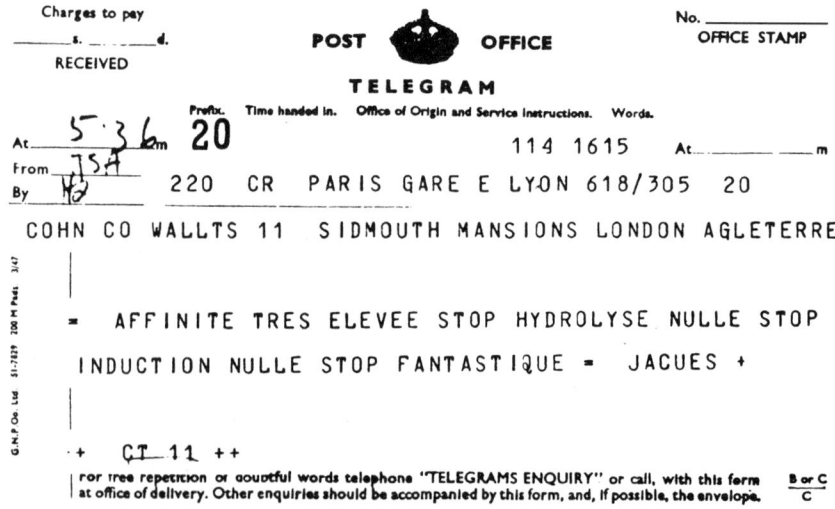

FIG. 1. Very high affinity stop hydrolysis negligible stop induction negligible stop fantastic. Jacques.

I include this telegram to illustrate the pleasure which Jacques Monod derived in proving that his favorite idea was wrong: "Fantastique" was the exact word. He was one of Karl Popper's greatest admirers and, like Popper, he insisted that scientific advance consisted in the falsification of hypotheses. I wish now that I would have realized that the Monod hypothesis on subunit complementation, which proved wrong for induced enzymes, was later to prove correct for induced antibodies.

The existence of nonsubstrate inducers had a profound philosophical impact, for, like Ionesco, Monod had created a theatre of the absurd. A bacterium growing on succinate was producing a useless enzyme, β-galactosidase, in response to a substance it could not metabolize. Monod, with great humor, invented the renowned Scottish philosopher MacGregor (his mother's maiden name), whom he quoted in all of his later writings. This time he attributed to MacGregor the following quote: "Each of science's conquests is a victory of the absurd." The vitalist Hinshelwood–Michurin–Lysenko position which irked him had been answered with experimental vengeance. For this reason he decided to drop the term "enzymatic adaptation" and use instead "induced enzyme synthesis," a term which was adopted eventually in an encyclical (*Nature*, **172,** 1096–1098, 1953) issued by the Adaptive Enzyme's College of Cardinals: Monod, Pollock, Spiegelman, and Stanier.

These four findings provoked Monod to toy with an idea which was very daring for 1951. The inducer had to be recognized by a stereochemically specific molecule which was *not* the induced enzyme itself. However, this idea left unexplained the autocatalytic nature of the response to lactose, a fact which now pointed strongly to a self-replicating gene product, the "plasmagene," postulated by Spiegelman.

In 1951, Seymour Benzer, François Jacob, and Elie Wollman (returning from sabbatical leave) joined the laboratory. Jacob and Wollman viewed adaptive enzymes with great suspicion and by exploring elsewhere paved the way for the era of the operon. It was only in 1953, when Max Delbruck visited Paris and demanded accountability, that the suspicion was diffused and our endeavors became respectable. Seymour Benzer, on the other hand, nettled by Stanier's published statement that it could never be done, decided to tackle the question of the cause of the S-shaped autocatalytic induction curve by using Monod and Wollman's finding that certain *E. coli* bacteriophages could block enzyme induced by lactose as the sole carbon source under conditions in which only cells which contained enzyme could be lysed. It became obvious that the S-shaped curve was due to the heterogeneity of response of individuals in the population. A bacterium with one molecule of enzyme could metabolize lactose to make more enzyme and therefore had a great advantage. In other words, the postulated *E. coli* "plasmagene" turned out to be the bacterium itself. For Monod, the second paradox was resolved.

From these studies Monod now developed the concept of *gratuitous induction*. Under conditions where the carbon source and the inducer were separated, the heterogeneity and the S-shaped induction kinetics disappeared.

At this point Monod was ready to face his third basic fact, the competition between substrates. This implied competition for precursors, which had led him to the subunit hypothesis that preformed subunits were shared between different enzymes. It became inescapable that he know whether the enzyme was made de novo after induction, or from preformed precursor subunits.

The answer required an isotope experiment in a laboratory that had never seen even the shadow of a Geiger counter. Fortunately Monod captured the interest of a Canadian physicist, Lou Siminovitch, who had been working with Louis Rapkine and André Lwoff since 1947. Siminovitch had discovered ^{35}S and proposed its use as a general protein marker. Siminovitch scrounged through the physics laboratories of Paris collecting junked parts which he checked off on his scribbled wiring diagram. He handed the precious do-it-yourself kit to Monod, who, like a child with a tinker toy, put it together and made it work. At the Christmas party that year I joshed Jacques in a skit which cast him as a bicycle repairman (réparateur de vélos).

David Hogness, now in the laboratory, began the experiment, which required purification of very small amounts of β-galactosidase to greater than 95% purity. The only way to do this at the time was by immunologic methods. Six months later, Dave Hogness completed the definitive experiment, nervously counting each point on the tinker toy through the night, while Monod played his cello and I uncorked André Lwoff's best properly chilled Sancerre wine, which he had carefully hidden in the cold room.

The result was clear. The enzyme was made from amino acids de novo after induction, at a maximum rate, virtually without lag.

This led Monod to formulate a new parameter which we christened as Monod's law, symbolized by $\Delta Z / \Delta B$ (the differential rate of synthesis), the basic unit of which was physiological time.

With hindsight it is easy to appreciate taste in science. The three most important characteristics of induced enzyme synthesis formulated in 1949, misleading as they were, had led by 1953 to a clear definition of the problem, and Monod was prepared to pursue it, virtually alone.

However, why we were so insufferably sure of ourselves is not clear to me. Given what we know today one might say that we had not advanced very far. Justifiably annoyed by our arrogance, Martin Pollock produced a cartoon in 1953 (Fig. 2), which at the time was upsetting to me but brought pleasure to Monod.

Pollock's cartoon shows Monod standing over a starry-eyed American (myself), symbolized by an outlandish tie, to whom he is saying, "Bravo my fine fellow! You have made remarkable observations—naturally without having done or understood anything—but nevertheless spectacular. Bravo! Continue the good work."

In the wastepaper basket are the papers of Pollock on penicillinase; on the wall is a depiction of "Who killed cock robin (Sir Cyril Hinshelwood)?"; above that is Max Delbruck smiling approval; next to Max is plotted Monod's temperature as a function of Sol Spiegelman's publications (notice how normal it is after the Benzer experiment); Monod's law, $\Delta Z/\Delta B$, is inscribed on the French tricolor behind us; and on the left was Pollock's evaluation of our accomplishments: we had destroyed all existing so-called facts, replacing them with nothing he was willing to believe "faits confirmés," and we had produced nothing but wild theories. This is how Pollock saw us in 1953. (He had a personal piece of advice to me which did not escape my notice, symbolized by the mouse in the left corner. Go back to the study of antibody synthesis in mice! In fact, long before molecular biology could influence immunology, Pollock proposed as the key the study of the clonal distribution of antibodies [1 cell–1 antibody].)

Today, I understand Monod's reaction of pleasure because such understanding could only have been the consequence of profound friendship.

Just before the modern era of the operon, one striking fact which we had generated had been ignored. With Georges Cohen and Germaine Stanier, Monod had shown that the end product of a biosynthetic pathway, in this case, tryptophan and methionine, repressed the synthesis of the corresponding enzymes on that pathway. Not only was function inhibited, as Novick and Szilard had shown, but constitutive enzyme synthesis itself was also repressed by its end product—a remarkable energy-saving device.

In his Nobel lecture, Monod muses about this:

I had learned like any schoolboy that two negatives are equivalent to a positive statement. Mel Cohn and I debated this logical possibility, which we called the "theory of double bluff," recalling the subtle analysis of poker

by Edgar Allan Poe. How blind I was not to take this hypothesis seriously sooner, above all since several years earlier we had discovered that tryptophan inhibits the synthesis of tryptophan synthetase. I had always hoped that the regulation of constitutive and inducible systems would be explained by a similar mechanism. Why not suppose that induction could be effected by an antirepressor rather than by repression of an anti-inducer? This was precisely the thesis which Leo Szilard proposed to us in a seminar. The preliminary results of the injection experiment (PaJaMa experiment) confirmed Leo Szilard's penetrating intuition and my doubts about "the theory of double bluff" were removed.

In a parallel world next door to us were Elie Wollman and François Jacob creating the basis for genetic analyses, which was soon to merge with induced enzymes to reveal what we know today as "operon theory."

I did not participate in the merger, which began in 1956, after I left Paris. This period is modern operon history: the discovery of the permease and transacetylase; the PaJaMa experiment; operator constitutive and promoter mutations, coordinate induction, polarity, and that remarkable insight "messenger RNA," all part of the 1961 Jacob–Monod Cold Spring Harbor paper. It was another great classic, written like Monod's 1947 Cold Spring Harbor paper in that simple and direct Anatole France style. It took only one more concept formulated in 1965, that of allosteric interactions, to round out the story of regulation at the physiological level.

The key to the power of these Monod theories, 1947, 1961, or 1965, was simply that they were physiological-level theories capable of reductionism; that is to say, they were capable of an analysis at the level of chemistry. They were truly theories of molecular biology. This was the basis of their elegance and their parsimony.

Monod and I never finished our 1974 discussion on the Torrey Pines beach. What was the next problem of regulation to be? Monod was concerned with the universality of the elements used in the regulation of the *lac* operon. Was there a limited number of elements which required minor rearrangements or was the number going to be large? Did we have to search for new generalizing rules on how they had to be organized? Were there any new laws which would come from the wiring diagrams, the logic of the circuitry? Were both positive and negative regulation fundamental to the integrated organism or could individuals have been constructed using only one or the other switch?

I believe that Jacques Monod had one of the most creative minds of our time not because he was a leader of righteous causes, not because he was a creator of molecular biology, not because he founded and directed institutes of learning. He had one of the most creative minds simply because he thought deeply, ascetically, and in a Socratic way about how knowledge is acquired; and it is this process that he insisted should be the only basis for a system of ethical and aesthetic values.

Addendum

André Lwoff asked me to write of "The Great Adventure in Lactose"; yet I have already accomplished that in the first part of this article and now wish to add small vignettes to that story in order to give another dimension to that adventure and to Jacques Monod.

It was Dr. A. M. Pappenheimer, Jr., who steered me to Paris to work with Jacques Monod. It took one of the great teachers of our times and someone who cared deeply about me to have thought that move out so carefully. I am surprised, still today, at the way in which it turned out.

In 1949, my desk in the laboratory at the Pasteur Institute was too poorly lighted to work in the evenings. I soon learned that the tomb of Pasteur was well lit. Thanks to an understanding concierge, the gate to it was left unlocked in the evenings. I spread my papers out using the low tomb as a desk and blissfully studied into the wee hours, smiled upon by garish mosaics depicting Pasteur's great discoveries. Monod surprised me one late evening while he was showing visitors around and found the idea so congenial that we spent many evenings there chatting across the death mask of the Institute's patron saint. Imperceptibly we changed from English to French and from "vous" to "tu." This is how we became friends with a love so deep that often without words we shared the many dramas of our lives.

From these earliest days, in spite of overriding problems, Jacques was preoccupied with the failure of French science and in particular the failure of the Pasteur Institute to have any credibility, much less world leadership. He decried the mediocrity of the Sorbonne, Académie des Sciences, and the Collège de France. Further, no improvement (as he jealously noted to be taking place in Germany and England) was in sight. His puritanical devotion to this theme throughout his life and his romantic dream about the Pasteur Institute not only were misunderstood and unappreciated, but ruined much of the creativity he might have expressed in the later years of his life when he became Director.

The political climate of the 1950s in the States and the bleakness of postwar Europe sent many leading intellectuals, poets, writers, painters, sculptors, and scientists to work in Paris. At first it was the darkness of the hour rather than the renown of our laboratory that brought together in the "City of Light" so many outstanding minds. Paris was the world center of creativity. All of us lived in it: in the laboratory, at the lunch table, in the theater, at concerts, in art galleries, in book stores, at the *chansonniers, la cuisine, le salon metaphysique*; we were in the middle of history and politics; we were *les hommes engagés*.

So, Seymour Cohen, Bernard Davis, Michael Doudoroff, Stephen Fazekas, David Hogness, Niels Kjeldgaard, Margaret Lieb, Aaron Novick, A. M. Pappenheimer, Jr., Lou Siminovitch, Roger Stanier, Leo Szilard, and Annamaria Torriani all came from afar, and along with one French student, Germaine Bazire, made up the pre-operon era of our laboratory. This was unusual in France; at that time foreigners could not compete with the French for positions in other academic centers. Further, aside from maintenance support from the Pasteur Institute, both public and private American grants supported the laboratory—a "debt" Jacques never forgot.

Michael Doudoroff brought "heart" into our work. Experiments worked when he was around because he cared about experimenters, somehow making the goals and aspirations of others his own.

Together we worked a night shift starting early evening, then going to dinner and resuming in the laboratory toward midnight. Usually we were a bit high after a few bottles of beautiful French wine and our work seemed to sing.

It was late one night that we were carrying our preparation of amylomaltase through its last step of purification, a process that had taken two long months from the moment we grew a kilo of bacteria. We were distinctly drunk, happy, and loquacious.

As we were preparing to centrifuge the precipitate, I hit the beaker on the side of the lab bench and the preparation went all over the floor. I sobered up and became ill. Mike rushed to get a dust pan and large window wiper. He carefully gathered up the liquid, rinsed the floor with buffer, and gathered the rinsings. He reprecipitated the enzyme and recovered it from the floor in 70% yield. I watched in deep depression. He bragged often about the best preparation we had ever made. We never dared tell anyone.

During a short visit to the States, I attended by chance a seminar in New York by the famous Wisconsin biochemist H. A. Lardy, who described his work on the cation activation of β-galactosidase. He pointed out that the enzyme was Na^+ activated, despite contrary reports by two unnamed French scientists that it was K^+ activated. As I already pointed out, the credibility of French science in the outside world was low. I was too neophyte and upset to dare defend our work.

On hearing my story, Jacques reacted by showing his great sense of proportion and led us to the discovery that both were right (but as he put it, only we had understood it!), for with lactose as a substrate β-galactosidase was K^+ activated, and with orthonitrophenyl-β-D-galactoside as a substrate it was Na^+ activated. This was my first lesson that science was an adversarial system and it took Jacques' kind of internal security to cope with it. On numerous occasions, faced with similar situations, this incident has flashed across my mind.

In 1951, I had the theory that an inducible and constitutive enzyme differed only in that the latter was internally induced. So I decided to isolate the internal inducer (obviously a galactoside) from a constitutive β-galactosidase mutant. I painstakingly grew a kilogram of bacteria, made extracts, and discovered immediately that they contained a potent internal inhibitor (not an inducer). Jacques said immediately, "Drop it, the finding makes no sense given what we know of physiology." Stubborn, six months later, I crystallized 10 mg of the inhibitor and identified it as glucose. This was too embarrassing and I expected an "I told you so." Instead, I received an admiring comment about my tour de force and a bit of advice, "Once you have gone this far, find out how it works."

The experiments on memory in bacteria were devilishly long. Briefly, if glucose was added five minutes before inducer the enzyme never appeared; five minutes after inducer and the enzyme was made forever. So, we had two stable states of a given bacterium growing in a given medium depending on its memory of the order of addition of glucose and inducer several eons ago.

These experiments required that we dilute the cultures every hour to keep them growing continuously, and Jacques and I took turns sleeping in the laboratory.

I decided one evening to simply set up an automatic system for feeding the removal of culture. Since I had a liter of culture which I diluted with a liter of medium every hour, I simply fed in a liter per hour of fresh medium and siphoned off a liter of culture per hour continuously. To my surprise, the bacteria could not keep up and the density of the culture fell. In fact, to maintain it I could not feed more than 690 ml/hour. As I was wrestling with this paradox, obviously upset, Jacques sat down with me and asked if I had any idea why I could not feed more than 690 ml/hour when I expected 1,000 ml/hour. "It may sound wild to you, Jacques, but I think I have discovered that bacteria, like men, have a biological need for rest." He smiled patiently and said, "You have discovered that the ln 2 = 0.69. Think about that."

The next day, both he and I had the detailed theory of continuous culture. However, I had been told the answer and it took the pleasure away.

We named the thing the "bactogène."

Annamaria Torriani was central to our laboratory. We were an ensemble (at first a trio) made possible by the aristocratic greatness of this woman. She had been through years of underground activity during the period of Italian fascism, and with Luigi Gorini was trying to build a new creative life. This extraordinary duo greatly influenced my life and I loved them.

Annamaria was struck by the mutation lac^- to lac^+. When isopropyl-β-D-thiogalactoside became available, we knew that both lac^- and lac^+ strains were inducible for β-galactosidase. Further, constitutive strains for β-galactosidase which were lac^- had been isolated, and Annamaria proposed that the mutation affected a lactose transport system (today called permease) which had to be inducible.

Jacques' reaction surprised me. He insisted that the problem was hopeless if one had to assume that two enzymes were induced coordinately; all of the elegant indirect experiments which Annamaria did could not convince him that her hypothesis was worth considering. Only years later when direct experiments on this transport system became possible did Jacques change his mind in a way which led to the operon theory.

Almost two decades later, at a *sympathique* moment, Jacques, slightly drunk, bragged that we had never been wrong. Today, I am wary when such a thought crosses my mind, for I can see how Jacques' memory was totally false. Almost no aspect of the operon theory was formulated correctly the first time: from the confusion of promoters and operators, to the postulate that the repressor was an RNA, to the denial that an operon ever existed.

This was not the important memory for Jacques to have had. What is important was the operon theory came uniquely as a result of correction from within, not from without. It was this self-correction that made Jacques unique.

We strolled on the beach at La Jolla talking about Jacques' desire to become Director of the Pasteur Institute. He knew that he could do it better than any other candidate. He felt that he was the only one who could get the loyal support of a scientific staff riddled with a sense of insecurity and inferiority. I tried to talk him out of it because, even successful, his creativity was more important than the revitalization of an institute. In the last year of his life, Jacques came to the realization that he had failed but so had the Pasteur Institute, for he left it as he found it—without leadership in the world.

Jacques tried to convert his failures at home into successes abroad. Everything he wanted for the Pasteur Institute he tried to achieve for the Salk Institute. He played the major role in writing an enlightened set of bylaws which with time and his absence have succumbed to the erosion of expediency. Often at our faculty meetings, he warned us to learn from the mistakes of the Pasteur Institute. Today, without his vigilance, we commit the very mistakes Jacques warned us to avoid. I wonder if such mistakes are inevitable (a sign of the times), or simply that only sages learn from history.

Jacques was far more gifted and creative than most of us who write about him today. His uniqueness came from the rare quality that his internal image of himself was accurate; he knew when he was derivative and when he was original, when he was honest and when he lied, when he used his heart and when he used his head.

To me he revealed, uninhibited, his puritanical accuracy in dealing with himself. He was in constant conflict between internal accuracy and external inaccuracy; and this drove him to leadership and tore him apart. One evening, he summed it up to me, "You can be rational and immoral, or irrational and moral. If I don't resolve this for myself, I will remain in pain either way." He came close in *Le hasard et la nécessité*.

M. Cohn, J. Monod, (unidentified).

Melvin Cohn

Melvin Cohn, a founding and resident fellow of The Salk Institute for Biological Studies since 1962, is a professor in the Conceptual Immunology Group. His investigations are theoretical and deal with the evolutionary selection pressures that shape the vertebrate immune system. A major thrust of his recent work has been the creation of a computer simulation of immune responses using the principles of nested cellular automata and adapting them to programs that can be run on typical desk-top computers or on a remote Internet server.

Dr. Cohn received his M.A. in Chemistry of Proteins from Columbia University, New York, N.Y., in 1941, and his Ph.D. in Biochemistry from New York University in 1949. From 1949 to 1955 he was a Fellow of the National Research Council at the Institut Pasteur in Paris. In 1955 he left to become Professor of Microbiology at Washington University School of Medicine, St. Louis, Mo., then (1959) Professor of Biochemistry, Stanford University School of Medicine. In 1961–1963 he returned to the Institut Pasteur as a National Science Foundation Fellow, and in 1963 returned to California to join The Salk Institute. Since 1964 he has also held the position of Professor in Residence, Department of Biology, University of California-San Diego, and he is currently head of the "Cell Distribution Center," a facility created in 1972 for supplying immune-related cell lines to the scientific community.

Dr. Cohn has held Visiting Professorships at different institutions: in Melbourne, Australia; in Paris, France; and in Madrid, Spain. He has lectured at many universities and institutes all over the world. He held the position of Scientific Advisor to CSIRO, Delhi, India, 1965–1970, and has been an Advisor to the National Institute of Immunology, New Delhi, India, since 1978. Dr. Cohn has held Fellowships at the Basel Institute for Immunology, Basel, Switzerland (1970–1990), and Churchill College, Cambridge, England (1971–present) and held the position of Fogarty Scholar-in-Residence at the National Institutes of Health, 1996–1998. He received the Eli Lilly Award in Microbiology and Immunology in 1956 and the Sandoz Prize in Basic Immunology in 1995. He is a member of numerous American, French, and international scientific societies.

Permeability as an Excuse to Write What I Feel

Georges N. Cohen

I met Jacques for the first time in September 1944. He was still wearing the uniform of a French Army major. We were just emerging from the hellish four years of Nazi occupation. I was totally ignorant of microbiology and, at the age of twenty-four, I had but limited laboratory experience: six months at the Laboratory of Animal Physiology at the Sorbonne (1939–1940), a few months in the Laboratory of Pharmacology of the Pharmacy School in Montpellier (1941), about six months in the biochemistry department of the medical school in Marseille (1942), and a year in the Service de Chimie Biologique of the Institut Pasteur (1943–1944), then headed by Michel Macheboeuf—the whole thing interrupted by military service, some weeks of captivity, escape, underground life, marriage, the birth of our first child. This, I should hope, conveys my lack of training and my profound need for guidance.

From the very first day, I was attracted by the radiating personality and the scientific and human qualities of Jacques, who rapidly became my mentor. Germaine Cohen-Bazire (the present Dr. Germaine Stanier), who had joined me at the Garches branch of the Institute to work on the mechanism of anaerobic fermentations, and I visited him every time we had a problem.

Very soon, Monod organized informal meetings at which we presented our work, subjecting it to the criticism of all the attendants. Later, I discovered that such seminars were common in the United States, but to my knowledge, this highly commendable habit had not yet been established in France. Germaine and I attended all of these seminars, which were held either in a large laboratory of the Service des Fermentations or in André Lwoff's office.

I had the ardent desire to join Jacques Monod's research team, but the space he was occupying in Lwoff's unit was too exiguous to allow my transfer. My purgatory at Garches lasted until the spring of 1954. However, I was enjoying Jacques' moral support and critical mind; his door was always open and he always listened with extraordinary patience to the report of our experimental results, subjecting them to benevolent but harsh censorship.

Following André Lwoff's suggestion, I spent a few months of 1948 at Oxford, in the Microbiology Unit headed by Donald D. Woods, where I learned the fundamentals of microbial nutrition. Returning to Garches, I became interested in the antagonisms between exogenous amino acids during the growth of *Escherichia coli*, in the hope of using this approach to elucidate certain biosynthetic pathways. During this period, Marie-Louise Hirsch and I observed that the growth of a leucine-requiring mutant of *E. coli* was inhibited by valine or isoleucine, whereas its growth on leucyl-glycine or glycylleucine was unaffected by the antagonists. I brought the manuscript to Jacques. In the discussion of the paper, one of the hypotheses put forward was the existence of a selective permeation system in *E. coli*, stereospecific for the three branched-chain amino acids. He struck out the corresponding paragraph with a choleric red pencil and told me, "Every time a microbiologist has no clear explanation for a nutritional puzzle, he calls upon permeability to conceal his ignorance." The paper appeared in 1953 with alternative explanations, which turned out to be entirely wrong (*Biochem. J.*, 1953, **53**, 25–29). The irony of fate was that in 1955, in Jacques' laboratory, Howard Rickenberg and I demonstrated the existence of a stereospecific permeation system for the three amino acids, not active on their peptides (*Comptes Rendus*, 1955, **240**, 2086–2088; *Ann. Inst. Pasteur*, 1956, **91**, 693–720). Jacques and I enjoyed telling this misadventure to younger people.

This brings me to my arrival in the Service de Physiologie Microbienne in 1954. Monod had been appointed Chef de Service in 1953, but was still working in Lwoff's laboratory, under the roof of the Institute—the famous attic. When I arrived there, the team consisted of Jacques, Melvin Cohn, David Hogness, Germaine Cohen-Bazire, and Annamaria Torriani. In Lwoff's group were Marguerite Lwoff, Elie Wollman, François Jacob, Dale Kaiser, Cyrus Levinthal, Julius Marmur, and Pierre Schaeffer. According to my recollections, the month of October witnessed the arrival of Bernard Davis, Aaron Novick, and Howard Rickenberg, and the departure of Cohn and Hogness to Saint-Louis, in the Department of Microbiology headed by Arthur Kornberg. It is during those few months between May and October that I discovered what was to be known as the β-galactoside permease.

I had decided to abandon for a while the study of amino acid biosynthetic pathways and to join the research actually done in my new environment. Mel Cohn and Jacques had just discovered the gratuitous induction of β-galactosidase by thiomethylgalactoside (TMG); Melvin had synthesized a small quantity of radioactive TMG. Jacques proposed that I see whether after addition of this labeled material to E. coli cultures, the radioactivity could be found linked to one of the cellular macromolecular components: DNA, RNA, or protein. In retrospect, this naive experiment had no chance of succeeding. It required the elaboration of the concept of a specific lac repressor by Pardee, Jacob, and Monod to enable Gilbert and Müller-Hill to succeed in answering the question asked by Jacques. However, the experiment brought interesting fringe benefits: the amount of radioactivity that could be found intracellularly was negligible in noninduced cultures, but very high in cultures that had been preinduced by growth in the presence of a galactoside.

In ten days' time, I found that the induced "accumulation" of TMG was energy-dependent, reversible, and stereospecific. Internal radioactivity could be displaced by cold TMG and by other thiogalactosides but not by the corresponding glucosides. The material extracted from the cells behaved chromatographically as authentic TMG; if the exposure was prolonged, another spot became increasingly important. I called it TMG-X. A year later, Len Herzenberg (who was to become a renowned immunologist) identified it in our lab to be 6-acetylthiomethylgalactoside, the product of an enzyme also inducible and characterized by Zabin and Kepes as thiogalactoside transacetylase. When I returned from summer vacation, Howard Rickenberg had arrived and Jacques sent the two of us to Helferich's lab in Bonn to learn how to prepare thiogalactosides from 1-bromotetraacetylgalactose and mercaptans. After returning from Germany, we synthesized radioactive ^{35}S-thiomethylgalactoside from 50 millicuries of radioactive methylmercaptan. Since the latter boils at $6°C$, we had to take all sorts of precautions, but we did not contaminate Lwoff's laboratory. Once the ^{35}S-TMG was made, we resumed the work and found that strains constitutive for β-galactosidase were also constitutive for TMG "acceptors," the i^- mutation being pleiotropic. Some strains were devoid of acceptors but could make galactosidase, thus being cryptic. Others could not synthesize galactosidase, but the specific acceptors could be induced.

Galactosidase and acceptors were thus independently genetically determined. Howard and I presented our preliminary results in 1955 (*Comptes Rendus*, 1955, **240**, 466–468). Monod later joined us, and he is responsible for the concept of a catalytic *permease*, as opposed to stoichiometric acceptor sites. Our results had been calculated so far in cpm/dry bacterial weight. Jacques made us recalculate the data in more sensible units; it appeared immediately that the molarity of the TMG incorporated into the cell was very high, reaching 2 to 4% of the bacterial dry weight. This excluded the hypothesis of a stoichiometric fixation on stereospecific receptors and led us to the concept of a catalytic permease, part of an "active pump" system. We developed this concept in two long papers in French and in a review in English, extending the concept to other small molecules, such as amino acids (*Ann. Inst. Pasteur*, 1956, **91**, 693–720 and 829–857; *Bacteriol. Rev.*, 1957, **21**, 168–194).

In retrospect, I am rather proud of this series of experiments, which established the existence of a galactoside permease and laid the grounds for the discovery of thiogalactoside transacetylase. The fact that more than one activity is controlled by the same pleiotropic mutation did in an indirect manner help the development of the operon concept.

In 1955, Monod, Rickenberg, and I went down two stories when the untimely death of Michel Macheboeuf liberated the Service de Biochimie Cellulaire, which was renovated thanks to the generosity of Madame Bethsabée de Rothschild, the baroness Edouard de Rothschild, and the Rockefeller Foundation of New York. During the renovation, Jacques was present every day, keeping an eye on the works, watching, that all the details were according to his plans. The Service, which I head presently, looks pretty much today like what it became after this remodeling (the first to have taken place since the building construction in 1900). Jean-Marie Dubert, David Perrin, François Gros, Alain Bussard, and later Adam Kepes joined our group. The foreign visitors I remember very vividly in the period 1955–1957 were Leonard Herzenberg, Frederick Neidhardt, Dean Cowie, and Harlyn Halvorson. With the two latter, a strong personal friendship was established, and I spent eighteen months in their laboratories during 1957 and 1958.

After that period, I went back to my studies on the regulation of amino acid biosynthesis in collaboration with François Jacob and Earl Stadtman. I earned the nickname of "Saint Georges l'Aminosaure," bestowed upon me by André Lwoff, for my insistence on working on something that was neither DNA, phages, colicins, nor β-galactosides. The discoveries we made justified the creation of a small team of workers. In Monod's lab, I occupied

80 square feet. The laboratory was establishing the existence of messenger RNA, discovering operators and operons. I was at a loss with my problems among all this intellectual revolution. Although I had strong ties with the Institute and with Jacques, I decided to cut the umbilical cord (at the age of forty!) and emigrate to the United States, where the National Institutes of Health was offering me ample space and facilities. I went to Bethesda in 1959 to make contacts with interested scientists and to examine the possibilities of sending my children to school. I accepted the position which was offered. I returned to France to prepare for our departure to find that Monod had obtained a promotion for me to Director of Research, and a new laboratory at Gif-sur-Yvette, near Paris. If I had thought for a while that Jacques had lost interest in me, I had been grossly mistaken. I stayed in France and, ten years later, I reentered the Institute, as the successor of André Lwoff. In 1972, when Monod became the director of the Institut Pasteur, he appointed me as his successor in the Service de Biochimie Cellulaire, where I am to this day, in the very laboratory where I started thirty-five years ago.

Others may relate in this book how Jacques' lab was associated with the discovery of the repression of biosynthetic enzymes by their end products and with the generalization of the negative-regulation model in biosynthetic systems. Others may describe the rationale by which Jacques

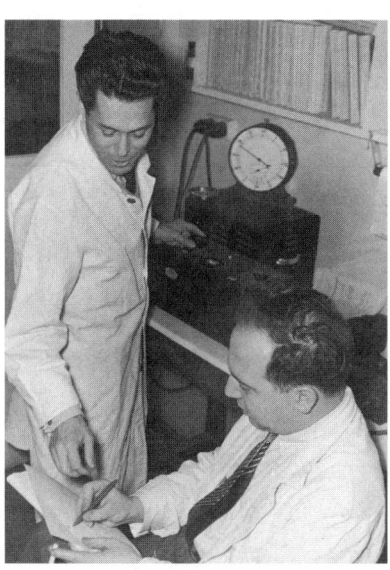

J. Monod and G.Cohen, 1954.

arrived at the concept of allosteric enzymes. I was physically present in his laboratory for a short period only (1954–1960), but very few have witnessed the birth of so many fundamental concepts in such a short period in a single place. This was due above all to the unusual intelligence of Jacques Monod.

As all human beings, he was not exempt from shortcomings, but the balance is overwhelmingly positive. His former students and colleagues occupy leading positions in France in the biological sciences, which Monod has greatly revolutionized not only in the Institute, but also in universities. He was a great scholar and a man of great charm, creating around him an atmosphere in which discoveries had to be made. His absence is and will continue for a long time to be severely felt.

Georges Cohen

Georges Cohen obtained his Ph.D. degree at the University of Paris in 1945. Among his many scientific achievements have been studies on the mechanism of anaerobic fermentations by clostridia; elucidation of the biosynthetic pathway leading from aspartate to threonine in microorganisms; discovery and study of the bacterial β-galactoside permease and amino acid permeases; and discovery and study of the incorporation of amino acid analogues into proteins (norleucine, p-fluorophenylalanine, and selenomethionine). The incorporation of selenomethionine is now used by crystallographers as a built-in isomorphic replacement. Dr. Cohen has in recent years isolated, characterized, and established the amino acid sequence of the methionine repressor.

Dr. Cohen has also conducted an enzymological study of the biosynthetic pathways leading to threonine and methionine in bacteria, including the genetic structure of the DNA regions coding for the enzymes involved. His work has helped establish the existence of multifunctional enzymes, that is, of polypeptide chains carrying more than one catalytic activity, and these studies have laid the theoretical grounds for amino acid overproduction by fermentation industries.

In parallel to his research with prokaryotes, Georges Cohen has been interested in the structure of cDNA. Investigators in his laboratory have established the structure of the gene coding for human transferrin and have also conducted studies of the genome of Tacaribe virus, a close relative of viruses causing severe hemorrhagic fevers in Africa (Lassa) and in South America (Junin).

Dr. Cohen was director of the Laboratoire d'Enzymologie of the CNRS from 1960 to 1969. At the Institut Pasteur, he headed successively the Units of Microbial Physiology and of Cellular Biochemistry. He has been a member of the Scientific Council of the Institute for 10 years, has chaired it for 4 years, and has been a member of its Board of Regents. He has served as executive secretary of the International Cell Research Organization, a nongovernmental organization, since 1987.

After his retirement in October 1989, Georges Cohen began a new career in bioinformatics. He has annotated the genome of the hyperthermophile *Pyrococcus abyssi*, is presently annotating the genome of *Acinetobacter* sp., and is collaborating with laboratories at the U.S. National Institutes of Health (NIH), at the Genoscope at Evry, and at the University of Brussels. Dr. Cohen has to his credit about 240 publications including six books translated into many languages (English, German, Italian, Spanish, Japanese, and Chinese). He has been named a Fellow of the Carnegie Institution of Washington (1957, 1960) and a Fogarty Scholar in Residence at the NIH (1977, 1978, 1979) and has received the Charles Leopold Mayer Award of the French Academy of Sciences (1974) and the UNESCO Carlos Finlay Prize (1989).

Honorary doctorate, University of Chicago, June 11, 1965. Ernst Chain, Jacques Monod, George W. Beadle, and François Jacob.

The Switch

François Jacob

O ne September afternoon in 1958, I walked into Jacques Monod's office. I was both extremely tired and extremely excited. Tired, because the previous night I had flown back from New York, where I had delivered a Harvey Lecture. Excited, because while preparing this lecture at the end of July, I had found what seemed to me a new way of looking at both lysogeny and induced enzyme synthesis. During the summer there had been no opportunity to see Jacques. I was, therefore, especially eager to discuss these ideas with him and to examine their relevance to the lactose system. I spoke with the glib tongue that a mixture of fatigue and excitement often produces. Jacques barely listened to me. He began to smile and soon started to roar with laughter—that well-known laughter which almost filled the whole building and enabled one to locate him. He thought that the idea was simply childish, and he was ready to provide at least five arguments against it. My fatigue prevailed over my excitement, and I decided to postpone the discussion until the following day. I went to bed.

The collaboration between Jacques and myself had begun only in 1957. Until that time, like all members of the Pasteur group, we had frequent discussions, mainly at lunchtime. But only occasionally did we perform experiments together. In fact, we were working on what everybody then regarded as two completely different systems. In 1957, Jacques had at last brought order into a problem that had been long confused. The old phenomenon of "enzymatic adaptation" became transformed into "induced enzyme synthesis," after Jacques and his group had shown that the increase in enzyme activity observed upon addition of inducer resulted not from the conversion of a preexisting protein, as then frequently believed, but entirely from de novo protein synthesis. Furthermore, a remarkably ingenious analysis of the properties of lactose analogues had revealed the separate roles of β-galactosides as inducers and as substrates. A posteriori, these two points might have appeared sufficient to reject what Joshua Lederberg called the "instructive," as opposed to the "selective," mechanism of enzyme induction. Nevertheless, in a paper that Jacques wrote during the winter of 1957–1958 for a symposium in The Netherlands, a flavor of instructionism still persisted. In what he then considered *the* theory, the amino acids were assembled in the correct peptide sequence on a specific template, the ribosome, resulting in the formation of a precursor, the "preenzyme," which remained attached to the template. Specific and reversible combination of the preenzyme with β-galactosides then resulted in folding and detachment of the molecule to produce the active enzyme. Either spontaneously or under the inductive influence of another, endogenous substance, the preenzyme could undergo a different type of folding to produce another protein configuration, endowed with another, unknown enzymatic activity. Jacques' main concern at that time was to explain by a unitary hypothesis two phenomena apparently worlds apart: enzyme induction and the inhibition of the synthesis of biosynthetic enzymes by an end product of the pathway. The latter phenomenon, first detected by Jacques' group and by D. D. Woods, was subsequently further investigated in several laboratories and termed "repression" by Vogel. The necessary unity could be attained either by a "general induction" or by a "general repression" hypothesis; the former was favored by Jacques.

However, the most important aspects of the paper written by Jacques were not the theoretical underpinnings, but rather the clarity and the precision with which the questions raised by the lactose system were defined. It had become clear that a new methodology was required for further advances. The most promising approach to the study of the lactose system appeared to lie through genetic analysis.

It was just at this time that Elie Wollman and I had made bacterial conjugation a workable tool for such an analysis. In 1953, a very close collaboration with Elie had started on lysogeny. This field had been completely revived by André Lwoff. The main questions which then required analysis were of a genetic nature and could be approached through bacterial conjugation. As it turned out, conjugation did not simply help with the analysis of lysogeny; for lysogeny proved to be a major factor in understanding the mechanism of conjugation. One of the first experiments we did together revealed the phenomenon of "zygotic induction," i.e., the production of phage when a chromosome bearing λ prophage entered a nonlysogenic recipient cell, but not when the reciprocal cross was performed. This proved that the immunity of lysogenic bacteria is caused by a cytoplasmic factor preventing prophage expression; and it also showed that in conjugation, genes could be expressed without undergoing recombination. A second series of experiments involving separation of the happy couples in a Waring blender, a torture conceived by Elie, showed that the male chromosome was injected into the female at a constant rate, following a precise time schedule. Although Elie did not like the metaphor, this became known in the lab as the "spaghetti experiment" because the female appeared to swallow the male chromosome like spaghetti. It then became relatively easy to dissect the main events of bacterial conjugation; to construct a map of the bacterial chromosome; to demonstrate its circularity; and, finally, to detect a class of genetic elements, the episomes, which moved back and forth between chromosome and cytoplasm. In short, conjugation had become a useful instrument for the analysis of any bacterial function.

In September 1957, the decision was reached with Jacques to use conjugation for a genetic analysis of the lactose system. Jacques had isolated a number of mutants affected in the production either of β-galactosidase or of permease, as well as constitutive mutants—the z, y, i trinity. It was easy to insert them in various combinations in either male or female bacteria. Arthur Pardee, who had come to spend a sabbatical year with Jacques' group, was interested in the project. This led to the so-called PaJaMa (Pardee, Jacob, Monod) experiments. In brief, these experiments showed that β-galactosidase was synthesized at maximal rate within two to three minutes after entry of the gene in a cell; that the i gene determining the inducible versus constitutive character of enzyme synthesis was distinct from the z and y genes, which controlled the synthesis of β-galactosidase and β-galactoside permease, respectively; and, finally, that inducibility was dominant over constitutivity. It was therefore the inducible i^+ allele, and not the constitutive i^- one, that was expressed in the cytoplasm by an active product. This finding did not support Jacques' earlier hypothesis, according to which constitutivity resulted from the synthesis of an endogenous inducer. However, it nicely fitted a proposal made by Szilard, who happened to spend a few days in Paris at that time. Like Jacques, Szilard wanted to explain the induction and the repression of enzyme synthesis by a common mechanism. In contrast to Jacques' "general induction model," he preferred a "general repression model," for several reasons. The dominance of i^+ over i^- was clearly in agreement with the latter model, which furthermore could easily accommodate the immunity of lysogenic bacteria. These new, abstract entities, the products of the i^+ gene in the lac system and of the C_I^+ gene in phage, were called "repressors."

The results of the PaJaMa experiments strongly influenced the thinking of the whole Pasteur group in two respects. One was, obviously, the nature of the regulatory mechanism involved in enzyme synthesis. The other was, somewhat less obviously, the nature of the "template" involved in protein synthesis. Proteins were known to be synthesized on ribosomes. Ribosomes were known to contain stable RNAs. And the central dogma then proposed by Francis Crick was epitomized as DNA $\rightarrow$ RNA $\rightarrow$ protein. At that time, it was tacitly admitted that the stable RNA components of ribosomes represented the specific templates for protein synthesis; in other words, a gene produced one or several specific ribosomes, which in turn manufactured the corresponding protein. However, the notion of stable intermediate RNAs was not in good agreement with the results of the PaJaMa experiments, which had shown that, in a mutant bacterium unable to synthesize β-galactosidase, the transfer of the gene almost immediately led to a maximal rate of synthesis. During the long discussions with Arthur and Jacques, we examined some unorthodox ideas: perhaps there was no RNA intermediate and the protein was synthesized on the gene itself, or perhaps the RNA intermediate was very unstable. We did not like the former hypothesis; as for the latter one, there were very few arguments pro or con. We finally decided that the presumed instability of the intermediate RNA could be determined by transferring a ^{32}P-labeled lac^+ male chromosome into a lac^- female, allowing enzyme synthesis for a short time, and then destroying the gene by ^{32}P decay. The amount of residual enzyme synthesis following gene destruction would provide a measure for the stability of the template. In the fall of 1958, Arthur Pardee returned to Berkeley armed with the design of this difficult experiment, which he later very skillfully performed with his student Monica Riley.

During that year, 1957–1958, I came to work much more closely with Jacques and to know him better. At least twice a week there were long meetings in his office where various aspects of the work in progress were discussed. He had an unusual feeling for the interplay between theory and experiment, and was a virtuoso of the hypothetico-deductive method. Not only did he very rapidly perceive the experiment necessary to check a particular point; he also squeezed the results to the very limit of their significance. He had the gift, mainly possessed by poets, of seeing signs that others did not. At the same time, his attitude toward theories always amazed me. I have a certain taste for changing fixed ideas, for throwing old idols onto the garbage heap, even if I have contributed to setting them up. Jacques, in contrast, did not like to get rid of his theories. He had a strong tendency to stick to his model, sometimes slightly beyond the point of reason.

As is frequent with so rich and strong a personality, several different and sometimes contradictory individuals coexisted within Jacques Monod: two at least, if one considers only the scientist. Each took over in turn, depending on his mood and on the circumstances. The first of these individuals—let us call him Jacques—was a very warm and generous man of great charm; a man interested in people as well as in ideas, constantly available to his friends, ready to discuss their problems and find a solution; a man of great rigor and insight, always to the point, asking cogent questions, and sharply self-critical. The second individual—let us call him Monod—was incredibly dogmatic, self-confident, and domineering; a person unceasingly in quest of admiration and publicity, demanding to be the focus of attention; a person making definitive black-and-white value judgments on everything and everybody, fond of teaching fellow scientists the *real* meaning of their own work but sweeping away as nonsense any objection they might timidly offer. Jacques was able to bring all his personal activities to a halt and go out of his way to help a friend in a difficult situation. Monod could quite easily turn a friend into an enemy with a few words. In private, one dealt almost always with Jacques. In larger gatherings, one sometimes had to deal with Monod. Working with the former was an exceptional pleasure. Arguing with the latter could be a difficult experience. Fortunately, at that time, it was mainly Jacques who was in command.

The afternoon that I returned from New York, I was received by Monod. The next morning, however, when I again entered his office after an eighteen-hour sleep I found Jacques. I started once more to tell my story, and the discussion between us went on until the end of the afternoon.

My main point was the following: We all had noticed the analogies between the results of zygotic induction with lysogenic bacteria and those of the PaJaMa experiments with the *lac* system. But only when trying to produce a reasonable picture of lysogeny for my Harvey Lecture on "Viral Functions" did I realize how far the parallel could be pushed. In both cases, a group of normally silent genes could be triggered and become expressed at will; in both cases, this silence was due to a single, distinct gene: C_I in phage λ, i in the *lac* system; in both cases, genetic analysis showed that the wild-type allele of this gene was expressed by a cytoplasmic product, a repressor blocking in some way the expression of the other genes. These analogies appeared so great that the postulate of an identical mechanism seemed to me inescapable.

If so, each system with its particular advantages and disadvantages should aid in the analysis of the other. In lysogeny, genetic analysis was especially easy, while gene products were barely known and difficult to assay. Conversely, in the *lac* system, the proteins were easy to assay, but genetic analysis was not so precise. If one assumed an identical mechanism and a complementarity of the two systems, then lysogeny already set some constraints on possible models. In order to produce λ phage particles, some 50 to 100 different proteins had to be synthesized, all these syntheses being blocked by the C_I repressor. It seemed to me extremely unlikely that the phage did produce 50 to 100 different species of stable templates, of stable ribosomes, each one being blocked by the same repressor. Repression had to operate on one element common to all functions. It could prevent the synthesis of one peptide chain common to all proteins, an idea which I did not much like. Alternatively, it could work on a single lock, a master switch simultaneously controlling the production of several proteins at some level—then unknown—of the protein-synthesizing machinery. Since the only molecular species at that time known to contain the genetic information for the structure of several proteins appeared to be DNA, it seemed most likely that repression operated at the level of DNA. Starting from the viewpoint of the phage system, I therefore wanted Jacques' opinion about the implications for the *lac* system of two basic concepts: (1) Repression (or induction) operates not progressively, but like a switch, by a yes-or-no, an on-or-off mechanism that involves only two states. (2) Genetic units of an order higher than the gene must exist: "units of activity" that contain several genes subject to unitary expression, such expression probably being regulated at the level of DNA.

During my long presentation, Jacques offered only a few remarks designed to make me clarify certain points. I had the impression that he was interested, although somewhat reluctant to swallow the whole story. After a long silence, he began to argue carefully, point by point, attempting to explain as clearly as possible the implications of the model for the *lac* system. There were some aspects that he just didn't like, for instance the notion of repression acting at the genetic level. "However," he said, "this might simply reflect the bias of a scientist formed by classical genetic training." For him, the gene is a noble, intangible entity. "But bacterial geneticists," he added, "have shown that episomes can be introduced into or removed from a chromosome almost at will, a situation unthinkable ten years previously. Actually," Jacques concluded, "there is no direct evidence either for or against the idea of repression at the level of DNA, and we should keep this possibility in mind."

His objection to the switch, the on-or-off concept of protein synthesis, appeared more serious to me. Differential β-galactosidase synthesis was always linear, but the rate of synthesis varied as a function of the nature and concentration of inducer. This, Jacques believed, could not be reconciled with an on-or-off system of synthesis. I myself have never been much of an expert in reaction kinetics. My main argument was the simplicity of a yes-or-no system for phage induction in lysogenic bacteria. A few weeks before, however, I had perceived that an on-or-off mechanism could in principle account satisfactorily for different rates of synthesis. This insight had come to me while I watched one of my sons playing with a small electric train. Although he didn't have a rheostat, he could make the train travel at different but constant speeds, just by turning the switch on and off more or less rapidly. It seemed to me that a similar mechanism could govern protein synthesis, provided that the system had sufficient inertia. This argument did not appear very convincing to Jacques!

When considering the role played by polygenic units of activity in the *lac* system, Jacques immediately picked out an important implication of the model to which I had not paid attention. If both β-galactosidase and permease genes formed part of a single unit which could only be expressed as a whole, then the syntheses of the two proteins should be completely coordinated *under any condition*; in other words, irrespective of the type of synthesis, the two activities should always occur in the same ratio. Indeed, if one knew more about the permease, the two proteins should even be always present in the same relative amounts. And Jacques firmly stated that this was not so. The glucose effect—i.e., the decrease in rate of synthesis in the presence of glucose—did not affect β-galactosidase and permease to the same extent. Furthermore, several compounds, such as inositol-α-galactoside, induced β-galactoside permease but not β-galactosidase.

For me this was a hard blow. In fact, it was exactly the type of argument I was both expecting and fearing to hear from Jacques. After a lag, it occurred to me that the glucose effect should not be perhaps taken too seriously since nobody knew what it meant, where it acted, or what the mechanism was. However, the uncoupling of induction by inositolgalactoside worried me since it could turn out to be fatal to the whole hypothesis, unless one wished to argue that the status of permeases, their nature and specificity, were not yet very clear.

It was late in the afternoon; I was completely exhausted while Jacques remained as fresh as in the morning. We stopped. We had drinks and started to talk with other members of the group. We soon began a new argument in which Monod replaced Jacques for a while. He always insisted on being purely logical. He decided that I was mainly intuitive. This was all right. I had no particular objection to being intuitive, until Monod claimed that intuition did not exist, and he never understood what it meant. Then I replied with the usual argument, the comparison between the computer and the human mind: while the former has to scan all possible situations before reaching a conclusion, the latter uses shortcuts. The discussion soon cooled down and turned to the respective virtues of cognac and whisky.

From that day on, a particularly close and friendly cooperation began between Jacques and myself. The next few years remain in my mind as among the most active and exciting times of my scientific life. Very critical of the model to begin with, Jacques rapidly became more and more interested in it. His criticisms were more and more constructive. Some of his objections disappeared. Others, such as the uncoupling of induction by inositol-galactoside, remained as skeletons in the closet. We agreed, however, to leave them there for a while. Almost every day we had long working sessions, each bringing new results, each trying a new concept on the other, as if we were tennis players. As ever, Jacques continued to focus his attention on the *lac* system. I myself tried to consider both systems, fascinated as I was by the analogy between them. The detection of a mutant in one system immediately led to the description and the isolation of a symmetrical mutant in the other. If our collaboration became so close and our work so exciting, it was mainly because the more results we obtained, the greater the similarities became. In the course of our dialogue, each of us could proceed with his own internal monologue. I think that this may well be one of the keys to intellectual pleasure.

In the fall of 1958, during one of our sessions which I believe Georges Cohen attended, Jacques pointed out that if the switch considered as the acceptor site of the repressor did exist, it should be specific, therefore genetically controlled, and in turn liable to alteration by mutation. Immediately we began to discuss the properties expected from such a mutant, drawing arrows on the board to represent the elements of circuits. Such an acceptor mutant (called A^- at that time, later to become O^c) would no longer be sensitive to the repressor produced by the i gene; it should be constitutive. Furthermore, in diploid cells, the chromosome carrying an A^- mutation should continue to be expressed in the presence of another chromosome carrying the wild A^+ allele. The hypothetical A^- mutation accordingly had very clear-cut properties. An A^- mutant should be constitutive; but in contrast to i^- constitutives, which are recessive to i^+ in diploids, A^- should be dominant over A^+. In addition, only β-galactosidase and permease genes located *cis* with respect to A^- should be expressed in A^-/A^+ diploids.

During all this discussion, we had become very excited. Suddenly, I realized that a mutant presenting exactly these characteristics had been known in phage λ for several years. This was the so-called "virulent" (v) mutant that Elie Wollman and I had isolated and analyzed. It grew on bacteria lysogenic for λ. It was dominant, since it multiplied in lysogenic bacteria mixedly infected with v and v$^+$ phages. Finally, the use of suitable genetic markers had shown that only those genes located in position *cis* to the v mutation were expressed. This mutant accordingly possessed every property anticipated for the acceptor of the repressor. How stupid I had been not to have thought of it before! From that day on, our joined confidence in the model increased by several orders of magnitude.

During the winter of 1958–1959, we were both busy testing experimentally different aspects of the model. Jacques wanted first to check the coordinated expression of β-galactosidase and β-galactoside permease under the most varied possible conditions. This he did with Madeleine Jolit, Carmen Sanchez, and David Perrin. They showed that the ratio of the two activities was constant under all conditions, except when inositol-galactoside was the inducer. We were again forced to lock the closet. Two additional features in favor of the model also subsequently emerged from this work. Certain mutants in the β-galactosidase (z) gene turned out to produce permease in lower amounts than did wild type, while mutants in the permease (y) gene always produced a normal amount of β-galactosidase: polarity of expression soon became an additional argument for interpreting the *lac* region as a unit of activity. A new protein, β-galactoside transacetylase, was discovered by Adam Kepes and Irving Zabin. This added a new gene to the *lac* region, a new activity to measure, and a new site of polarity. The induction of transacetylase was completely coordinated with that of β-galactosidase and permease, again except when inositol-galactoside served as inducer. Like β-galactosidase, transacetylase was not synthesized in response to inositol-galactoside, which induced only the permease. This result began to weaken our belief in the strength of the inositol-galactoside argument.

As my contribution to the work, I wanted to isolate the dominant constitutive A$^-$ mutant, homologous in the *lac* system to the v mutant of phage λ. This would have been easy, provided that stable diploids for the *lac* region had been available. Unfortunately, the zygotes obtained by conjugation and used in the PaJaMa experiment were only transient and did not allow mutant selection. However, a few months earlier, before Elie Wollman had left to spend a year in Berkeley with Gunther Stent, we had shown that in Hfr strains the sex factor F is transferred to females at the tail end of the chromosome. Furthermore, Edward Adelberg, who had spent a sabbatical year with us, had found upon his return to Berkeley a strange F⁺ derivative of an Hfr: the sex factor F had become autonomous, while retaining the memory of its former location in the Hfr chromosome. This property suggested that, when returning from the integrated to the autonomous state, the F factor had carried with it a small adjacent segment of the Hfr chromosome, just like transducing phage. If this were true, a way to obtain stable diploids for any chromosomal segment had been discovered, provided suitable Hfr strains were available. During the winter and spring of 1958–1959, Martine Tallec and I isolated a series of Hfr strains that ended close to the *lac* region, and we tried to derive F-*lac* from them. The first F-*lac* was obtained and characterized in June 1959. It covered the whole *lac* region, recombined with the chromosome, so that any *lac* mutation could be transferred to it, yielding stable diploids. From such a strain, it was easy to derive a constitutive dominant mutant. This was achieved three weeks later.

The cooperation with Jacques was closer than ever. In the large lab adjacent to his office on the first floor, people were adding inducers, assaying activities, and measuring syntheses. In my lab on the third floor, in André Lwoff's laboratory, mutants were being accumulated and mapped by conjugation. I spent a large part of my time going up and down the stairs. Almost invariably, every day ended in long discussions in Jacques' office, more and more arrows being drawn on the board. His logic, his imagination, and his tenacity were amazing. When a discussion was beginning to get off the track, he immediately stopped it, focusing the attention on what was the real point. He was always ready to invent some new experiment and to bet a bottle of whisky on its outcome.

During the fall of 1959, we were extremely busy checking the properties of dominant constitutive A⁻ mutants. The A⁻ mutation was first coupled with most other *lac* mutations on the chromosome and on the F-*lac* factor, to obtain all possible combinations of diploids in *cis* and *trans* positions. The properties of these strains were carefully determined and turned out to be exactly as predicted: the A⁻ mutation resulted in dominant constitutive expression of the three *lac* genes but only of those located in *cis* position. Furthermore, as expected of a structure that could switch the activity of the whole *lac* unit, the A⁻ mutation mapped at one end (z) of the unit. That part of the model which dealt with polygenic units of genetic activity had consequently received complete support. After a long debate on the respective virtues of Greek and Latin words, the *i* gene became a "regulator" gene, as opposed to "structural" genes which specify the amino acid sequence of proteins; the unit of activity was called an "operon" and the switch an "operator." The A^- mutation consequently became an "operator constitutive" (O^c). The general terminology applied, of course, to phage λ as well as to the *lac* system.

There was another aspect of the model in which the symmetry between λ and *lac* proved to be important: the site of action of inducers. When the PaJaMa experiments had shown the existence of a cytoplasmic repressor, the site of inducer action had been transferred from the preenzyme to the repressor: β-galactosides were considered to inhibit the inhibiting effect of the repressor on β-galactosidase synthesis—what Jacques liked to call "double bluff"—thereby releasing enzyme synthesis. There was, however, no direct proof of that postulated interaction between repressor and inducer. During the winter of 1958–1959, while Alan Campbell was spending a sabbatical year with me, we isolated a new mutant of λ. This mutant was noninducible (ind$^-$) by UV light or other inducers. In the λ version of the model, the regulator gene C_I was considered to produce a cytoplasmic repressor blocking the expression of one, or more probably several, operons. Ultraviolet irradiation was assumed to result in the production of some substance able to inactivate the repressor, thereby allowing expression of these operons and consequently phage multiplication. The ind$^-$ mutation turned out to be located in the C_I gene, and to be dominant over the wild allele: in lysogenic cells multiply infected with ind$^-$ and ind$^+$ phages, the ind$^-$ repressor prevented UV-induced expression and multiplication of all λ chromosomes. The most likely interpretation of the ind$^-$ mutation was a loss of sensitivity of the repressor to the unknown inducing substance produced by UV irradiation.

According to the rule of symmetry between the λ and *lac* systems, a homologous mutation had to be found in the *lac* system. This should be a mutation of the i gene that would be dominant over i^+ (or i^-) in diploids and resulting in noninducibility of the whole *lac* operon. At first, Jacques had some doubts about the possibility of obtaining such a mutation: were it to exist, he assumed that it would have already been isolated among the many *lac*$^-$ mutants he had produced. However, this mutant could conceivably be rare, and obtainable only from a diploid strain. Actually, it was with some difficulty that in the spring of 1960, I succeeded in isolating such mutants from a diploid i^+/i^+ strain. Located in the i gene, the mutations called i^s (for super-repressed) were dominant over both i^+ and i^- alleles. For an i^s mutant to produce the three Lac products in small amounts, inducer concentrations 100 to 1,000 times higher than those required for wild type proved necessary. It seemed therefore that the mutated repressor was altered in such a way that it could scarcely be inactivated by inducers. The properties of such mutants were later worked out in greater detail by Clyde Willson and Mel Cohn in Paris.

Thus the pieces of the puzzle were being rapidly assembled. In the fall of 1959, Arthur Pardee sent us the first results of the experiment he had performed with Monica Riley: after destruction of the z gene by ^{32}P decay in zygotes which had already produced β-galactosidase, the enzyme was no longer synthesized. The conclusion that protein synthesis did not involve *stable* intermediate templates was thus inescapable. Since it was not possible to consider DNA itself as the template, an unstable intermediate RNA had to exist. Despite many discussions, in particular with François Gros, who was the RNA specialist in the group, we did not get much further. At a meeting in Copenhagen attended by most molecular biologists during the fall of 1959, I discussed our ideas, in particular the idea of an unstable RNA as a template in protein synthesis: nobody paid the slightest attention. At Easter 1960, I went to a symposium on bacterial genetics in London and spent a few days in Cambridge. During a small meeting in Sydney Brenner's room at King's College, I described the latest results obtained in Paris and Berkeley on the regulation of protein synthesis and mentioned once again the unstable RNA hypothesis. Francis Crick and Sydney reacted immediately; they made the crucial connection between the unstable template hypothesis and RNA with a rapid turnover, which had been previously described by Hershey, and by Volkin and Astrachan in bacteria infected with T-even phages. This was a day of great excitement, which ended up in a party at the Cricks'. In spite of the presence of many pretty girls, Sydney and I spent the whole evening planning the details of an experiment to be done at Caltech, where we both happened to be invited at the same time. In June 1960, we thus met in Pasadena and performed this series of experiments with Matt Meselson in Max Delbrück's laboratory. Mainly thanks to Sydney's exceptional skill and quickness of mind, it could be shown that, upon infection with phage T4, the RNA with a rapid turnover—then called "tape" and later "messenger"—became associated with ribosomes formed prior to infection, and produced phage-specific proteins. In Jim Watson's laboratory at Harvard, François Gros and Walter Gilbert simultaneously demonstrated the existence of a similar RNA with a rapid turnover in uninfected bacteria.

With the discovery of messenger RNA, the process of protein synthesis was shown to occur in two successive steps: transcription from DNA to messenger RNA, and translation of messenger RNA into polypeptide chains on ribosomes. This gave a new impetus to the whole field. For the

Pasteur group, it raised a number of specific questions, which were endlessly debated: Is there a unique messenger for the whole operon? Does repression act to prevent translation, or as it appeared more likely in view of messenger instability, does it act directly on DNA to prevent transcription? On the whole, however, the messenger concept gave strong support to our model. The repressor existed in two alternate states: active or inactive, resulting from interaction with the inducer. When active, it reacted with the operator, thus blocking the expression of the entire operon, either at the transcriptional or at the translational level. The whole system could be viewed as an on-or-off switch alternating between two states, just like the switch of the small electric train. The rate of protein synthesis was determined by the relative times spent in the on and off positions. A few years later, the notion of on-and-off regulation was used again by Jacques together with Jeffries Wyman and Jean-Pierre Changeux, when they explained the allosteric properties of proteins by their transitions between alternate conformations.

Our simple regulatory circuit—which an engineer could have easily designed—applied not only to *lac* and λ but also to certain biosynthetic pathways. With Georges Cohen, we had shown the existence of a similar system acting in the regulation of the synthesis of enzymes involved in the production of trytophan. Even our old enemy, the uncoordinated induction of permease by inositol-galactoside, had vanished when it was realized that the permease thus induced was not β- but α-galactoside permease, whose structural gene was not located in the *lac* operon! This was the last skeleton to disappear. When the true role of inositol-galactoside as an inducer was discovered by Adam Kepes and Claude Burstein in June 1960, I was at Caltech with Sydney Brenner, hunting for the messenger. About half an hour before I was going to give a seminar, I received the following telegram: "Be careful. New compounds give uncoordinated induction galactosidase, permease, acetylase. Regards. Jacques." I took it as a practical joke, which it was, and sent back a telegram: "Model already destroyed. Proof inducer folds each protein. Regards."

At that time, we had both become convinced that the model gave a reasonably good description of the regulatory mechanism involved in the lactose, λ, and tryptophan systems. We usually called it *negative* regulation, meaning that in the absence of the regulatory gene product, the operon is expressed at high rate. We often discussed the possibility of the symmetrical situation, or *positive* regulation, where in the absence of the regulatory gene product, an operon would not be expressed at all. Our attitudes about positive regulation were opposed. I liked the idea of the existence of both positive and negative regulations, because it seemed to me that the combination of both types offered a much greater flexibility for complex systems and differentiation. Jacques, on the other hand, was not very fond of positive regulation, because he liked nature to provide unique solutions. Since a combination of two negatives was equivalent to one positive, he did not see the logical necessity of adding another, distinct mechanism. He made the clear-cut point that a decision could be reached experimentally only by deletions of regulatory genes: in negative systems such deletions should result in a constitutive expression of the structural genes, while in positive systems they should lead to a complete lack of expression of these genes. For a long time, however, Jacques remained refractory to positive regulation, even when, later in the 1960s, evidence in favor of it began to accumulate in several laboratories. It was shown, for instance, that regulatory genes N and Q of phage λ were likely to act in a positive way. Ellis Englesberg, then working on the arabinose system, gave several seminars at Pasteur. His results also required some positive effect of the regulatory gene product. After each seminar, however, he received a severe lesson in regulatory genetics from Monod, who always insisted on a notion "that even a schoolboy cannot ignore: $- \times - = +!$"

In the fall of 1960, Jacques and I decided to assemble the various pieces of information then available into a story, mainly written by Jacques. The paper was sent in December to the *Journal of Molecular Biology*. On the whole, the model described in this paper has withstood the test of time and of deeper biochemical analysis, with one major exception. For rather poor reasons, we had decided that the repressor should be RNA rather than protein. There were several arguments, the main one being that of specificity. We visualized the operator as a short string of DNA, a few nucleotides long. In the bacterial chromosome as a whole, the specificity of several hundred operators therefore had to be determined by short base sequences. The recognition of such sequences by proteins appeared to us to be difficult. On the other hand, recognition by RNA through base pairing would be easy. This notion had a major weakness: the interaction between a repressor RNA and a β-galactoside inducer required a protein, for which no mutation had ever been detected. During the winter of 1961–1962, however, together with Raquel Sussman, who was spending a year in Paris, I isolated a large series of new regulator C_I mutations in λ. Some of these mutations turned out to be thermosensitive; others were *amber*, i.e., corrected by suppressors known to act at translation level. This was unambiguous proof that the λ repressor was—or contained—a protein. It was then easy to look for and to isolate similar mutations of the *i* gene in the *lac* system.

In both λ and *lac* systems, the time was thus ripe to hunt for repressors. They were captured at Harvard some years later.

However rigorous its planning, a piece of research always contains a good deal of sound and fury. Only a posteriori, as fixed by the flashes of memory, does it become organized into a story. By then, the story has lost much of the flavor of life. I am fully aware that the story I have told here is only one among many possible accounts that might be given of the same events. It does not reflect the random motion and background noise, the exaltation and boredom, the small pleasures and large disappointments which, in various combinations, make everyday life in a laboratory. Nor does it convey the richness of Jacques Monod's unusual personality and inimitable style. It is now more than one year since he has disappeared. It is still difficult to realize that he is no longer present in some room of Pasteur, ready to talk about anything and to bring you back to the real world: the lactose system.

Announcement of the Nobel Prize: the three laureates in the garden of the Institut Pasteur: A. Lwoff, J. Monod, and F. Jacob, October 14, 1965.

Afterthoughts: François Jacob

In the beginning of the 1970s, I shifted from work on *Escherichia coli* to work on mouse embryo and mouse teratocarcinoma. Around 1975–1976, I realized that it had become difficult to study evolution without a good knowledge of embryonic development, a relationship which later became a new field under the name Evo-Devo.

It had long been believed that every organism was made of particular molecular types. There were cow molecules, sheep molecules, snail molecules, etc... And it was assumed that a cow was a cow because it was made of cow molecules. Around 1975, it began to appear progressively that there was no such things as cow molecules but that, in contrast, all organisms were made of similar molecular species. Evolution was building new molecules with old ones through a process similar to tinkering ("bricolage"). Tinkering means that old molecules were transformed by adding or removing or recombining or transferring pieces of DNA sequences, or that some molecular types were given new functions. In many instances, changes in regulatory circuits controlling gene expression were likely to be involved. And, since that time I have essentially been interested in this aspect of biology.

The Editor's suggestion for reading some of François Jacob's books to learn more about him:

The Logic of Life, Pantheon Books, New York, N.Y., 1973

The Possible and the Actual, University of Washington Press, Seattle, 1982

The Statue Within, Cold Spring Harbor Laboratory Press, Cold Spring Harbor, N.Y., 1995

The PaJaMa Experiment

Arthur B. Pardee

I n this memoir I recall what I can of my association with Jacques Monod during the sabbatical year I spent in his laboratory from 1957–1958. During this period we carried out the bacterial mating experiment, which was subsequently named the "PaJaMa experiment" because it was done by Pardee, Jacob, and Monod (*J. Mol. Biol.*, 1959, **1**, 165). This experiment provided the fundamental basis for the regulatory phenomena of enzyme induction and repression.

Prior to 1957 I did not know Monod very well, having met him only briefly in 1952 when I gave a lecture on enzyme changes following bacteriophage infection, at the Pasteur Institute in Paris. Later that summer we both attended the Conference at Royaumont, at which many of the fundamental discoveries of molecular biology were discussed. During the next few years (1953–1957), work in my laboratory often was parallel to publications from Monod's and Jacob's groups.

My interests in enzyme induction arose from several studies I had made—on enzyme changes following viral infection, on regulatory interactions between protein synthesis and nucleic acid synthesis—and from our discovery of feedback inhibition of enzyme activity, and of enzyme synthesis repression. All of these researches led me to questions regarding the means a cell used to regulate its metabolism. Evidently the regulation of enzyme *production* was at least as important as the regulation of enzyme *activity*, and these thoughts led me to study enzyme induction and repression. During the mid-1950s both Monod's laboratory and mine came up with discoveries of inducible transport systems (permeases) and the repression and derepression.

During that period Gunther Stent, my colleague at Berkeley, visited the Pasteur Institute yearly and always came back with glowing reports of the progress in science there. I recall that I often told him about our most recent results and heard him respond that the Pasteur group had "done it already." It was a glorious day when I told him about something new that we had done and, after a long pause, he responded, "Well, they haven't done that, but they are thinking about it!"

The convergence of my research interests with Monod's led me to apply to spend my sabbatical leave at the Pasteur. At that time Monod and Jacob had been collaborating in studies on the transfer, between bacteria, of the genes for production of β-galactosidase and galactosidase permease. After these genes were transferred by conjugation from a galactosidase-positive (lac^+) donor to a lac^- recipient, the mated cells were capable of forming colonies on lactose plates. These results demonstrated transfer of the genes for lactose utilization, as measured by their ability to permit growth. Monod suggested that I do the opposite experiment, i.e., to see if deletion of the transferred genetic material would arrest growth. His proposal for a method to delete the genes was based on some experiments done by Elizabeth McFall under guidance by Gunther Stent and myself. These studies showed that decay of ^{32}P that had been incorporated into bacterial DNA destroyed the capacity of the unmated cells to make β-galactosidase. So the initial project that I was to work on was to transfer ^{32}P-labeled genetic (lac^+) material by conjugation into lac^- cells, and then to ask whether decay of the ^{32}P would destroy the ability of the mated cells to grow on lactose plates. Although I did not do these experiments in Paris, they were later done by Monica Riley in my laboratory.

Upon arriving at the Pasteur Institute in the fall of 1957, Monod proposed that I first become acquainted with the mating system by asking whether transfer of the lac^+ genes into a lac^- cell would permit the mated cells to form β-galactosidase. This direct measurement of enzyme activity was a dramatic departure from the previous techniques for studying mating, which depended upon the appearance of colonies on selective plates a day or so after the mating event. The immediate or delayed appearance of an enzyme after conjugation could provide much more direct data than growth studies as to the timing and mechanism of gene transfer and expression.

My first technical problem was to distinguish between production of the enzyme β-galactosidase by mated cells and the unmated donors. (The unmated recipient in a cross between a lac^+ donor and a lac^- recipient is genetically incapable of producing the enzyme.) Monod's suggestion was that I obtain chloramphenicol-resistant lac^+ donor cells; when mating was done in the presence of this drug the donors' protein-synthesizing ability would be blocked, but the mated cells might be resistant. Therefore only the mated cells would be capable of making the enzyme. But chloramphenicol-resistant *E. coli* are hard to produce. I suggested that we could use the streptomycin-resistance property of the recipients for our kinetic work; this marker had been used in all the previous mating and plating experiments to distinguish resistant mated cells from sensitive, unmated donors. Within a few days of my arrival (September 25, 1957) I had shown that streptomycin-resistant recipient cells provided a means for studying enzyme production as well as colony formation by mated bacteria. I think that this rapid initial success provided the mutual confidence that was essential for our later progress.

We soon demonstrated that the enzyme β-galactosidase appeared at maximal rate in mated cells within minutes of the time that the corresponding lac^+ genes were transferred, timing of transfer being measured by the interrupted mating technique of Jacob and Wollman. This result in itself was of great interest, since it demonstrated that the gene becomes active virtually as soon as it enters an appropriate environment. Also it showed that the techniques for mapping genes according to mating times were almost surely determining the time of entry of the gene into the cell. The discovery of messenger RNA had a major root in this experiment.

I recall noting with interest how technical approaches of our laboratories differed. It was a custom in the Pasteur Institute to make each measurement of β-galactosidase as a function of duration of reaction, with the result that each measurement was relatively laborious. Perhaps a dozen experimental values could be obtained in a day's work. I decided that a cruder approach would suffice, and that only one time point was needed for each measurement. In this way I could increase the acquisition of data by a factor of about 10 and yet have sufficient reliability to work out the main features of the mating system. I recall Monod coming into the laboratory and shaking his head in amazement at the way that I was setting up the experiments with a dozen cultures and a hundred or more sample tubes. This "rough-and-ready" approach was important in the rapid progress we made in the following months.

Following our success in demonstrating that enzyme production accompanies gene transfer, Jacques (as he preferred me to call him) suggested that I look at the transfer of the β-galactosidase (z^+) gene into a constitutive z^- recipient cell. It will be recalled that wild-type *E. coli* need to be fed a low-molecular-weight inducer related to lactose to make them produce

β-galactosidase. The Pasteur group had discovered constitutive mutants ($lac\ i^-$), which produce β-galactosidase at a high rate independent of the presence or absence of an inducing compound. Did the constitutive cells make the enzyme because an inducer is synthesized within them? We could transfer the z^+ gene into a cell that is z^- (and hence cannot itself make the enzyme) and that also has the constitutive property $lac\ i^-$ as well. The proposed internal inducer would be expected to act upon the transferred gene; hence the mated cell should produce the enzyme in the absence of externally added inducer.

The first experiment of this kind was done on December 3, 1957. To our great pleasure, the enzyme appeared promptly upon gene transfer, in the absence of added inducer. Thus, conditions within the recipient cell were "constitutive" because the introduced genetic material found an environment that permitted its expression.

Although this result was qualitatively what we had hoped for, there were some problems. First, the rate of enzyme production in the absence of external inducer was lower than in the presence of the inducer at the end of an 80-minute period following mating. Second, transfer of the i^- gene into a $z^+\ i^+$ cell did not permit enzyme formation.

It was not until several months later that we worked out the kinetics of enzyme production in the original $z^+\ i^+ \rightarrow z^-\ i^-$ system. We showed that initial rates of β-galactosidase production were very similar with and without inducer. But within two hours the culture without inducer had stopped making β-galactosidase, while the culture with inducer continued to make the enzyme at an ever-increasing rate. The mated cells thus had switched from constitutivity to inducibility.

These results were considered in relation to the apparently converse phenomenon of repression, in which removal of a small molecule permits specific enzyme synthesis to occur. A common basis was sought by Monod, Jacob, myself, and Leo Szilard, who visited the Institute. Our thinking led to the now-familiar idea that the inducible cells contain an inhibitory substance, a protein named a "repressor," which prevents induction unless it is antagonized by the added low-molecular-weight inducer. The constitutive cells have genetically lost ability to form the repressor (in some cases by deletion of genetic material), and hence their production of β-galactosidase is not blocked. Following mating, the gene ($lac\ i^+$) which codes for the repressor molecule is introduced along with the β-galactosidase gene ($lac\ z^+$). Expression of the repressor is delayed for an hour or more, in contrast to the almost immediate expression of β-galactosidase, a problem later studied further by Dr. Stephen Barbour in my laboratory. Immediately after mating, the cells do not contain the repressor, and the enzyme is formed; only after the repressor has accumulated during an hour's further growth does the requirement for inducer appear, to counteract the newly formed repressor.

I think it is safe to say that this PaJaMa experiment provided the basis and frame of reference for further studies on the mechanism of enzyme regulation. It led to the isolation of the repressor protein by Walter Gilbert and Benno Müller-Hill, and their elaborate studies of repressor structure and action in their laboratory, and by Suzanne Bourgeois, Arthur Riggs, and others. The PaJaMa experiment is the basis of a historical-philosophical study of the nature of discovery by Kenneth Schaffner, who interviewed the individuals involved and reconstructed a composite of what happened (*Stud. Hist. Phil. Sci.*, 1974, **4**, 349).

Contacts between Jacques and me did not end with my departure from the Pasteur Institute in the fall of 1958. We both studied induction and repression for a few years longer. Then we both became involved in determining the mechanism of control of enzyme activity (as distinct from enzyme formation). Prior to my sabbatical year (between 1953 and 1956), Richard Yates and I had discovered another major mechanism of metabolic control: feedback inhibition of enzyme activity as seen in the pyrimidine biosynthetic pathway (by cytidine triphosphate). This control mechanism was independently discovered by Edwin Umbarger, for the isoleucine pathway.

After my return to Berkeley in 1958, I took up the question of how feedback inhibition works—how enzyme activity can be controlled (activated or inhibited) by compounds that do not at all structurally resemble the enzymes' substrates. I decided we needed a pure enzyme for these studies, so Margaret Shepherdson and I completely purified the regulated enzyme, aspartate transcarbamylase, which Yates and I had previously shown to be under feedback control. My student John Gerhart and I during 1958 to 1962 then did some kinetic and physical chemical studies which led to our discovery toward the end of 1960 that an enzyme can have special regulatory sites designed for control. We showed by selective denaturation that regulatory sites are separate from the classical catalytic sites. We concluded that feedback-inhibitable enzymes are designed to be regulated, by possessing special regulatory sites to which effectors attach and modify catalytic activity.

From the PaJaMa studies on the molecular nature of induction— proposed to be the effects of interaction of a lactose-related inducer with the repressor protein, which initiates a quite different process of gene expression—Monod became interested in how a small molecule can modify a protein's function. He and I had discussed feedback inhibition during my sabbatical in 1957–1958. A little later he and Jean-Pierre Changeux started to study feedback inhibition of threonine deaminase by isoleucine, the system pioneered by Umbarger. We were in contact from time to time, and I gave a seminar about our results on the regulatory site of aspartate transcarbamylase at the Pasteur Institute in April of 1961. We recognized similarities between our two studies. Changeux's observations, closely similar to those of Gerhart and myself, and made independently and simultaneously, led them to conclude, as we did, that regulatory sites exist in addition to catalytic sites.

Jacques coined the word "allostery" to signify the function of these other, regulatory sites (allosteric sites). He and François Jacob in 1961 made the far-reaching hypothesis that allostery applies beyond the control of enzyme activities and hemoglobin binding of oxygen, and is perhaps the major mode for control and coordination of metabolism. Through allosteric interactions, a compound produced by one metabolic sequence can affect another chemically dissimilar, but functionally related process. In later papers, Monod, Changeux, and Jacob integrated the few solid pieces of data on allostery, principally our work and theirs on feedback-inhibitable enzymes and classical studies on hemoglobin. Wyman, Monod, and Changeux developed an elegant mathematical model for the role of subunits in simple cases of allosteric regulation, based on a symmetry principle.

Gerhart and I proceeded with the molecular side of the problem during this time, aimed at providing definitive proof for the existence of allosteric sites. Our further kinetic, binding, and physical chemical studies on control of aspartate transcarbamylase culminated with Gerhart's beautiful experiments in which he separated allosteric sites physically from catalytic sites and showed that the enzyme consists of two sorts of protein subunits, one carrying catalytic and the other regulatory activity.

Jacques and I took different paths after the mid-1960s. We were invited to give the main lectures, on metabolic regulation, at the National Meeting of the Japanese Biochemical Society in October of 1966. I looked forward to our being together for a few days of leisure talk and sightseeing. But Jacques was awarded his Nobel Prize that month and remained in Europe. We saw little of each other after that.

A few personal footnotes might be amusing. I recall one experiment in which I intended to measure the appearance of β-galactosidase, while François Jacob was to determine the time of entry of the corresponding lac^+

genes by separating the mated cells with a Waring blender. I routinely did my mating experiments in 2-liter flasks using a very thin layer of liquid culture so as to give maximum aeration with minimum perturbation of the cells. When it came time for Jacob to take his samples, his pipette would not reach to the bottom of my flask, so there was a momentary crisis during which there was some rapid-fire French that never got published; the mated culture was rapidly poured into a smaller flask and the protocols of both laboratories became amalgamated.

My experimental results were discussed frequently and exhaustively whenever something novel happened. Many of our ideas were generated in these exciting discussions. I sometimes found myself at a disadvantage when both Jacob and Monod were in the conversation; as soon as the subject became really intense, they switched from English to French, and at a rate far too fast for me to follow. I had to interrupt and beg them to return to English, the laboratory's lingua franca. In fact, my French was so inadequate that I finally was assigned a French-speaking technician who I'm sure was under orders not to use English; thus I was forced to expand my pitiful French vocabulary.

I close with a few more personal observations and memories of Jacques Monod.

He was a man of broad talents. One of our most pleasurable activities outside of the laboratory involved music. Jacques was an excellent cellist, and I, a much more humble student of the cello. I spent a fair amount of time visiting the luthiers in Paris looking for a cello; on a number of occasions he joined me in my search. I finally found an instrument to my liking (which I still possess and treasure). I recall taking it to his apartment and comparing it with his own instrument, a fine old cello, though dull brown and quite undistinguished in appearance. Mine, on the contrary, was a gleaming mahogany red and quite spectacular. Another scientist commented that we really should exchange cellos because the instruments, although both very fine, had just the opposite personalities from their owners!

Another interest of Jacques' was rock climbing. On one occasion he took us to the rocks in Fontainebleau, where he introduced me to the fine art of scaling the local boulders. He was an extremely skillful and agile climber, and he could get up vertical faces that looked impossible to climb, without hand- or foothold. One of the few occasions on which he expressed strong emotion occurred when I stepped on his hand!

In science he expressed a number of fundamental approaches and ideas that remain with me. He frequently said that science is like art. He looked upon a finished piece of science like a finished painting. Perfection consists of doing just enough, not one stroke too many or one too few. His writings reflect this choice of a clean line over completeness.

He also had strong views on what should be accepted as evidence and what should be omitted. In particular, I recall a discussion of the operon idea, that a group of genes such as the ones for β-galactosidase and galactoside permease must be induced simultaneously, owing to the linkage of these structural genes to the same regulatory system.

I raised the objection that galactinol induces the permease but not β-galactosidase, in apparent contradiction to his hypothesis. He told me that in view of all the other facts, he was prepared to omit this one. I was horrified at his willingness to suppress valid data that did not fit his conceptions, to the extent that later I investigated this matter. He turned out to be right, since the permease that is induced by galactinol responds to the same assay but is different from the one linked to β-galactosidase. Hence the different permease can be induced by galactinol separately from β-galactosidase although the linked one cannot. Sometimes an ugly fact only seems to destroy a beautiful hypothesis. In initial stages of an investigation, where very little is known about a system, one should not expect every fact to fit perfectly.

Another facet of our conversations that remains with me concerns hypotheses. I was constantly suggesting hypotheses, some good and some less so, regarding the nature of our experimental results and their fundamental basis. Monod would reply that a hypothesis, no matter how clever, is only useful to the extent that it can be put to a crucial experimental test. Nevertheless, I feel one first needs a hypothesis, and then one tries to think of ways of testing it.

These recollections have a bearing on Jacques' ability to bring scientific ideas to general scientific notice. He had a tremendous talent for sifting and assembling information, for selecting the most important pieces, eliminating much that was logically secondary (including actual data), and constructing a logical edifice that was most compelling. He also had a remarkable ability to coin word-labels for major ideas: *permease, operon, promoter*, and *allosteric* come to mind at once. These traits were coupled with a forceful personality, striking appearance, and great clarity of speech and writing. His generalizations probably have had as much effect on science as have his discoveries.

His impact on an audience is beautifully illustrated by the following "appreciation," written just after having heard Monod's Harvey Lecture in New York (1961–1962) by a colleague who prefers to remain anonymous.

A Night at the Operon

Opening night at the Harvey Society featured an impeccable French scientist in one of the great performances of his career. Professor Jacques Monod captivated his colleagues by the eloquence and simplicity with which he pleaded his case. No other Harvey Lecture in our time has been characterized by such a brilliant exposition of a logical line uncluttered by experimental detail. One felt the great heritage of French literature and philosophy in every phrase. Descartes would have loved it. The audience sat breathless through the last extrapolation and at the end, the burst of wild applause surpassed any ovation heard in our town north of 57th street. Whatever the fate of the operon theory, Professor Monod has achieved a special kind of immortality tonight, in his creation of a new art form which transcends both science and literature.

Jacques had a remarkable combination of personal traits: brilliant, polished, self-possessed, dramatic when necessary, and always on display. He could be kind and thoughtful to his friends, but arrogant and distant to those in whom he was not interested. Once a colleague remarked that Jacques thought of himself as a Renaissance prince; indeed he acted like one. Truly he was a man to respect and, in many ways, to admire.

My year in Monod's laboratory was certainly one of the most remarkable of my career. I learned there of the great power of genetics in combination with biochemistry. The day-to-day interaction with first-class minds, including Monod, Jacob, Lwoff, and Horecker, was enormously stimulating and it reinforced the high standard set by my associations with various previous mentors and colleagues. And far from least was my opportunity in that year to become acquainted with and to gain as a friend Jacques Monod, a man exceptional in both his intellectual ability and his strength of personality.

Arthur B. Pardee

Arthur Pardee returned from Jacques Monod's laboratory to resume his research on feedback inhibition at the University of California at Berkeley. He became Professor and Chairman at Princeton University and then Professor at Harvard University, where he is presently Professor Emeritus. He is a member of the American Academy of Arts and Sciences, the National Academy of Sciences (United States), Institute of Medicine, and the American Philosophical Society, and a Fellow of the

American Association for Microbiology. He has been President of both the American Society of Biological Chemists and the American Association for Cancer Research.

Dr. Pardee has received numerous honors including the Paul-Lewis Award; Harvey Lecture; Sir H. A. Krebs Medal; Rosensteil Medal; Honorary Member, Japanese Biochemical Society; FASEB 3M Award; Princess Takamatsu Award, Japan; CIIT Award; Docteur Honoris Causa, Paris; Boehringer Bioanalytica Award; Outstanding Alumnus Award, California Institute of Technology; Honorary Faculty Member, Nanjing University; and Fellow, International Institute for Advanced Studies, Nara. Dr. Pardee has published about 500 articles, most regarding regulatory processes in biochemistry and molecular biology.

The Messenger

François Gros

T his book is meant to depict, by means of successive patches (as an impressionist painting), a man through the epoch he lived in, and this epoch through the man. The epoch, that of molecular biology, may in its turn be seen as a composition of various "scenes," as those which are threaded along a play by Bertolt Brecht, where a multitude of protagonists move about and spread their energy with passion. At first, we see no connection between their actions, which even seem superficial and disorderly; but, as the play unrolls, we begin to perceive the paradigm, to recognize the hero who is the lordly inhabitant of the place, to understand the meaning and the logical organization which emerges from the confusion. Jacques Monod has been, and will be remembered as, one of those very great men around whom the main events of contemporary biology have synthesized and harmoniously gathered, as the large tableaux around Brecht's hero. These events have indeed, most of the time, sprung from the powerful ideas which he himself conceived, or at times integrated with profit in the logical train of his thoughts.

The history of the messenger is particularly meaningful in that respect.

The Heroic Period

If I try to remember the history of RNA through the prism of my memories, I can rediscover several images, warped by time, which are nonetheless of great emotional content, probably because they date back to the awkward beginning of my scientific research.

The history of RNA, such as I lived it before the great hypothesis of the messenger, is that of a mysterious substance which did not interest anybody except a few cytologists. It was the time of an inventory without passion or excitement when one observed in the cytoplasm the preeminence of a molecule "whose properties were comparable to, although slightly different from those of the deoxyribonucleic acid (DNA)." For a time, the exclusive presence of the pentose nucleic acid in yeast cells—where it was first identified—was stressed, but the idea was soon given up (1947–1949). Actually, with the sophistication of cytochemical techniques, the "pentose-rich compound" was soon shown to be present in all cells including bacteria, which, for strange reasons, were for a long time not considered as normal cells! (The fact that they have no nucleus was at the origin of this way of thinking.) The time had not yet come to think of the special part played by RNA! I remember, retrospectively with extreme surprise, a theory according to which the "zymonucleic" acid was a reservoir of energy, yielding its phosphorus atoms to ADP, a sort of "phosphagen." It should be noted that until 1950, biochemistry was the science of degradation or energy-yielding processes, the archetypes of which were "muscular contraction" and "glycolysis." Many enzymes were known which were able to "dissect" the pentose nucleic acid: pancreatic ribonuclease crystallized by Kunitz, nucleotidases on which Hermann Kalckar was so keen, various phosphatases. This RNA, progressively liquefying into its elements, was for all—including the apprentice biochemist that I was then—a cause of profound satisfaction, conscious as we were that the secret of life did reside in the covalent organization of nucleic acids.

I remember my exhilaration when, carrying out experiments that were highly significant at that time, I mixed bacterial suspensions, previously "washed" with ribonucleotides, and studied the oxidation of pentoses into acetic acid and CO_2 with a Warburg manometer, convinced that I was about to find the clue. (Warburg, how many holocausts were offered in thy name!) If I dare return to that maze of experiments it is because, in my opinion, they illustrate what I call the dynamism of "dead-end approaches." By considering the RNA as a mere *metabolic* substrate we bypassed the reality. This took place around 1950.

The Template Hypothesis and the Role of RNA in Protein Synthesis

Three "groups" of concepts, which developed after the Second World War, have apparently allowed RNA to surge from limbo: these concepts ensued from several observations on the surprising size of this strange molecule as well as from the first results obtained by Brachet and Caspersson establishing a relationship between the protein-synthesizing capacity of the tissues and their RNA content. Last but not least, the "template hypothesis" also played its part in the understanding of RNA function.

The finding (Bawden and Pirie, Bonfenbrenner et al.) that plant viruses contain, apart from proteins, ribonucleic acid has not only led us to perceive the functional universality of the RNA in biological systems, but has also helped in the analysis of its physicochemical composition. Considered at the beginning as a tetranucleotide (Levenne model) composed of the four usual bases (adenine, guanine, cytosine, uracil), RNA was shown to be present in viruses as a large molecule, a *polymer*. As ultracentrifugation techniques developed, molecular parameters became more precise; but as molecular weights of one million or more were first reported, doubt began to hover. Could it be that the size of RNA was equal to or even exceeded that of the already known proteins? Possible artifacts, discontinuity, and the existence of repeated patterns united by linkers were imagined. And what else was not thought of! After some time, evidence had nonetheless prevailed: RNA proved to be very big—a macromolecule. We were still far from imagining that DNA could be yet larger, but a notion was diffusing of a molecule whose constituents are colinearly assembled according to a sort of *code* which might have a deep physiological significance to transmit genetic information. Indeed, new ideas developed in a parallel direction concerning the biosynthesis of proteins, and models were imagined. Dounce, Lipmann, and Borsoock formulated their first template hypothesis: the colinear sequence of amino acids, which is the quasi-immutable mark of the species and predetermines the properties and functions of proteins, is not just the result of usual enzymatic catalysis. By analogy with what was known about the biosynthesis of polysaccharides, it was proposed that the protein sequence is, in some way, predetermined by some "primers" or "templates" whose chemical nature nonetheless still remained unknown.

But the works of Caspersson in Sweden and J. Brachet in Belgium operated as real trampolines in the study of the role played by RNA in the economy of the cell. Basing their judgments first on microspectrophotometric measurements, and second on quantitative cytochemical analyses, these two biologists reached independently the conclusion that a direct relationship exists between the proteosynthetic activity of a tissue and the amount of RNA it contains.

It was to the merit of A. Claude and G. Pallade, then of Schachman, Pardee, and Stanier, to demonstrate that the RNA is mainly present in tiny cellular corpuscles, later named "ribosomes" by scientists of the Carnegie Institute in Washington. This finding of the particulate characteristic of

RNA is at the origin of outstanding progress in molecular biology. It nonetheless led scientists to the wrong track.

At first, it was not difficult for the scientists of the Carnegie Institute (Bolton, Cowie, Britten) to support with precise quantitative data the fact that in vivo the protein chains begin to appear on ribosomes before they accumulate in the soluble part of cytoplasm, the cytosol. In 1957–1958, the group of Zamenick and Hoagland, then that of Tissières and Watson, established (after Lipmann) the main stages of the biopolymerization of proteins by using cell-free systems. Thus the mechanism of what was then called the assembly of polypeptidic chains was progressively made clear. And yet the role of ribosomal RNA still remained ambiguous. Works by Graham and Siminovitch, and Koch and Levy, had proved that this RNA was endowed with great metabolic stability both in bacteria and animal organisms. The template RNA would then have the stability of genes.

Surely, in 1949–1950, a decisive point in the evolution of concepts had been reached. But all the pieces of the puzzle had not yet been fitted into place. Little by little, without knowing exactly who was the first to formulate the proposition with exactitude, the idea arose that an RNA of great molecular weight, present in ribosomes, could operate as a template for the assembly of amino acids "activated" in a polypeptidic chain. It seemed that people were reaching their goal, and yet there still remained a long way to go.

The Genetic Code—the Hypothesis of the Messenger

Ideas on the genetic code and determinism had developed until then quite independently from the study of mechanisms of protein genesis. Along the same general lines, as a result of the template hypothesis, and because many examples tended to prove that genes played an active part in directing the assembly of amino acids for the formation of proteins (molecular diseases, enzymatic alterations connected with point mutations), several theories appeared concerning the possible nature of the code Gamov and Crick argued brilliantly about the hypothesis of combinations of nucleotides (or bases) which might *determine* the order of amino acid chains, "calling," as it were, each amino acid to its right place in the sequence of polypeptides. Crick explained, with particular insight, why in his opinion the combination used was a triplet code, without overlapping or "commas." But no experimental demonstration was available to support those beautiful ideas.

I think that it was about that time (1959–1960) that Jacques Monod began to get seriously interested in ribonucleic acids. Until then, in fact, he did not believe, or refused to consider, that these substances might be of some importance. The discovery of the polynucleotide phosphorylase (Ochoa, Grunberg-Manago) had shaken him by making him realize that a biological system was able to fabricate polyribonucleotides in random sequences. Although he agreed with the ideas, then prevailing, of the template hypothesis, he could not understand how the RNA could fulfill its function of assembler or assembly line. The RNA came as an intruder in the organized world of his thoughts. I was about the only one in his group to deal with ribonucleic acids, and, upon my return from a period of training with S. Spiegelman, I had begun studying the effects that nucleic base analogues might have on enzymatic induction, first in yeast, then in *E. coli*. I have the feeling that Jacques recorded my results with great leniency and generosity but great skepticism. His reservation was based, in part, on rather indirect reasonings that would soon prove extraordinarily fruitful.

In 1960, the famous experiment (Pardee, Jacob, Monod) of transferring a galactosidase gene from a male bacterium to a female one had done more than simply prove the existence of regulatory and repressor genes; it had opened the way to entirely unexpected data concerning the *kinetics* of the expression of the transferred gene. Everything happened as if during the minutes which followed the transfer gene *z* started to function at a maximum rate, and not in an autocatalytic way. We have to remember that, if we were to believe the most trustworthy authors, the genes, essentially composed of DNA, determined the specificity of proteins through the mediation of ribosomes whose RNA, an integral constituent, was metabolically stable. And yet nothing was observed in the kinetics of galactosidase synthesis which might look like a period of latency corresponding to the accumulation of new specific ribosomes. Such a difference between the metabolic stability of ribosomal RNA and the rapid induction or "deinduction" of bacterial populations in exponential growth did not fail to surprise Monod. He was even more puzzled when I let him know of an experiment I had just carried out using 5-fluorouracil, an analogue incorporable in RNA, which *immediately after addition* stopped the synthesis of active galactosidase molecules.

I also remember a seminar which took place in the great lecture room of the Institut Pasteur (1960). During that seminar on the biosynthesis of proteins, Monod drew a conclusion which emphasized the paradoxes which I mentioned above. For him there were two possibilities: either the ribosomal RNA was the famous template bearing the genetic information whose existence was now suspected by everybody, or—and this seemed less probable—proteins were assembled directly on the gene proper, which then operated as a template. Unless, he thought, a specific form of RNA, not yet identified, existed and was endowed with great metabolic instability, which would operate as a messenger between the gene and the protein. But, since no such thing was known, all was not yet well in the kingdom of Denmark.

Jacques Monod shared his doubts with F. Crick, S. Brenner, and F. Jacob during many passionate talks. It may be in the course of their conversations that their attention was drawn to the RNA of Volkin and Astrachan. These authors had indeed already proved the existence of such a metabollicaly unstable RNA by analyzing newly synthesized ribonucleic acids in bacteria infected with T-even phages. The RNA, thus shown by ^{32}P labeling, had a very short lifetime, but it also had a total base composition very close to that of the phage DNA. Little attention had been given to this extraordinary observation, probably because the model system used in this experiment had, in a way, seemed "abnormal" and it was to be feared that metabolic deviation might be due to viral infection. Furthermore, the authors themselves thought they might well be dealing with a precursor form of the DNA.

Finding of Messenger RNA

One day in June 1961, I met F. Jacob on the steps of the main building of the Institut Pasteur. Both of us were about to leave for the United States and we exchanged a few ideas on the experiments we were planning to carry out there, each one in a different laboratory. François meant to study thoroughly the nature of Volkin and Astrachan's RNA in Meselson's laboratory in Pasadena, where S. Brenner was to join him. As for me, J. Watson had invited me to carry out in his lab work I had initiated some years before with F. Neidhardt on what was then called "the chloramphenicol particles."

What follows is well known. Jacob, Meselson, and Brenner brilliantly demonstrated that Volkin and Astrachan's RNA operated on *preexisting* ribosomes as a template biopolymer, organizing as a real "viral messenger" the proteins newly synthesized by the phage.

For me it was more painstaking! When I arrived at Jim's lab, I was not yet thinking of working on messenger RNA. To study the RNA accumulated in the presence of chloramphenicol one had to examine the newly synthesized RNA in control cells, without antibiotic. C. Kurland, who had already acquired great practice in using sucrose gradients to fractionate macromolecules, was at the lab. He was extremely interested in the properties of ribosomal RNA. Our experimental projects included labeling with radiophosphorus during increasing periods of time, from two to three minutes to one hour. We were very much surprised when we found out that for very short labeling periods the distribution profiles did not coincide at all with those of ribosomal RNA. As labeling went on, we could notice a coincidence between the profile of newly synthesized RNAs and that of preexisting RNAs. Jim was disappointed; the profiles were not regular; some heterodispersion could be observed. To cheer us up and forget the experimental results, which should have filled us with joy but, for the time being, left us downhearted, we used to take walks on the Harvard campus and talk about RNA and women (Jim was still single and extremely romantic). I don't remember how we came to talk of the effects of 5-fluorouracil and Volkin and Astrachan's RNA and to make comparisons with our own results. It was then we ventured to think that the "pulse-labeled" RNA might be worthy of interest. After many control experiments, we reached the conclusion that an important fraction of "rapidly labeled" RNA had the properties of a messenger RNA; its reversible association to ribosomes and the study of its composition strengthened our conviction, but we hesitated a long time before drawing a conclusion for fear we might be in the presence of a metabolic precursor of ribosomes.

In Jim's laboratory, in the severe Harvard University building, adorned on both sides with two huge bronze rhinoceroses, we lived epic moments! The heat was suffocating, the laboratory glassware reduced to nil, the radioactivity counters old, enormous, and noisy; many times experiments would trail along late into the night. Kurland and I would end our rapid labeling experiment at three o'clock in the morning. Immediately after, I would go to bed, fall asleep, then jump awake, and realize that, in order to make the tubes stand more evenly in the centrifuge, I had happily mixed the culture labeled for two hours with that labeled for two minutes—which did not seriously impair the results. We decided nonetheless that we had to get better organized: day shift, night shift. One day Jim introduced me to a Martian. I was supposed to give a briefing in chemistry to this eminent professor of physics. He used to follow me like my shadow and looked extremely intelligent (a ventured judgment, since I never heard him pronounce more than two words a day). This physicist with enormous glasses was W. Gilbert, who rapidly made himself famous in biology.

Back in Paris, F. Jacob and myself had the pleasant surprise of finding out that the hydrodynamic and kinetic parameters of the RNAs which we had studied 5,000 kilometers from each other were closely related. The characteristics of a messenger were plainly established for the RNA formed after infection with a bacteriophage, and it also appeared from our own experiments that an RNA with similar properties did exist in normal bacterial cells but that it accumulated in such a small amount (2 to 3% of total RNA) that it had not been possible until then to identify it.

The birth, at times difficult, of messenger RNA was received with great acclaim. We must admit that it is only the laborious and gradual finding of the main chemical, metabolic, and, particularly, template properties of the rapidly renewing RNA which succeeded in convincing everybody (1963). The finding of a strict complementarity in the sequences of the messenger RNA and the homospecific DNA (Spiegelman) was also a major contribution. At the same time, the first experiments on the role of artificial messengers played by polymers also supported the hypothesis of Monod and Jacob as well as helped the study of the chemical nature of the code (Nirenberg, Ochoa).

During a symposium on molecular biology which took place in the Salle d'Iéna in Paris, upon my return from Watson's laboratory, I introduced the first results I had obtained on messenger RNA. And I will always remember Jacques Monod saying to me, "This time, François, we have a beautiful story. We are going to have fun." (That was one of his favorite expressions.) I can see again his penetrating look and this sentence still resounds in my ears.

This is how the story of messenger RNA began. At least this was how I lived its beginning. As in many other circumstances, Monod's genius made him doubt the reality of current ideas at given stages of the evolution of science, and made him clearly foresee transitory situations. He then supported the hypothesis with a network of facts, thus replacing doubt with certainty.

François Gros

After Jacques Monod's death in 1976, François Gros became Director General of the Institut Pasteur, of which he was in charge until 1981. He then served as science adviser in the Prime Minister's cabinet (1981–1985), while continuing to give lectures at the Collège de France as Professor of Cell Biochemistry (1973–1996). In 1991, he was elected Permanent Secretary at the Académie des Sciences for the divisions of Chemistry, Biology, and Medicine.

The year 1978 saw an important shift in his research interest, from molecular biology of bacteria and phages to the study of somatic cell differentiation. As head of the Biochemistry unit at the Pasteur Institute, Dr. Gros has concentrated on gene expression patterns during muscle development, with special emphasis on myogenic regulatory genes and on sarcomeric proteins. At the Collège de France, his research group has made contributions in various aspects of mouse neurogenesis, particularly on polymorphism, molecular properties, and roles of neurocytoskeletal components.

1962.

Between 1960 and 1965.

Origins of Molecular Biology: a Tribute to Jacques Monod

From Lactose to Galactose

Gérard Buttin

In 1954, the curriculum of a student in natural sciences, at the École Normale, included a one-year period during which he had to become acquainted with experimental work in a laboratory. I was such a student, more interested in knowing about the molecular mechanisms underlying cell physiology than hunting insects and flowers in their natural environment, convinced in addition that the most exciting field in which to tackle problems of cell physiology was the bacterial world. For biological students bacteria were part of the plant kingdom. This classification was perhaps based on more administrative than taxonomical considerations, but it led me to seek the advice of Roger Buvat, the head of the Department of Botany at the École Normale. My good luck was that, in contrast to most biologists, he was interested in bacteria. He was somewhat disconcerted about helping me because the only place in Paris where such research topics were actively investigated was the Pasteur Institute, an institution which seemed to have had no interest at all in maintaining any connection with the University; yet, he managed to arrange an interview for me with Jacques Monod.

Hence, one morning in the summer of 1954, I went to "Pasteur" and met with Monsieur Monod. He asked me to explain precisely what I expected from a training period in his laboratory. I had to confess that I would not be free for research on a full-time basis. He expressed some concern but did not object. He suggested a few reading references, which were all American journals. Due to my complete ignorance of the English language, I had to ask him to repeat the name of the magazines slowly. At this stage in our talk, I was almost prepared to forget about the idea. He looked a little surprised indeed that somebody should find this a problem, but his only frightening comment was that I would have plenty of opportunity to "improve" my English. As I understood it better later on, his acceptance had some flavor of a challenge, which compensated for the difficulties to be expected.

I started working after the summer holidays, under the direct guidance of Germaine Cohen-Bazire and with a lot of advice from Annamaria Torriani. At that time, the *lactose* inducible system comprised only one enzyme: the β-galactosidase. Lactose analogs—the thiogalactosides—which were not substrates for the enzyme had recently been synthesized by Turk and Helferich. Some of them exhibited inducing properties, while others competitively inhibited the induction process. These "gratuitous" inducers—a term coined by Jacques Monod to express their property not to contribute to the overall cell metabolism—opened a new dimension in the quantitative analysis of induction kinetics. I undertook a systematic screening of the inducing capacity of all available thiogalactosides, of their affinity to the purified β-galactosidase, and of their "apparent affinity" to the intracellular enzyme. Day by day, the complete lack of correlation between the values of these three parameters became more apparent. This further argued against the simple so-called "instructive" models of enzyme synthesis (viewing the active site of the β-galactosidase molecule as the target for the inducer), which a more restricted offensive, using the same weapons, had just started to put into question. Besides, constitutive strains, which synthesized β-galactosidase in media devoid of inducer, had been isolated. The prevailing hypothesis was that they accumulated an "internal" inducer. The availability of powerful inhibitors of induction prompted us to check their ability to block the constitutive synthesis, with entirely negative results. We were so disappointed that we shelved the constitutive strains, a fortunate decision because several years elapsed before it became clear that the inhibitors could not be more active in this system than a key trying to open a door which has no fitting lock. Germaine got over these frustrating results by concentrating attention on the cultures of photosynthetic bacteria, which together with E. *coli* monopolized her interest. I had no such refuge, but realize today how exceptional was this training period, when Jacques Monod was free enough to devote the necessary time every evening to criticizing each experimental curve. When he left, it was customary that dinner service at the École Normale was over.

A good deal of these talks in the "attic," where the laboratory was located, focused on possible interpretations of a simple observation: partially "preinduced" cultures synthesized almost immediately β-galactosidase at its maximal rate when exposed again to low inducer concentrations, which, when supplied to uninduced cultures, triggered synthesis only progressively and after a marked lag. Obviously, the identification of the stable "factor," distinct from β-galactosidase, which accumulated during—and promoted—induction, would be a major clue to understanding the induction process.

The enlightenment came from the ground floor. There, Georges Cohen and Howard Rickenberg already occupied a room of the renewed Department of Biochemistry, of which Jacques Monod had been appointed head. Using radioactive thiogalactosides, they directly analyzed the uptake of these compounds by a variety of bacterial strains. It soon became clear that the inducible factor which puzzled us was a specific transport and accumulation system for β-galactosides: the β-galactoside permease. The properties of the permease accounted for most peculiarities observed in the determination of the "apparent affinity" of the galactosides for the intracellular enzyme. It was also clear, however, that the inhibition exerted by some thiogalactosides on the induction process could not be explained on the basis of a competition between inducer and inhibitor for the transport system. Besides the permease and the β-galactosidase active sites, a third intracellular structure which recognized the β-galactosides was unmasked, at the level of which these compounds had to exert their property to control induction.

I had no further chance to contribute to sketching out the target for the inducer. The school year was over. It had been rich with excitement; I had enjoyed the informal friendship of renowned scientists. I felt bitter when I thought of the near future: the preparation of the Agregation competitive examinations, and then military service!

Three years went by before I could return to the laboratory to prepare my Ph.D. thesis. Jacques Monod and his colleagues had all moved down from the attic to the Department of Biochemistry, which spread along both sides of a long corridor on the ground floor. This was now a busy avenue, channeling a flow of information which everyone could glean from better-informed colleagues. Some encounters were recurrent: every Monday morning Georges Cohen and Alain Bussard would exchange jokes and puns heard during the weekend. Some encounters were unpredictable: for example, literally bumping into David Perrin, who was always eager to sum up the content of the latest issue of *Proceedings of the National Academy of Sciences* as he simultaneously dissected the package or innards of a new laboratory apparatus.

Science had progressed. The analysis of the *lactose* system remained the research topic for most people in the department, but the genetic approach had been fruitful enough to shift interest to a more precise understanding of the relationships between genetic material and inducible protein molecules. When I showed up, Jacques Monod was excited by the tremendous experimental possibilities offered along this line by the inducible enzymes of galactose metabolism. The Lederbergs had just shown that the chromosomal segment carrying the genetic information for these enzymes could be carried by the phage λ and thereby inserted in duplicate into the genome of a recipient bacterium. A manuscript by W. Arber described that more than one prophage genome could be integrated in this way. Obviously, the possibility was open in this system to carry out dominance and complementation studies for bacterial genes involved in the control of inducible enzyme synthesis; and piling up a variable number of galactose genes did not seem to be out of reach. This would allow a quantitative estimate of the influence of gene dosage on the rate of synthesis of a well-defined protein. I considered the transduction processes as the most amazing phenomenon in the microbial world and enthusiastically accepted tackling this problem.

Very little was known on the regulation of the three enzymes—galactokinase, transferase, and epimerase—which had been identified in this system, other than the fact that they were all induced by galactose. Well informed of the advantages presented by "gratuitous inducers," I wished to characterize galactose analogs with properties mimicking those of the thiogalactosides in the lactose system. After we discussed this project, Jacques Monod came back holding a little pine box full of odd tubes, each containing an uncommon sugar in trace amount. This treasure was a unique collection of rare chemicals, a gift from Gabriel Bertrand, who despite his advanced age still frequently visited a laboratory piously preserved from the overall refurbishing of the department. The magic box did contain the key to the solution of my problem: a sample of pure D-fucose, which came to light as a nonmetabolized inducer devoid of any significant affinity to galactokinase. When it turned out that the lactose inducer TMG (thiomethylgalactoside) behaved in this system as a powerful specific inhibitor, my weapons were ready. With Agnes Ullmann's help, in one evening I synthesized enough TMG for several years of research work. The kinetics of galactokinase induction were analyzed; constitutive strains were obtained by exploiting their property to grow on galactose in the presence of TMG; a set of galactose-negative mutants was isolated; and the phage λ fulfilled its expected role in allowing the analysis of the complementation pattern between these mutants. Week after week, it became more obvious that the galactose genes expressed themselves as an integral unit, and obeyed a negative type of control. Genes governing induction were identified, with properties impressively similar to those of the so-called regulator (*i*) and operator (*o*) genes of the lactose system, to which "the model" attributed respectively the emitter and receiver function for a "repressor" signal. The galactose regulator mapped unambiguously on the bacterial chromosome at the antipodes of the supposedly "structural" genes of the three enzymes. This remote location was consistent with the properties expected from the emitter of a diffusible signal, and indeed it pleased everybody. The first attempts to map the operator gene were less successful. My data made it difficult to avoid the conclusion that it was located right in the middle of the enzyme genes, a result which then made it difficult to understand how at the molecular level the operator could control the expression of the whole structural sequence. Jacques Monod was skeptical. He could indeed suggest interpretations, but they were not straightforward. François Jacob, who paid a growing interest to the development of this work, was more than skeptical: an operator might sit at either end of an operon, but not in the middle. He advised me to go back to my bench and to look for some bias in the mapping procedure. I did this. The bias was that the strains utilized were heterogeneous in the activity of a galactose transport system. The operator moved back to one end of the galactose operon, supporting one of Jacques Monod's favorite statements, according to which "a good model deserves more confidence than one conflicting experiment . . . "

The striking similarity exhibited by the regulatory elements of the lactose and galactose systems strengthened both the repressor hypothesis and the notion that the unit of genetic expression can comprise several genes. I could not help expressing some surprise at how perfectly my results fitted the theory: "You man of little faith!" was the first comment from the boss, soon followed by a straightforward laugh in which one could discern more complicity than boldness.

This docility of the galactose system was indeed reassuring. In the neophyte, it might just as well have given rise to the slightly disappointing feeling of repetitive work, if the puzzling results of experiments involving the transducing phage λ had not initiated a different line of investigations. Clearly, the regulation of the galactose genes obeyed the same regulation, be they in their natural environment or part of a λ transducing prophage inserted in the bacterial chromosome. But, during the vegetative development of the transducing phage, an intense synthesis of galactokinase was observed, the rate of which was essentially independent of the presence of an inducer. The galactose genes now seemed to obey the regulation common to the phage genes. We had in the past discussed the possibility that the expression of a "structural" gene might be influenced by the overall regulation of the genome in which it is inserted, but this vague hypothesis had now to be reconciled with our evidence that the galactose repressor somehow acted at the level of an operator which remained associated to the galactose structural genes in the transducing particles. We proposed two explanations for this paradox. The first one considered the derepression of the galactose enzymes as the consequence of an imbalance in the ratio of repressor molecules to their operator target sites when multiple copies of the latter were generated by the replication of the transducing phages. The other one postulated a real change in the regulation of the galactose enzymes, arising from the synthesis of a phage transcription enzyme which could displace the bacterial repressor. The first interpretation was strongly supported by the work of Salvador Luria, who, at the same time, analyzed a very similar phenomenon, using a lactose-transducing p_1 phage. Taking advantage of the fact that the lactose regulator gene i is close enough to the genes it controls to also remain associated to them in the transducing phages, he could establish that the cointegration of an active i gene was necessary and sufficient to prevent the derepression of the lactose enzymes during the phage vegetative growth. Unfortunately, the structure of the galactose system did not enable one to decide if the same explanation accounted entirely for the derepression of galactokinase.

Yet the most surprising observation arose from the simplest control experiment. When I checked whether the induction of a wild-type λ prophage in a wild-type bacterial strain had any influence on the expression of the nearby galactose genes, an abundant synthesis of galactokinase—weakly stimulated by galactose or D-fucose—was again detected.

A very friendly competition to explain this phenomenon was engaged in with Herbert Wiesmayer and Michael Yarmolinsky, who had just observed that epimerase synthesis also escaped its normal control upon λ induction. The naive speculation that the altered expression of host genes caused by the turn-on of prophage functions might have some relevance to the mechanism of animal cell transformation by oncogenic viruses was a strong stimulus. A variety of experiments were carried out, which all showed that a close physical association of the galactose bacterial genes to the λ prophage was a prerequisite to the manifestation of this effect. The kinetics of the constitutive enzyme synthesis suggested that, at least in some cells, the linkage between the viral and host genome was maintained. We were back to considering either phage-dependent replication or transcription of the bacterial genes as the basis for this peculiar effect. The conflicting results of experiments devised to clarify this question could be reconciled only much later, when the availability of appropriate phage mutants made it possible to show that both replication and transcription processes initiated on the phage genome could proceed uninterrupted up to the bacterial galactose operon. Even more recently, refined biochemical experiments established that if λ does not code for a new transcription enzyme, it governs the synthesis of a regulatory factor which makes the bacterial RNA polymerase blind to its stop signals and able to "read-through" a nearby operon.

While I attempted to assemble the pieces of the galactose puzzle, five years had flown by, consisting of days of great excitement, followed by weeks of questionable progress. In the laboratory, a few rites resisted time, like the seminar program, which attracted an increasing number of people from various institutions. I scrupulously attended the seminars, which Jacques Monod considered, with excellent reasons, the necessary basis for the education and recycling of a scientist. But, because most of the talks were delivered in English, for a long time I missed so many of the speakers' points that I left the room more often discouraged than stimulated. Strengthening my feeling of being handicapped was the ease with which the boss could join the game ten minutes late and ask within the next ten minutes a first pertinent question about problems very remote from his actual field of investigation. Another permanent institution was the lunch ceremony, which from two floors drained André Lwoff, Jacques Monod, François Jacob, and their colleagues. Over the bread and cheese—or the more sophisticated masterpieces of a nearby *traiteur*—we talked about scientific news, political topics, or artistic events. These discussions showed that whatever the topic, the very same people were once again the most competent.

Some changes were discernible during this period. The *tour d'ivoire* was infiltrated by a growing number of students who brought their own originality to a renowned and experienced community which was constantly being enriched by long-term visits of foreign scientists. New rites had also been established, like François Jacob's daily morning visit to our laboratory. On his way to Jacques Monod's contiguous office, he would stop and either banter or urge us to speed up the program, but would always make invaluable comments on our experiments in progress.

The growing fame of Jacques Monod and his appointment to a professorship at the University exerted increasing and convergent pressures— which he resisted remarkably well—to take him away from the laboratory. Then, Saturday afternoons became the best time to join him. The silence of the building contributed to generating an informal and relaxed atmosphere for discussions which he liked to pursue while carrying out—perhaps as an antidote to administrative duties—some experimental work.

The former student today faces, in a very different context, the duties of a mature scientist. Among the questions which are often raised—even if there are no answers—on a university campus, two remain for me a major source of worry. How is one to instill in young people the necessary passion for research? How can one evaluate a scientist's individual responsibility in the success—or in the lack of success—of a program with certitude? Perhaps my special concern when it comes to these problems is nothing but the ransom to be paid for a very fortunate first contact with the scientific community; and does it perhaps also illustrate the sly danger of an example?

Gérard Buttin

Born in 1931, Gérard Buttin joined the laboratory of Jacques Monod in 1954 to perform a short experimental work requested by his cursus at the Ecole Normale Supérieure. Under the guidance of G. Cohen-Bazire, he analyzed the kinetics of induction of β-galactosidase, which contributed information on the mechanisms of sugar internalization studied by G. Cohen and H. Rickenberg. Following a three-year break to prepare the "Agregation de Sciences Naturelles" and accomplish his military service, he returned to the laboratory and prepared his Ph.D. thesis under the supervision of Jacques Monod, with training in microbial genetics by François Jacob. The switch "from lactose to galactose" induction analysis also supplied unexpected insights on the deregulation of host enzyme synthesis by viral induction.

As a Rockefeller Fellow, Gérard Buttin faced a new switch—from gene expression to gene synthesis—during a postdoctoral stay in Arthur Kornberg's laboratory at Stanford University (1963–1965). Recruited by François Jacob upon his return to Paris to work on the biochemical part of his "replicon" project, he identified with his first student, M. Wright, the first bacterial "recombination enzyme" (exonuclease V).

Elected in 1970 as Full Professor of Genetics at the University of Paris (Paris 6), Dr. Buttin created the Unit of Cellular Genetics in the newly opened Institute for Research in Molecular Biology (presently the Institut Jacques Monod), where he directed research on both microbial and mammalian cell division before focusing on two major topics of somatic cell genetics: gene amplification and monoclonal antibody production. He transferred his laboratory to the Pasteur Institute in 1981, organized there the biotechnological Hybridolab laboratory, and developed research on genomic instability (with M. Debatisse) and exploitation of anti-DNA antibody-derived peptides for cell internalization of macromolecules (with S. Avrameas).

Dr. Buttin was affiliated with the CNRS (1958–1965), the Collège de France (1965–1970), the University Paris 6 (he retired as Emeritus Professor in 1999), and the Pasteur Institute, where he has served since 2000 as Emeritus Professor in charge of ethical questions. He was awarded the Silver Medal of CNRS (1963) and elected in 2001 as the president of the French Society of Genetics.

The Wonderful Year

David Perrin

T he circumstances in which I met Jacques Monod already reveal something about the man; it was in 1954, when he gave a lecture on Louis
Rapkine's esthetic theory at the Philosophical Society. Since childhood a
friend and admirer of Louis, I went to listen and was more fascinated
by the man than by the subject (the president slept through most of the
talk and only woke up to give a brilliant conclusion). Having just finished
my "license," majoring in zoology, I went to the not-yet-famous attic of the
Pasteur, to ask Monod if I could work with him. He inquired whether I had
any biochemistry—I did—and whether I played a musical instrument—I
did not, but admitted to singing in the university chorus (to my relief he
did not ask for a demonstration). He told me to go learn genetics; I did and
that started it all. I worked with him for fifteen years, and discovered later
that he too started as a zoologist.

Monod was very conscious that students coming from the French
university knew very little. When he designed his lab, he set aside a large
room with six benches, which was to be used for a two-month lab course,
to be given each year to twelve students. The ideal candidate was defined
by Monod as being infinitely ignorant and infinitely intelligent. Actually,
the course was only given for two years, in 1956 and 1957. It was intended
to breach the gap between university teaching and living research. It was
our real introduction to what was to become molecular biology—covering
DNA, proteins, physiology, and biochemistry of bacteria and phages, of
which we had been taught practically nothing. For France, the system was
completely new. The students helped design the experiments; anything feasible was tried beyond the basic canvass of the course. The staff of the lab
was mixed with the students and everybody learned together, discussions
being general.

This atmosphere prevailed in the lab: the corridor in the evening was the meeting place where the day's experimental results were submitted for discussion. The weekly seminars were at the same time an informal but important event. The room was uncomfortable and crowded. Slides were forbidden, since Monod did not want data "hocus-pocused"; curves had to be plotted, tables laid out, everything in micromoles and milligrams of something, not just in cpm or activity per milliliter. Slides were introduced only when every speaker was lost without them, and wanted to show about twenty different sucrose gradients of the same profile. Monod sat in front. His questions were not only sometimes quite aggressive but some people could not stand his brilliant and very deliberately logical approach to all subjects. What amazed me most was that he often embarked on a far-fetched idea, but always managed to land back on his feet by proposing a feasible experiment that would provide a test for it.

In the lab things were more relaxed and the atmosphere was probably more like that of an American lab than a French one. We practically lived in, all having a joyous lunch seated at a table. But it was more like a big family. True respect for Monod's obvious superior ability was mixed with a deep affection for him. We, the students, were his children, and he treated us as such, sternly when he thought it better for us. He wanted each of us to have our own private research project, and appreciated people who became completely involved. Once a brilliant scientist from another lab gave a seminar, after which I asked Monod what he thought of it, expecting scientific appreciation. His answer was that the man really seemed to be inhabited by his problem; it was probably his highest praise for a scientist.

So 1958 came. I was working on enzyme induction in *Pseudomonas*. One day Monod came to me and said, "Jacob and I have started a study of the genetics of galactosidase and permease. The system is ripe, we are getting a lot of new mutants that have to be analyzed and we need more hands. I would like you to study CRM [cross-reacting material]-producing mutants of galactosidase for which there is preliminary evidence." So I dropped *Pseudomonas* and mutants of catechol oxidase and joined this extraordinary team for what was to be a fantastic scientific adventure. It was a wonderful year, during which every experiment proved what it was supposed to prove.

The study of CRM was of course prompted by the need to map the structural gene for galactosidase so it could be distinguished from genes involved in induction. But it was also to understand the obscure relation between that old ghost Pz and CRM of galactosidase. The experimental approach is interesting in that it reveals how Monod worked. He wanted

Agnes Ullmann and David Perrin, 1970.

quantitative results that could be expressed in absolute values. We devised a scheme of competition between galactosidase and CRM in which the only thing measured was enzyme activity. Never did I use an Ouchterlony plate to detect CRM, even though it might have been a fast screening procedure. Likewise, we practically never looked at bacteria under the microscope. Monod did not have visual intelligence; he saw with his brain, and the only result that really pleased him was a linear plot. His Cartesian logic prevailed over everything. Experimental conditions had to be strictly controlled; bacteria were always grown in synthetic medium and had to be exponential. He often made fun of the British, who added beef extract and 10% tap water to synthetic media. Still, there was available to us as a last resort, when bugs did not grow, a mysterious tube labeled "oligo elements," to be added in case of disaster—I've never seen it work.

We worked following a certain number of Monod's practical aphorisms such as: "Bacteria never make mistakes" or "There are no bad experiments." Monod had a knack for extracting from "muddy" data what went wrong and what had to be done to have a significant experiment. Statistics were not used; effects had to be all or none or so great that you only wondered whether they were artifacts, not whether they were meaningful. The last aphorism was the most potent: "Autosatisfaction is death."

Monod loved music so much that he could not stand a radio in the lab. He said that when he heard music he could not think. But during that wonderful year Jacob kept whistling a theme from Mozart's clarinet quintet and Monod walked through the corridors booming a trombone theme from a Brahms concerto. Music was with us anyway.

When the different CRM proteins of galactosidase came to be reasonably measurable proteins, Monod asked me to prove that their induction was quantitatively the same as that of galactosidase and that they had no affinity for the inducers. That resulted in a short paper that I have seen quoted in literature as "by Perrin and Coll.," which gave me a shock, for the "Coll." were Jacob and Monod. Now we had two new tools at our disposal: Jacob's Flac episomes, which permitted the study of dominance and complementation, and CRM, which made possible quantitative measurement of the expression of two alleles of the structural gene for galactosidase in a diploid. During that year every experiment added a significant piece to the puzzle. From Flac episomes came O^c mutants. By measuring CRM and galactosidase in diploid O^c/O^+ mutants, O^c was shown to be not only dominant but cis-dominant, which was unequivocal proof of the existence of the operator. The i^s mutant which became a cornerstone of the theory of induction by negative control appeared by luck. It was a strange lac negative, giving numerous revertants which were all constitutive. It could only be understood if you were prepared for it; it proved to be $trans$-dominant and fit nicely with the theory.

One day shortly after, some Flac heterozygotes were made. There were strange results that could be interpreted as compartmentalization of proteins inside the bacterial cells. For two days "vesicular biology" was rampant in the corridor, then fell flat when a mistake in strains was discovered. Two weeks later, a visitor from the States arrived and excitedly asked about vesicular biology, which was already forgotten. This shows how fast news traveled and the anxious interest of our foreign colleagues in the work in progress. We felt that the theories, the techniques, and the strains at our disposal really put us ahead of many other labs. Therefore we were not involved in a "rat race" and could be somewhat relaxed about publication of results. This resulted in papers that could really be constructed around an idea, with proper controls, and not the hasty unloading of notebooks that is now common.

The wonderful year passed, the chase for the repressor was started, and allostery came. The lab was crowded; on a central bench Changeux and I had only half as much space as Buttin, more advanced and swamped with galactose plates. On the shelf above me stood a large flask with Changeux's "allosteric buffer"; no other buffer worked, and when it was half empty he used to replenish it, neglecting the slime at the bottom which may have been responsible for the miracle.

More years passed. Monod became Director; Jacob moved to mammalia; bacteria seemed less an exciting subject as molecular biologists moved to differentiation and oncogeny. But we still tried to do meaningful work with bacteria following one of Monod's last aphorisms: "Remember there is always plenty of room at the bottom."

David Perrin

When recombinant DNA technologies became available in the mid-1970s, Jacques Monod, a year before he died, decided to create a new unit at the Institut Pasteur to develop this new technology. In 1977, David Perrin was nominated head of this unit, called the Unité de Génie Génétique, a title which he coined with Jacques Monod instead of "engéniérie génétique," the awkward phrase translated from English ("genetic engineering") and used in French at that time.

In 1979 Perrin joined the unit headed by Maurice Hofnung (who passed away in 2001), where, at the beginning, he participated in the study of structure-function relationships of the membrane proteins of the maltose transport system, mainly by using an in vitro system of protein synthesis to facilitate the biochemical characterization of those proteins. Some time later, Hofnung's laboratory discovered the first class of complex repetitive DNA elements in bacterial genomes, the BIMEs (bacterial interspersed mosaic elements), thought to be related to genome plasticity and evolution. From then on, David Perrin worked on the functional and structural diversity among these BIMEs.

In September 1995, David Perrin retired from the Institut Pasteur.

From Diauxie to the Concept of Catabolite Repression

Boris Magasanik

I do not remember when I met Jacques Monod for the first time. I was certainly very much aware of his work when I began to study enzymatic adaptation to *myo*-inositol in 1950. I recall that shortly after that time André Lwoff visited J. Howard Mueller, the head of my department at Harvard Medical School, who asked him to talk to me about my work. André Lwoff kindly suggested that my results would interest Jacques Monod and that I should write to him. I was quite abashed at the idea that such a great man would see anything of interest in the modest efforts of a beginner and could not bring myself to take up André Lwoff's suggestion.

It turned out that our efforts to discover, by the use of amino acid auxotrophs, whether the formation of an adaptive enzyme involved synthesis de novo of protein paralleled similar efforts of Jacques Monod (1, 2). I am quite certain that we became personally acquainted in 1953 or 1954, but it was a meeting in Boston in the spring of 1956 that has remained most vivid in my memory.

Our studies of inositol metabolism in histidine-requiring mutants of *Aerobacter* (now *Klebsiella*) *aerogenes* had yielded a most unexpected result: the histidine requirement for growth on *myo*-inositol was approximately 25 times greater than that for growth on glucose (1). We then found that when grown on *myo*-inositol or most other energy and carbon sources, but not when grown on glucose, the cells produced a series of enzymes capable of degrading histidine to ammonia, glutamate, and formamide. Glutamate could be further degraded to serve as a general source of energy and carbon. These enzymes, whose synthesis was induced by histidine, caused the loss of the exogenously supplied histidine, and therefore increased the requirement of the histidine auxotroph for this amino acid (3). Our observation that glucose prevented the formation of these enzymes was thus a rediscovery of the glucose effect identified by Jacques Monod thirteen years earlier as the cause of the diauxic growth of *Escherichia coli* on mixtures of glucose and other carbon compounds (4).

One of my students, Fred Neidhardt, continued the study of this phenomenon. He could show convincingly that glucose did not interfere with the uptake of histidine and did not limit the cell in substances necessary for protein synthesis. He made the completely unexpected, but to us very exciting, observation that *K. aerogenes* could grow on glucose with histidine as the only nitrogen source and produced in this case the histidine-degrading enzymes in spite of the presence of glucose; addition of ammonia immediately arrested the synthesis of these enzymes (5).

At this stage, Fred Neidhardt, having completed his dissertation, expressed the wish to be accepted by Jacques Monod for postdoctoral study at the Pasteur Institute; but unfortunately, we were informed that there was no space for him in the coming year. The only hope was that Jacques Monod, who was about to visit Boston, would reverse this verdict upon meeting Fred.

My wife, Adele, and I were planning a small reception for Jacques at our house. We decided to take him and Fred to dinner at a well-known seafood restaurant before the reception in the hope that good food and pleasant conversation would soften Jacques' heart toward Fred. During this dinner, Fred told Jacques about our *Klebsiella aerogenes*, which, when faced with the dilemma of obeying the command of glucose to stop histidase production, or of disobeying this command, chose to disobey when this disobedience was essential for growth. It became obvious that Jacques considered this an unlikely, perhaps incorrect, but certainly unattractive view. Somewhat crestfallen, we finished dinner and drove quickly to our house in order to arrive there before our other guests. While they gathered, Jacques sat quietly in a chair, apparently absorbed in thought. Then suddenly he turned to Fred with a smile and said: "You have proven the existence of God." And he accepted Fred for postdoctoral work at the Pasteur Institute, where Fred eventually worked with François Gros.

My own study in Jacques Monod's laboratory as a Guggenheim Fellow in 1959 made me appreciate even more the blend of uncompromising critical rigor and of human concern in Jacques' personality. I began at the Pasteur Institute to study the regulation of the histidine-degrading enzymes in *Salmonella typhimurium*, where the combination of genetic and biochemical techniques developed by Monod and Jacob for the *lac* system of *E. coli* could be used to advantage.

Although Jacques initially did not like Fred Neidhardt's apparently teleological explanation of the escape of histidase from the effect of glucose in cells deprived of ammonia, this explanation was actually based on an earlier discovery by Monod: the inhibition of the formation of an enzyme essential for tryptophan biosynthesis by the addition of tryptophan to the growth medium (6). The *repression* of an enzyme by its product appeared as a rational counterpart to the *induction* of an enzyme by its substrate.

In our analysis of the escape of histidase from the effect of glucose, we made the assumption that the formation of an enzyme may be subject to *both* substrate induction and end-product repression. We assumed that such an enzyme would only be produced when the intracellular level of its substrate was high and that of its product low. The ultimate products of histidine degradation are the catabolites, common products of the degradation of all carbon compounds capable of supporting growth, as well as ammonia and glutamate. In a cell growing with glucose as source of carbon and ammonia as source of nitrogen, the intracellular level of catabolites, glutamate, and ammonia would be high enough to repress histidase, overriding the induction by histidine. However, limitation of one of these products by depriving the cell either of ammonia or glutamate or of glucose, an excellent source of carbon catabolites, would permit induction to prevail and histidase to be produced. It is quite easy to see that this hypothesis would also explain why enzymes such as β-galactosidase and the enzymes responsible for inositol degradation would be subject to the effect of glucose, but not be able to escape from this effect when starved for ammonia. The only products of the degradation of lactose or *myo*-inositol are the carbon catabolites more readily available by the degradation of glucose, which would therefore repress the enzymes responsible for the degradation of lactose and of inositol under all conditions. The hypothesis of catabolite repression predicts that energy sources other than glucose should repress glucose-sensitive enzymes in cells grown in media that limit their ability to utilize the catabolites rapidly for the synthesis of macromolecules; and in fact, partial amino acid, purine, pyrimidine, or phosphate starvation has this effect (7).

As the names imply, in a physiological sense, induction is a positive control and catabolite repression a negative control. Nevertheless, for *lac* and *hut* (histidine degradation) and many other systems, induction actually reflects negative control at the molecular level, and in all cases catabolite repression appears to reflect positive control at the molecular level. The postulation by Monod of the existence of a specific *lac* repressor, capable of preventing transcription of the *lac* operon unless neutralized by the specific inducer (8), has been verified completely by the isolation of a repressor protein with these exact attributes (9). The observation by Ullmann and Monod (10) and independently by Perlman and Pastan (11) that the addition of cyclic AMP can overcome the repressive effect of glucose led to the discovery that the transcription of genes coding for enzymes subject to catabolite repression requires activation by the catabolite-activating protein (CAP) charged with cyclic AMP (12). Glucose appears to lower the intracellular level of cyclic AMP by an as-yet-undiscovered mechanism.

The discovery of a relatively nonspecific molecular mechanism for catabolite repression (CAP recognizes a site located near the promoter of every catabolite, sensitive operon or gene) clearly raised anew the question of how the transcription of the *hut* system of *K. aerogenes* could be activated when cells were grown in an ammonia-free medium with glucose as source of energy. Cells growing in such a medium cannot be induced to form β-galactosidase, a clear indication that CAP is not present in its active form. An explanation for the escape of histidase from repression by glucose was finally found, twenty years after Neidhardt had first observed the phenomenon. The transcription of the *hut* genes and of other genes coding for enzymes whose activity can provide the cells with energy, as well as with ammonia or glutamate, can be activated either by CAP charged with cyclic AMP or by nonadenylylated glutamine synthetase (GS) (13, 14). In the presence of excess ammonia, GS is present at a low level and partly in an adenylylated form. Starvation for ammonia results in the deadenylylation of GS and in a rise in its cellular level. This increased amount of nonadenylylated GS is responsible for the activation of the transcription of the *hut* genes in ammonia-starved cells.

In short, the apparent escape from catabolite repression of histidase really reflects a phenomenon distinct from catabolite repression: the activation by GS of the transcription of genes coding for enzymes that can supply the cell with ammonia or glutamate. As expected, the synthesis of an enzyme such as urease, capable of supplying the cell with ammonia but not with catabolites, is activated by GS but not by CAP (15).

In retrospect, Fred Neidhardt and I were stimulated to formulate the hypothesis of catabolite repression by our analysis of the apparent escape of histidase from the effect of glucose. It is this analysis which failed to win Jacques Monod's approval when we first presented it to him. His sensitivity and taste made him aware of a flaw in our analysis that only much more work could elucidate: the apparent escape of histidase was unrelated to the effect of glucose, but reflected a new, unsuspected regulatory mechanism.

References

1. Ushiba, D., and Magasanik, B. 1952. *Proc. Soc. Exp. Biol. Med.*, **80**, 626.
2. Monod, J., Pappenheimer, A. M., Jr., and Cohen-Bazire, G. 1952. *Biochim. Biophys. Acta*, **9**, 643.
3. Magasanik, B. 1955. *J. Biol. Chem.*, **213**, 557.
4. Monod, J. Recherches sur la croissance des cultures bacteriennes, Paris, 1942.
5. Neidhardt, F. C., and Magasanik, B. 1956. *Nature*, **178**, 801.
6. Monod, J., and Cohen-Bazire, G. 1953. *C. R. Acad. Sci.*, **236**, 530.
7. Magasanik, B. 1961. *Cold Spring Harbor Symp. Quant. Biol.*, **26**, 249.
8. Pardee, A. B., Jacob, F., and Monod, J. 1959. *J. Mol. Biol.*, **1**, 165.
9. Gilbert, W., and Müller-Hill, B. 1966. *Proc. Natl. Acad. Sci. USA*, **56**, 1891.
10. Ullmann, A., and Monod, J. 1968. *F.E.B.S.*, **2**, 57.
11. Perlman, R., and Pastan, I. 1968. *Biochem. Biophys. Res. Commun.*, **30**, 656.
12. Zubay, G., Schwartz, D., and Beckwith, J. 1970. *Proc. Natl. Acad. Sci. USA*, **66**, 104.
13. Magasanik, B., Prival, M. J., and Brenchley, J. E. 1970. p. 9. *In* S. Prusiner and E. R. Stadman (ed.), *The Enzymes of Glutamine Metabolism*. Academic Press, New York, N.Y.
14. Tyler, B., DeLeo, A. B., and Magasanik, B. 1974. *Proc. Natl. Acad. Sci. USA*, **71**, 225.
15. Friedrich, B., and Magasanik, B. 1977. *J. Bacteriol.*, **131**, 446.

Boris Magasanik

Boris Magasanik was born in Kharkoff, Ukraine, in 1919. He received his preliminary and secondary education in Vienna, Austria, and studied chemistry at the University of Vienna. He continued his studies at City College, New York (B.S., 1941). From 1942 to1945 he served in the United States Army in England and France. He obtained his Ph.D. degree in biochemistry from Columbia University in 1948. A faculty member in the Department of Bacteriology and Immunology at

Harvard Medical School from 1951 to 1960, he then continued his career as Professor of Microbiology at MIT and he is still there, serving as the Jacques Monod Professor of Microbiology.

The main subject of his research since 1978 has been the regulation of nitrogen utilization in response to nitrogen availability (nitrogen regulation) in enteric bacteria and in *Saccharomyces cerevisiae*. This work is reviewed, respectively, in *Escherichia coli and Salmonella typhimurium: Cellular and Molecular Biology* (edited by F. C. Neidhardt et al., ASM Press, Washington, D.C., 1996) and in *Gene* (B. Magasanik and C. A. Kaiser, *Gene*, **290**, 118, 2002). He is the author of over 300 publications dealing with microbial physiology and the regulation of gene expression in bacteria and yeast.

Dr. Magasanik became a member of the National Academy of Sciences in 1969 and received the Selman A. Waksman Award in 1993. He closed his laboratory a few years ago but continues to teach the subjects of microbial physiology and biological regulatory mechanisms.

Permeases and Other Things

B. L. Horecker

E arly in the 1950s Jacques Monod visited Washington, D.C., and delivered a lecture at the National Academy of Sciences on the subject of enzyme induction in *Escherichia coli*. His earlier careful measurements of the dynamics of growth in microorganisms had led him to describe a phenomenon, which he called diauxie, that characterized growth of *E. coli* on pairs of substrates such as glucose and lactose. He observed that after a period of rapid growth during which the glucose was consumed, there was a leveling-off of the curve before growth resumed on the second substrate. Pursuing this observation and fortified by his early training in genetics in T. H. Morgan's laboratory at the California Institute of Technology, Monod discovered that the second phase of growth was dependent on the synthesis of a specific enzyme, β-galactosidase, whose synthesis was inhibited by the presence of glucose and induced by lactose, the second substrate.

Monod was an inspiring lecturer and his presentation at the National Academy of Sciences was a revelation to those of us who were hearing him for the first time. He described an elegant series of experiments, carried out with his colleagues David Hogness and Melvin Cohn, in which they demonstrated that the appearance of β-galactosidase activity was not due to activation of a preexisting proenzyme but rather to the de novo synthesis of the enzyme protein from amino acids. They also observed that the decrease in specific activity of the enzyme, after removal of the inducer, was due to simple dilution by new protein synthesis and not to the degradation of the enzyme protein. Indeed, the turnover of *E. coli* proteins under exponential conditions of growth was negligible, so that no radioactivity from prelabeled proteins was used for the synthesis of the induced β-galactosidase.

The one-gene, one-enzyme hypothesis was by then firmly established and the challenging problem of the day was to describe the mechanisms whereby individual genes were turned on and off in response to changes in the environment or in the requirements of the cell. Monod sensed that the phenomenon of enzyme induction in microorganisms such as *E. coli* provided the key to this important question. His studies had already led him to suggest that the inducer did not act directly on the gene, but instead indirectly via a gene product.

During the course of an analysis of the ability of compounds related to lactose to act as inducers of β-galactosidase, Howard Rickenberg, working with Georges Cohen and Gérard Buttin in Monod's laboratory, discovered the inducible transport system ("permease") for β-galactosides. To quote from their classic paper (H. V. Rickenberg, G. N. Cohen, G. Buttin, and J. Monod, *Annales de l'Institut Pasteur*, **91**, 829–857, 1956), this discovery:

donne une solution à de nombreux problèmes que posaient le métabolisme des galactosides et l'induction de la β-galactosidase chez *E. coli*, et apporte une confirmation expérimentale à l'hypothèse, souvent envisagée, que des systèmes catalytiques stériquement spécifiques et fonctionnellement spécialisés, distincts des enzymes metaboliques proprement dits, gouvernent la pénétration de certains substrats dans les cellules microbiennes.
[provides a solution to numerous problems posed by the metabolism of galactosides and the induction of *E. coli* β-galactosidase, and confirms experimentally the often envisaged hypothesis that stereospecific and functionally specialized catalytic systems, distinct from strictly metabolic enzymes, control the penetration of certain substrates into microbial cells.]

While this work was in progress, Monod visited Bethesda and described these exciting and novel observations in a lecture at the National Institutes of Health. My group had just completed its investigations into the pathway of pentose fermentation and acetic acid production in *Lactobacillus*, and I was already much impressed with the use of microorganisms as models for the study of basic biological phenomena. I was convinced that the discovery of the permeases in Monod's laboratory had opened the door to the analysis at the molecular level of the mechanisms of cellular transport, and I decided to take advantage of a Rockefeller Public Service award to spend a year at the Pasteur Institute to work on this problem. In September 1957 my family and I sailed for Paris on the liner *Statendam*.

I could not have chosen a better time. The new science of molecular biology was emerging at the borders of microbial genetics and biochemistry, and the Institut Pasteur, with André Lwoff, François Jacob, and Jacques Monod, was at the center of the exciting developments in this field. When we arrived, the annual course, "Le Cours," had just begun, and the laboratories were filled with students learning the new techniques of molecular biology, microbial and phage genetics, adaptation, protein synthesis, and immunology. Every day in the tiny library off the refectory, there was a lecture by one of the staff members or by a visitor.

It was during that year that Arthur Pardee, on a sabbatical at the Pasteur Institute, carried out the classic experiment with Monod and Jacob that established the role of the regulator gene, the i gene, in enzyme induction. Pardee, who was one of the discoverers of the phenomena of feedback inhibition and enzyme repression, had come to Paris to work on this problem, and the idea, reinforced by Szilard during his frequent visits to the Institute, quickly developed that *induction* of catabolic enzymes by the substrate and *repression* of biosynthetic enzymes by the end product of the biosynthetic pathway were fundamentally similar processes, and that induction was really antirepression. The critical experiment, carried out with Hfr strains constructed for the purpose by Jacob and Elie Wollman, was finally conceived after long and excited discussions in Monod's little office. The experiments of Pardee, Jacob, and Monod also provided evidence that the messenger RNA for β-galactosidase in *E. coli* was unstable.

Monod's earlier studies had shown that the genetic regulation of the galactoside permease was coordinated with that of the β-galactosidase, and indeed that the gene for β-galactosidase (the z gene) and that for the permease (the y gene) were part of the same genetic unit, or operon. Thus both galactosidase and permease were induced coordinately. It was clear from these observations that the product of the permease gene was also a protein. Monod proposed that the permease would have the property of binding its substrate and suggested that we try to identify it by this property. Dietmar Türk, a young organic chemist from Germany, and I, working in a small hood in the hallway next to the refectory, set out to synthesize the lactose analogue, β-thiodigalactoside, labeled with very hot ^{35}S. To measure the binding, we used a simple and clever apparatus for equilibrium dialysis borrowed from Jean-Marie Dubert, designed for measuring the binding of antigen by antibodies. The experiment was negative, although there were a few more counts in the chamber with the extract from the induced cells. When Monod and I recalculated the number of counts expected, it became obvious that our labeled substrate was not of sufficiently high specific activity to allow us to detect the number of permease molecules that might reasonably be expected to be present in the extract. My notebook on this date, October 29, 1957, carries the title from Proust's monumental work *À la recherche du temps perdu*. Years later, with much more refined techniques available and with genetically enriched mutants, Walter Gilbert and Benno Müller-Hill used a similar approach to detect and isolate the *lac* repressor. The *active* galactoside permease has not yet been isolated, but Fox and Kennedy have isolated and characterized the *y* gene product, which they called the M-protein, after modifying it with the *N*-ethylmaleimide, using a clever modification of the binding idea based on Adam Kepes' observation that the *β*-galactosidase permease contains essential sulfhydryl groups that are protected by the substrate.

The rest of my year at the Pasteur Institute was spent analyzing the properties of another specific permease that was responsible for the transport of galactose. I was assigned a technician, a charming young lady named Janine Thomas, and from that point my notes were written in French, corrected each evening by Janine, who became my teacher as well as my able and hard-working assistant. The parting greeting each evening became "Est-ce que vous avez mis les souches?" The constitutive galactose permease, measured in a galactokinase-less mutant of *E. coli*, proved to be remarkably specific and able to detect the presence of very low concentrations of galactose in the external medium. Our work also led to the discovery that a specific mechanism, enhanced by 2,4-dinitrophenol (DNP), was responsible for the exit of the substrate, which had previously been thought to be a process of passive diffusion. Monod displayed his remarkable versatility by developing a logical, elegant, and simple mathematical formulation that permitted us to calculate the *exit* rate from the kinetics of the initial rate of *uptake*. This was published in the *Journal of Biological Chemistry* in 1958. This simple idea permitted us to analyze the activity of the exit process, which, to our surprise, turned out to depend on the constituents of the growth medium. Thus the galactose exit rate was enhanced in cells grown in the presence of galactose, despite the fact that the galactokinase-less cells could not metabolize galactose and the permease was constitutive. The exit of galactose was also found to be inhibited by substances, such as *α*-methyglucoside and succinate, that were not substrates for the galactose permease. Adam Kepes had already proposed that specific transporters are involved in the movement of substrate across the cell membranes, and his evidence and ours was later supported by the finding of specific binding proteins. Monod's laboratory proved to be a fruitful and stimulating environment for these early studies on transport mechanisms.

Monod was intimately involved in everything that was going on in the laboratory, from sporulation to permeation, and in those days he was always in the laboratory, available for discussion of the work. He was equally ready to talk about music, or art, or the Russian sputnik, the news of which he brought me one Saturday morning. He was full of ideas and he possessed a vast store of information. This always amazed me, because apparently he did not read the literature. Somehow he managed to hear about every important event in science and in that year I came to appreciate the value of the grapevine as a source of information about important new developments.

Jacques Monod's approach to science was the essence of the inductive method, including (1) careful analysis of the facts in hand, (2) development of a suitable model or hypothesis, (3) design and careful execution of appropriate experiments to test the model, and (4) revision of the model where necessary and then on to the next round of experiments. It was rare for him to make a serendipitous discovery. It was equally rare for an experiment to fail, and when this happened it was likely to be for lack of adequate methodology, rather than because the model was incorrect. Thus, it was fascinating to observe the series of hypotheses and experiments that led his laboratory to the elucidation of the role of the *i* gene in the control of expression of the structural gene. Behind the final conclusion was a carefully constructed logical edifice based on precise genetic and biochemical measurements and broad biological concepts. Another example of his logical approach to the problem of biological regulation was his elaboration of the concept of allosteric proteins. The earlier literature contained many observations that suggested that enzymes and other proteins might possess specific sites, distinct from the catalytic site, for the interaction with specific effector molecules, but it was the classic papers of Gerhart and Pardee that finally provided elegant and convincing evidence for this concept. With Monod's characteristic flair for developing physical and mathematical models, this time with Jeffries Wyman and Jean-Pierre Changeux, who also contributed new experimental evidence, and for articulating these ideas in a stimulating and popular manner, he quickly succeeded in making "allostery" a household word.

Monod's remarkable intellect may have tended to obscure the warm and sensitive human being. His loyalty and devotion to his family and his friends were constant and enduring, but these qualities were very private and were only evident to those who were close to him. His "good works," if made known, would fill many chapters. His qualities as a scientist were more widely recognized and he became the center of discussions at every scientific meeting that he attended. Those of us who had the privilege of working with Jacques Monod on a daily basis were indeed fortunate, and the inspiration that he transmitted will be our lasting legacy.

Bernard L. Horecker

Bernard Horecker was born on 31 October 1914 in Chicago, Illinois. He was educated at the University of Chicago, receiving his B.S. in Chemistry in 1936 and his Ph.D. in 1939. From 1939 to 1940 he was a Research Associate in the Department of Chemistry at the University, studying cellular electron transport and isolation of TPN-cytochrome c reductase. In 1941 he joined the Laboratory of Biochemistry and Metabolism at the National Institute of Arthritis and Metabolic Diseases (U.S. National Institutes of Health). His work there involved the enzymatic steps in the pentose phosphate pathway and their role in carbohydrate metabolism, including carbon dioxide fixation in plant photosynthesis and acetic acid formation in microbial fermentations.

In 1957 Dr. Horecker received the Rockefeller Public Service Award for study at the Pasteur Institute and worked for a year with Jacques Monod on galactose transport in *Escherichia coli*. Returning to the United States in 1959, he took the position of Professor and Chairman in the Department of Microbiology, New York University School of Medicine. During these years his studies concerned bacterial permeases; pentose metabolism in yeast and bacteria; Schiff-base mechanisms in catalysis by transaldolase and aldolases; and biosynthesis of *Salmonella* lipopolysaccharide. He continued investigations of *Salmonella* lipopolysaccharide, structure and mechanism of action of aldolases and transaldolase, and fructose-bisphosphatase and gluconeogenesis from 1964 to 1972 at Albert Einstein College of Medicine, New York, N.Y.

In 1972 Dr. Horecker joined the Roche Institute of Molecular Biology, investigating enzyme structure and mechanisms of action, particularly the isolation, structure, function, and biosynthesis of the novel polypeptides prothymosin and parathymosin. From 1984 to 1992 he served as Professor of Biochemistry and Dean of the Graduate School of Medical Sciences at Cornell University Medical College, continuing work on the molecular genetics of prothymosin and parathymosin and the regulatory role of cellular proteinases.

Dr. Horecker is currently Professor Emeritus of Biochemistry at Weill Medical College of Cornell University, New York, N.Y.

Early Kinetics of Induced Enzyme Synthesis

Adam Kepes

During the many years which I spent in his laboratory (1955–1967), I heard Jacques Monod enunciate a number of aphorisms and sayings. The most universally known and the most often misquoted is "Whatever is true for *E. coli* is true for an elephant." His faith in the universality of the laws and mechanisms of biology contrasting with his provocative attitude of apparent cynicism in front of the great problems of "the secrets of life" was fascinating to those of us who surrounded him.

But the cynical attitude was on the surface and I remember him saying that a real researcher must be more or less neurotic. I think he was emphasizing by this comment the necessity for the researcher to live with his doubt not only until "the experiments are confirmed by the theory," but even beyond that point. The necessity never to consider one's own published statement nor that of anybody else as an irreversible truth, and to be ready at any time to submit them to a revision, implies a strong feeling of insecurity.

One of his expressions which struck me the most at that time and which has since given me ample matter for reflection was the explanation of why he chose *E. coli* as the experimental subject. Somehow a liquid culture of *E. coli* reminded him of the perfect gas. Differences between individual bacteria in the culture, just like between molecules in the gas, are averaged out by the large number. An exponential culture remains "homothetic" to itself all the time; the relative increase of population in number, in mass, and in optical density remains identical to the relative increase of protein, DNA, and RNA, which can be compared in a gas during compression to the parallel increase in density, pressure, and concentration as well as in partial pressure of each molecular species (if the gas is a mixture). Upon a more detailed analysis of a growing bacterial culture, the relative increase of each individual stable molecular species, or for that matter each unstable molecular species (provided they are in a steady state), also remains identical. This rule is broken, of course, whenever an essential ingredient of the medium becomes limiting during the observation. One very illustrative expression of the impersonal existence of E. coli as a perfect gas is that "exponentially growing E. coli has no age." This explains the emphasis of the experimental routine in Monod's laboratory on exponential-phase bacteria and on conditions of "gratuity" in enzyme induction.

What happens during such an induction can be described by the analogy with one perfect gas, the uninduced E. coli, which is submitted to gradual dilution by the addition of another perfect gas, the induced E. coli; all increment in measurable properties is due to the latter. Thus, the total activity of induced enzyme is proportional to the increase of bacterial mass as shown by the classic "Monod plot": $\Delta E = f(\Delta B)$.

The formal linear relationship holds true with astonishing accuracy, in spite of the intuitively obvious fact that during the dilution there is no mixture of uninduced bacteria with induced bacteria, but the whole population is of individuals, partly induced and partly uninduced. This has been beautifully demonstrated by the experiments of Seymour Benzer. Thus it seems that the quality of E. coli as a perfect gas is not connected with the division of the matter into the relatively uniformly sized cells, but it is somehow a more profound property. This irrelevance of cell structure to the kinetics of enzyme induction explains (together with the brilliant achievements of phage research made without the help of any direct visual observation) another iconoclastic statement of Jacques Monod's, according to which, "had the use of microscopes been prohibited, biology could have leaped forward fifty years."

Not only was the biosynthetic process independent of the cell's age—its state of division—but the quality, the specificity of the biosynthetic product, was independent from the stimulus which brought about its synthesis, the inducer. Jacques Monod showed with Melvin Cohn that the kinetic parameters as well as the immunochemical properties of β-galactosidase did not change when a variety of inducers were utilized with the inducible strain or compared to the enzyme of the constitutive mutants, where no inducer was used. This lack of necessity for outside information led him closer to the idea of a master plan, the genetic information, but also inspired another attempt toward universality, the theory of generalized induction. If every enzyme is manufactured according to genetic information of a common kind, there would also be a common kind of signal to trigger the transfer of information. For constitutive enzymes, the inducer was present in the cytoplasm, presumably as a metabolic intermediate, while for inducible enzymes it had to be added to the medium. This purely intellectual postulate received confirmation with the example of a sequential induction in the mandelic acid metabolic pathway unraveled by Roger Stanier.

Several lines of reasoning started at this point and were solved in the spirit of simple logics, and of universality of mechanisms, by using precisely defined concepts without necessarily naming the underlying detailed cellular or molecular structures. The inducer is only the signal for biosynthesis; it happens to be a metabolite and a substrate of the inducible enzyme, even though these properties might be independent. A search was started to verify this independence and, effectively, nonmetabolizable inducers were found, namely, a whole family of thiogalactosides. Also discovered were metabolizable β-galactosides which had no inducing effect, e.g., phenyl-β-galactoside. Lactose itself turned out later to be of this kind would it not be for the transgalactosylating activity of β-galactosidase, which produces true inducers, as shown by Claude Burstein. This distinct function compels postulation that the inducer interacts with a receptor, distinct from the enzyme induced. For this, the inducer must first penetrate the cell. Hence the discovery of lactose permease by Howard Rickenberg, Georges Cohen, and Gérard Buttin, and generalization of the permease concept, one of the great moments of Jacques' intuition.

The definition of the structural gene of a transport system and the coining of the word *permease* was perceived as a challenge by the "transport-worker's union." The receptor of the inducer is also genetically determined, its gene being independent from the structural gene of β-galactosidase; it is the gene i. When it is damaged as a signal receptor, its function as a switch is frozen, and the cell is either constitutive i^- or super-repressed i^s. When genes are transferred by conjugation, the structural genes are first expressed constitutively, and only later when the i gene is sufficiently expressed does it become the inducer necessary for further synthesis (PaJaMa experiment). Therefore, the i gene codes for a cytoplasmic factor of negative-regulatory mode, the repressor. Induction, deinduction, and regulatory mutations of the i gene simultaneously affect the expression of the z gene, β-galactosidase, of the y gene, *lac* permease, and of the a gene, galactoside transacetylase, discovered in the meantime by Irving Zabin. This parallelism of expression, together with the more restricted coordination in *cis* in mutants of the operator o locus, led to the concept of the operon, the unit of genetic expression.

In the meantime it was demonstrated that DNA, in which genetic information is encoded (Avery, McLeod, et al.), assumed the configuration of a double helix with a base-pairing principle (Watson and Crick), and this enabled DNA to undergo self-replication (Meselson and Stahl), but apparently DNA was not the template for protein synthesis. Nascent proteins were found in ribosomal fractions containing no DNA, and in eukaryotes, DNA was confined in the nucleus while the bulk of protein synthesis took place in the cytoplasm. On the other hand, ribosomes did not fulfill the requirements for an intermediate between gene and its protein product; they remained the same in induced and noninduced, in uninfected and phage-infected bacteria, while they got involved in the synthesis of genetically unrelated proteins.

The rapid regulatory switches pointed toward an unstable intermediate embodying the genetic information between gene and protein. The intermediate, called "the messenger," soon became the messenger RNA, or mRNA. An unstable rapidly labeled RNA fraction, shown first by Volkin and Astrachan in phage T2-infected bacteria and soon in *E. coli* by François Gros, who went to Jim Watson's place at Harvard to fetch the fraction, appeared as the right candidate for this role, due to its base composition, its association to ribosomes, and its rapid chase.

Hypotheses were mushrooming about the use and fate of mRNA. Its "exceptionally high rate of turnover," with a half-life estimated at less than 15 seconds, suggested the possibility of a stoichiometry of one-to-one (that is to say that one molecule of messenger is destroyed for each molecule of protein synthesized). This possibility, expressed by Jacques at the 1961 Cold Spring Harbor Symposium, was still in the vein of the perfect gas analogy, but it seemed to meet with serious difficulties. Among other hypotheses which would permit a whole range of copy multiplicity was the possibility of a predetermined number of protein copies for each kind of messenger (fitted with a copy-counter) or a predetermined lifetime, which would set the range of the copy yield (messenger fitted with a timer like a time bomb). Rather seldom mentioned was the possibility of a random decay with a statistical life expectancy. At any rate, at the start of our kinetic study, nobody was ready to ask questions about the time course of elongation or of degradation of a single macromolecule.

The work outlined below could be qualified as the invention not quite of the time-microscope, but more modestly of a time-magnifying lens which focused attention to molecular events in the 10°- to 10^{2}- second time range and was one of the several pathways leading beyond the "perfect gas, perfect logic, perfect concept" era. Actually, the approach remained mainly on the conceptual rather than on the structural level; the main departure was the abandonment of the punctual molecule–punctual event approach. This work developed between 1961 and 1968. After the reports of Boezi and Cowie and of Pardee and Prestige on the 3- to 4-minute lag which elapses between the addition of the inducer and the appearance of β-galactosidase, the question arose, what happens if the inducer is removed before it brought about its effect? And I found that the enzyme synthesis occurred all the same, after a lag as usual, but it leveled off as a single wave of synthesis, its final yield being proportional to the duration of the presence of the inducer. I called this the "elementary wave," with the afterthought that at the limit an extremely short pulse of inducer would cause a short burst of messenger synthesis, substantially a single molecule of messenger per cell. Whatever protein synthesis is forthcoming then, its time course can serve to answer questions about the way the messenger gets inactivated and about its lifespan. Assuming that the rate of protein synthesis at any time reflects the amount of message active at that time, the time bomb model should result in a sudden slowdown of synthesis, whereas the progressive slowdown of the wave would be more consistent with a first-order, random inactivation of the messenger. Figure 1 represents the experimental and logical process. The top portion of the figure is the classic Monod plot of β-galactosidase induction taken from M. Cohn (*Bact. Rev.*, 1957). The bottom portion is a reconstitution of the first minutes of the same time course as a sum of consecutive inducer pulses, which result in the sum of time-shifted elementary waves. It shows that the contribution of each elementary inducer pulse is identical in amount and in its time program.

It turned out that the decay of the messenger was first order, i.e., exponential with time, with a half-life of 1 minute, at $37°$, a very reproducible result irrespective of growth rate and metabolic situations. The next question was why a 1.5-minute delay separates the addition of the inducer from the termination (translation) of the first molecule of enzyme. Soon it became clear that the whole of this period was occupied by the translation process, that is to say the elongation of the polypeptide chain. The synthesis of acetylase, the product of the third structural gene of the *lac* operon, was terminated only about one minute later. This was the first indication that the structural genes of an operon are expressed in sequence, and the use of rifampicin, and actinomycin permitted soon to ascribe this sequential program to the synthesis of a polycistronic messenger. Only one rifampicin-sensitive initiation event was followed by two distinct actinomycin-sensitive terminations, first that of the Z message and about

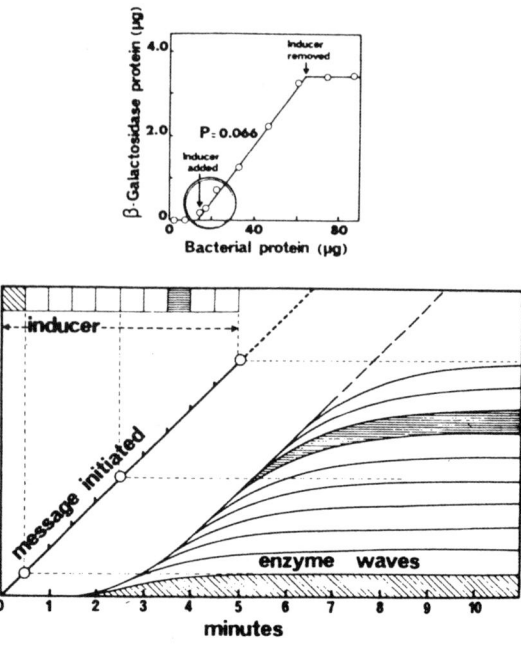

FIG. 1. Represents the experimental and logical process. See text for explanation.

one minute later of the *A* message. Knowing the elongation time of mRNA and its lifetime, the problem thus arose: what variants of the integral transcript have coding activity? This question was thoroughly explored by Michel Jacquet at the time, when I had already left Pasteur and started an independent group at the laboratory attached to Jacques Monod's chair of Molecular Biology at the Collège de France.

The picture which emerged from these kinetic studies was a highly coordinated sequence of events. Transcription, translation, and messenger RNA breakdown all proceed in the O-Z-Y-A direction. Translation proceeds, making on the average one peptide bond about every 65 milliseconds, and transcription produces three nucleotide bonds in the same time. The initiating end of messenger RNA has a half-life of about 60 seconds, or, in other terms, it has 1 chance out of 85 to be inactivated the next second, but it has no age; its life expectancy, does not decrease with time. The initiation of transcription is independent of previous transcriptions and does start usually before the previous transcription is terminated. The initiation of translation is independent of the length of RNA chain as soon as a sufficient length of RNA is available. The initiation of translation is also independent of previous translations, provided a minimum spacing; it only depends on the intact initiating end of mRNA. The progress of translation is independent of both the termination of the transcription and the survival of the initial end of messenger RNA. The survival of the initiating end of messenger RNA is independent of the intracellular concentration of inducer and largely, although not completely, independent of transcription.

The messenger RNA is polycistronic and stays probably as a single piece for the major part of its functional lifetime. At the steady state of all processes, the polycistronic messenger RNA is, however, seldom integral. The major part of it should be pieces: some unfinished, some already missing the initiating end, some devoid of both, still growing on one side while losing the other side at the same speed. The weighted average size is calculated as about half the integral size, the unweighted average one-third, and the frequency of integral size is about one-eighth of the total population (19% integral for the Z segment).

This detailed kinetic study followed the guidelines of a model largely elaborated in our lab. It gave more coherence to the picture; it helped to make a decision between alternate possibilities, but sometimes it also suggested the existence of steps not suspected before (e.g., a chemically distinct event of peptide chain initiation, inhibited by hydroxylamine, as shown by Simone Beguin). In other instances it brought the first experimental evidence in favor of a postulate, e.g., the polycistronic messenger, the colinear transcription-translation, the "instantaneous" accomplishment of events attributed to conformational changes like the interaction of inducer and repressor. Sometimes the kinetic demonstration preceded, sometimes lagged behind demonstrations by others with other methods, but it remained in permanent dialogue with every step of progress made in the laboratory and in "the world." But Jacques Monod, partly because he was disappointed that the permease work, which we started together in joy and excitement, did not lead to the expected triumph through the isolation of the protein, and especially that the acetylase did not keep its promises to be a part of the permease system, and partly because very soon he went further ahead to allostery and the "second secret of life," paid decreasing attention to this finicking work around the first secret (the base-pairing principle of genetic conservation and expression).

I personally feel that this piece of work is very much in the line of Jacques Monod's thinking. The elaboration of a rigorous methodology and a patient effort to try and answer all possible questions which that methodology can deal with, just as the formulation of a hypothesis followed by a patient exploration of all its possible predictions, slowly conducted to a new way to conceive pre-existing notions, in this instance the time dimensions of molecular events.

Adam Kepes

Profoundly humanist, Jacques Monod sought Adam Kepes for an introduction to Professor Augé of ICRO (International Cell Research Organization), associated with UNESCO. Professor Augé had identified cell biology as a special area for international scientific cooperation.

The choice of this upcoming discipline required outlets for the rapid diffusion of new discoveries, something greatly appreciated by scientists and international policy makers alike. Adam Kepes had the talent necessary to fulfil this undertaking: a global knowledge of the science as well as a profound solidarity with the young students and foreign researchers living in countries deprived of both outside contacts and foreign scholarships.

Animated by this cooperation, Adam Kepes organized, between 1965 and 1983, 191 practical laboratory courses in 59 countries, involving more than 4,000 individuals. These courses enabled the coming together of young scientists from countries sometimes in conflict with each other, creating solid friendships around the lab bench.

ICRO would not have been developed as well without Jacques Monod's essential contribution. I would like to pay tribute to his modesty and his ability to bring together human beings and their talents.

Suzanne Kepes

From Acetylase to β-Galactosidase

Irving Zabin

Since graduate student days, I had been working in lipid metabolism, and I wanted to do something different. Monod's name had some magic about it even in the 1950s; he worked in fundamental areas of biology, and he worked in Paris, one of my favorite cities. Therefore in 1957, well in advance of my first sabbatical leave, I wrote asking to spend a year with him at the Pasteur Institute. He answered yes, I obtained a fellowship, and in February 1959, I arrived in Paris.

The *lac* permease in *Escherichia coli* had been discovered a few years earlier (Rickenberg, Cohen, Buttin, and Monod, 1956) and an extensive review summarizing what was known about it had also been published (Cohen and Monod, 1957). *lac* permease was responsible for transport of lactose and other galactosides into cells. All experiments had been done using whole cells. To go further and get at the mechanism of transport required identification and isolation of the material or materials responsible. The reigning authority on transport in Monod's laboratory then was Adam Kepes, who had done some elegant experiments on the kinetics and energy requirements of *lac* permease (Kepes, 1957; Kepes and Monod, 1957). It had to be a membrane protein. Membranes contain lipids as well as proteins. And I had a background in lipid chemistry and metabolism. Therefore what was more logical than to ask me to work with Adam on the permease problem? So Adam patiently introduced me to *E. coli*, and to the mysteries of *lac* permease.

We had a fine time for some months dreaming up and doing experiments designed to help isolate the permease. For example, I remember trying to develop an assay by adding extracts from a permease-positive strain to cultures of a permease-negative strain in the hope of converting them to permease positive. Then, when this didn't work, the recipient cells were pretreated with certain organic solvents to try to pry them open so that the "permease" might be introduced. From time to time we got excited when it looked at first as if there was a small change in ability to transport, but nothing worked.

Monod listened and was available when necessary. He was interested and offered a suggestion now and then; but he let us proceed at our own pace. Then one day in May or June, he said to me, "Why don't you see if you can get the acetylation reaction to work in vitro?" The background was as follows: When radioactive substrates like isopropylthiogalactoside (IPTG) or thiomethylgalactoside were transported and accumulated by cells, a small amount of a radioactive derivative was also formed. About a year or so earlier, Herzenberg had been at the Pasteur Institute and had identified the derivative as a 6-O-acetyl compound. It was formed by permease-positive but not by permease-negative strains. It was a slow process. For any considerable amount of substrate to be converted to the acetyl form, cultures had to be shaken overnight in air, in the presence of a carbon source. This information was not published until later (Herzenberg, 1961), but it was available in the department. Did acetylation have anything to do with *lac* permease? A permease-negative cell would not transport IPTG into the cell, so IPTG would not be acetylated. Therefore acetylation of the substrate might be only an indirect consequence of the presence of the transport system in *E. coli*. It was foreign: perhaps it was acetylated like foreign amines were in animal tissues. Also, acetylgalactosides were not converted to galactosides, and were neither substrates nor inducers of *lac* permease. They were inert in *E. coli*. But even with all this, Monod's intuition suggested that acetylation should be examined anyway.

I was reasonably well-suited for the job. I had been a student of Konrad Bloch, who had done a great deal of work on acetate metabolism, and I had grown up in biochemistry while "active acetate" was finally identified as acetyl coenzyme A. So I prepared some necessary reagents, repeated some of the whole-cell experiments, explored part of the European countryside, and in September found that a cell-free extract of the constitutive strain ML308 would convert IPTG to acetyl-IPTG when supplemented with either acetate, coenzyme A, and ATP or with acetyl-CoA. That was fine, because now we could test for the ability of a strain to carry out acetylation and compare this ability with the presence or absence of permease in that strain. When we did, it turned out that an extract from a strain without *lac* permease could not make acetyl-IPTG, nor could a noninduced wild type make the derivative unless it had first been induced. The correlation was perfect. We were excited about these results. Within a few months we had enough information to publish a short note (Zabin, Kepes, and Monod, 1959). It ended with "The observed correlations constitute strong evidence that the acetylation reaction is carried out by a system closely connected with, or part of, the permease system." At the time, we thought we had a tag for the permease protein. The idea was that the protein which catalyzed the acetylation reaction also carried out transport, or some part of transport, in vivo. This statement is still literally true in the sense that acetylase is part of the *lac* operon.

Monod said to me one day in November or December, "Why don't you stay another year?" I was due to go back in a few months and the problem had just barely opened up. This was a fine suggestion to all except one small Zabin who missed his friends at home. And, because I felt a whole additional year would be too long to be away, I asked for and arranged a leave of absence for another half-year instead.

I worked pretty hard for the next period of time. There was a lot of biochemistry to do on acetylation. We wanted to examine the properties of the enzyme or enzyme systems responsible for the reaction. We wanted to see what we could find out, if anything, about the permease. And how did this fit in with induced enzyme formation? Now two apparently separate and distinct proteins were turned on by one signal, the addition of inducer to growing cells.

During this time, Monod was more than available. He was often in the lab, bringing up questions, suggesting experiments, waiting for the latest result. He was tremendously stimulating. I was too close then to think of him as a scientific giant or genius; this came later in retrospect, which was perhaps just as well. I have often asked myself what his special qualities were. He had an air of assurance, he was immensely articulate, he was very convincing. He was quick, of course, and he got to the point faster than most. He was imaginative and it always was a logical imaginativeness. But I think he was supremely a synthesizer; this was his great strength. He often looked at things from a quite new point of view; this was part of his ability to put things together. Also, I read somewhere that one quality of genius is the ability to focus intensely on one subject, then at a moment's notice do the same for another. He had that, too.

He had some favorite sayings. I remember one: "You don't do an experiment to prove something, but to disprove it." Of course that didn't prevent him from being highly persuasive. Another, about some experimental result or other, especially if it didn't fit an idea of his: "I don't believe it for a minute." I thought of these, privately, as Monod's "bon mots." He had great impact on people. At seminars in the small library room, speakers seemed to direct their efforts mostly to him. He was a charmer when and if he wished to be. Few were neutral about him, men or women. I once asked my wife why she thought he was attractive to women. He was certainly handsome, but that had nothing to do with it. Her answer was "He looks right into your eyes." An old long-time (male) friend of his said one never had a dull moment in his company. He was easy to be with. I was always on a first-name basis with him. And with all of this, there was a wall beyond which I, and I suppose all but a very few, did not go. Perhaps this was part of the mystery.

We measured the formation of acetyl-IPTG at first by a chromatographic method using radioactive IPTG. This was slow. Therefore I worked out a simple hydroxamate assay and things went much faster. I tested substrate specificities, affinities, and other properties, and I worked on the purification of the enzyme. We called it acetylase in the lab. It was also referred to as galactoside acetylase, galactoside transacetylase, and thiogalactoside transacetylase. Adam took part in some of these experiments and also busied himself with other work, including measurement of permease and acetylase in more mutant strains. The correlations we had done at first were with a few well-defined strains. There were many more available.

We got a lot of information during the next half-year but it soon didn't support the idea that permease and acetylase were the same. For instance, the substrate specificities for transport and for acetylation were different. Though IPTG worked for both, phenyglucoside was not transported by *lac* permease and was acetylated. Lactose was accumulated by the cells, but was not a substrate of the transacetylase. Then, the Michaelis constants were wildly different, high for acetylation and low for transport. Furthermore, acetylase was not a membrane protein, as presumably the permease was.

Meanwhile, measurements of permease in whole cells versus acetylase in extracts went on. From one permease-negative, acetylase-negative strain, a number of permease-positive revertants were picked out. Every single one was acetylase-positive. I remember Monod being very impressed with this data. The correlations also held up at first with other strains in the Pasteur Institute's collection. But then a number of strains were seen to be permease-negative and acetylase-positive. This didn't fit but it could be explained easily. Acetylase could be only part of the transport system. Worse, though, was one mutant strain (or maybe two) which was the reverse: permease-positive and acetylase-negative. This wasn't a well-characterized mutant and no one seemed to know its origin. We almost tried to ignore it on this basis.

I think Adam was the first to discard the notion that acetylase had anything to do with permease. His orientation was that of a kineticist, and the properties of the acetylase enzyme just weren't right for the permease. Monod hung on longer; how else could one explain change of two properties by a single nonregulatory mutation unless a single protein was involved in both properties? As for the rest, the properties of the isolated protein could be different inside the cell. This last argument was not terribly convincing to a biochemist who had learned that the way to find out what goes on in the intact cell is to take it apart. But I didn't know. I would have preferred to believe.

There was an excitement in the air. People stopped and talked in the labs and in the hallways of the old building. Monod's group was on the first floor. François Jacob, who was two floors up, came down often. I think it was he who was the first to ask me one day how much acetylase was in the cell. The operon model was being unveiled; it was a "groupe de gènes à expression coordonnée par un operateur." Some of the data to support it came from work on acetylase. The first paper, though, did not mention acetylase, nor was there any need to. In that paper (Jacob, Perrin, Sanchez, and Monod, 1960), data were presented on levels of β-galactosidase and permease in partial diploids.

In retrospect it seems likely to me that the operon model would have survived and flourished even if acetylase had not been discovered. But the discovery and availability of acetylase must have been an important psychological support to the architects of the model. It would have been hard to convince the outside scientific world that measurement of galactoside transport in whole cells was indeed as valid a measure of gene expression as a direct enzyme assay. Now, whether or not acetylase was permease, definite, quantitative, unarguable data on the expression of two different genes in the *lac* operon could be obtained. So, in the classic review on genetic regulatory mechanisms which appeared the next year in the *Journal of Molecular Biology* (Jacob and Monod, 1961a), acetylase had a prominent part. One important experiment showed the effect of different inducers on the formation of the two enzymes. With effective inducers, high levels of both β-galactosidase and acetylase were formed; with poor inducers, low levels of the two activities were seen.

Of course, some of the evidence for the operon was taken on faith. If assays of two different enzymes were done, then it was necessary to be sure they were direct measurements of quantities of two proteins, and that they were in fact different. By the summer of 1960, I had managed to purify acetylase about 25-fold. This was, though we didn't know it then, less than 10% pure. It was cleanly separated from β-galactosidase, and anti-β-galactosidase did not cross-react with acetylase, so they were two quite different proteins. But I'm sure this was after-the-fact support of the already mapped-out theory. We had no idea of the subunit composition of β-galactosidase or acetylase. I wondered (later) whether the two proteins might have a common subunit; they don't. But this would have explained coordinate induction in a quite different way.

I wrote a paper after I returned to UCLA in August 1960, but the manuscript went back to Paris once or twice before it was sent to the journal. We presented the enzymological and purification data on acetylase, pointed out some of the difficulties in believing that acetylase was permease or part of it, but left it an open question (Zabin, Kepes, and Monod, 1962). It was open just in case. Before the time of the 1961 Cold Spring Harbor meeting polarity had been discovered. It was evident that polarity (where a mutation in one gene may affect the expression of a gene or genes further down) could explain why acetylase reappeared in a permease-negative revertant. In their symposium paper, Jacob and Monod pointed this out and suggested that structural gene x (now a) specifies acetylase rather then gene y, the permease gene (Jacob and Monod, 1961b).

Well, if not, what was acetylase doing in the cell? We didn't think it was a detoxifier because the reaction was slow and affinities were low. Years later it was shown that $lac\ a^+$ strains had a selective advantage over $lac\ a^-$ strains when grown under certain conditions because they do detoxify (Andrews and Lin, 1976). This may be the raison d'être of the enzyme. When strains containing the y gene but no a gene were prepared, these carried out the transport process normally (Fox et al., 1966). Therefore there was no longer any reason to believe that acetylase was part of the permease system. However, I can't help but believe that acetylase is still unfinished business.

I left Paris with a feeling of some accomplishment and considerable enthusiasm. In fact, I talked with Monod about coming back to the Pasteur Institute, not on another sabbatical, but for an extended period. I didn't do so for a number of reasons both scientific and personal, but I closed down some work on lipids that was still going on in my lab, and switched entirely to the lac operon. It was pretty evident to me that the biochemistry of the lac proteins could be an important and useful area of work.

One of the first things to do was to get acetylase in pure form, find out how much was made in the cell, and measure its size. This took close to two years, but I isolated the enzyme in pure form and also crystallized it (Zabin, 1963). Gene Goldwasser spent a semi-vacation of several months in California and measured the molecular weight (about 60,000) in the ultracentrifuge. The surprise from the purification data was that there was very much less acetylase produced than β-galactosidase, 10 to 35 times less by weight, depending on conditions of growth. This seemed strange at first. There was some evidence by this time that *lac* DNA was transcribed into a single, polycistronic messenger, so approximately equal amounts of proteins might be expected to be formed.

I spent two or three weeks in Paris in the summer of 1963 and talked about these results. Monod called the finding that more β-galactosidase was made "wild-type polarity," and also, around the lab, "Zabinism." I didn't mind the publicity but it sounded like the name of a disease.

Wild-type, now often called "natural," polarity in the *lac* operon could be explained or rationalized in a number of ways. One of the most obvious had to do with the subunit composition of these proteins. It wasn't the relative amount by moles that was important in understanding operon function. Or in other words, one wanted to know the number, not the size, of polypeptide chains translated by a polycistronic messenger. To get that number it was necessary to find out something about the substructure, not only of acetylase but of β-galactosidase.

Acetylase was fairly straightforward. By physical and chemical studies it was clearly a dimer of two identical chains, each of about 30,000 daltons (Brown, Brown, and Zabin, 1967; Brown, Koorajian, and Zabin, 1967). Antibody to acetylase did not cross-react with β-galactosidase, nor with anything else in the cell (Berg and Zabin, 1964).

The subunit structure of β-galactosidase was a difficult problem. The molecular weight had been guessed to be 600,000 to 700,000 but this was revised by a number of physical studies to a value near 500,000. It was generally agreed that the protein was a tetramer; but what was the protomer, a single polypeptide chain or several smaller ones? There were measurements with the ultracentrifuge, end-group analyses, and, later, complementation experiments which favored the conclusion that there were small chains (cf. Zabin and Fowler, 1970). Also, a single polypeptide as large as 125,000 daltons was unheard of at that time. But when end-group and other studies were carried out, it became clear that β-galactosidase contains long, not short chains. In fact, the polypeptide contains 1,021 amino acids (Fowler and Zabin, 1977)!

This put a different picture on polarity of expression in the *lac* operon, because if the molecular weight of the β-galactosidase monomer is four times that of the acetylase monomer, the weight ratios of 10 to 35:1 must be divided by 4. These molar ratios are not so striking. Deviations from 1:1 ratios can then be explained by a variety of mechanisms including mRNA decay and/or ribosome loading and unloading. In any case, natural polarity stimulated a lot of interesting work in many laboratories.

β-Galactosidase has been a challenge, worthy of a lot of work. It still hasn't lost its interest. There were and still are interesting mutants, as well as the structure to wonder about. Though acetylase was and is revisited from time to time, β-galactosidase has been the focus of attention in my lab for many years. With the aid of many collaborators, work has been carried out on gene-protein correlations, on structure-function relationships, and on evolution in the *lac* operon. Some of this has been reviewed elsewhere (Zabin and Villarejo, 1975; Zabin and Fowler, 1978), and my purpose here is not to discuss this but to recall my association with Jacques Monod.

I've long since gone my own way, but I'll always be grateful to him for the chance to take part in a major and unique intellectual achievement. I saw Monod from time to time over the years and I spent a half-year sabbatical again at the Pasteur Institute in 1967. I remember, though not why or in what context it came up, he said then that the framework for the study of biology now exists. He meant this to include control of gene expression and allostery. He was right, I think. He set the stage.

References

1. Andrews, K. J., and Lin, E. C. C. 1976. *J. Bacteriol.*, **128**, 510.
2. Berg, A., and Zabin, I. 1964. *J. Mol. Biol.*, **10**, 289.
3. Brown, J. L., Brown, D. M., and Zabin, I. 1967. *J. Biol. Chem.*, **242**, 4254.
4. Brown, J. L., Koorajian, S., and Zabin, I. 1967. *J. Biol. Chem.*, **242**, 4259.
5. Fowler, A. V., and Zabin, I. 1977. *Proc. Natl. Acad. Sci. USA*, **74**, 1507.
6. Fox, C. F., Beckwith, J. R., Epstein, W., and Signer, E. R. 1966. *J. Mol. Biol.*, **19**, 576.
7. Herzenberg, L. A. 1961. *Arch. Biochem. Biophys.*, **93**, 314.
8. Jacob, F., and Monod, J. 1961a. *J. Mol. Biol.*, **3**, 318.
9. Jacob, F., and Monod, F. 1961b. *Cold Spring Harbor Symp. Quant. Biol.*, **26**, 193.
10. Jacob, F., Perrin, D., Sanchez, C., and Monod, J. 1960. *C. R. Acad. Sci.*, **250**, 1727.
11. Kepes, A. 1957. *C. R. Acad. Sci.*, **244**, 1550.
12. Kepes, A., and Monod, J. 1957. *C. R. Acad. Sci.*, **244**, 1550.

13. **Rickenberg, H. V., Cohen, G. N., Buttin, G., and Monod, J.** 1956. *Ann. Inst. Pasteur*, **91,** 829.
14. **Zabin, I.** 1963. *J. Biol. Chem.*, **238,** 3300.
15. **Zabin, I., and Fowler, A. V.** 1970. *In* J. R. Beckwith and D. Zipser (ed.), *The Lactose Operon*. Cold Spring Harbor Monograph, 27.
16. **Zabin, I., and Fowler, A. V.** *In* J. Miller and W. Reznikoff (ed.), *Molecular Aspects of Operon Control*. Cold Spring Harbor Monograph, in press.
17. **Zabin, I., Kepes, A., and Monod, J.** 1959. *Biochem. Biophys. Res. Commun.*, **1,** 289.
18. **Zabin, I., Kepes. A., and Monod, J.** 1962. *J. Biol. Chem.*, **237,** 253.
19. **Zabin, I., and Villarejo, M. R.** 1975. *Annu. Rev. Biochem.*, **44,** 295.

Irving Zabin

After my return to UCLA from my sabbatical year with Jacques Monod (1959–1960), I continued work in my laboratory on the *lac* operon, concentrating on the protein structures of β-galactosidase and transacetylase and on gene-protein relationships. Before retirement I was asked to do administrative work and I became for a three-year period Associate Dean in the School of Medicine of the University of California-Los Angeles, with a consequent decrease in research time. Some representative papers:

Fowler, A. V., and I. Zabin. 1978. Amino acid sequence of beta-galactosidase. XI. Peptide ordering procedures and the complete sequence. *J. Biol. Chem.*, **253,** 5521–5525.

Welply, J. K., A. V. Fowler, J. R. Beckwith, and I. Zabin. 1980. Positions of early nonsense and deletion mutants in *lacZ*. *J. Bacteriol.*, **142,** 732–734.

Zabin, I. 1982. Beta-galactosidase alpha-complementation, a model of protein-protein interaction. *Mol. Cell. Biochem.*, **49,** 87–96.

Fowler, A. V., and I. Zabin. 1983. Purification, structure, and properties of hybrid beta-galactosidase proteins. *Proc. Natl. Acad. Sci. USA*, **258,** 14354–14358.

Hediger, M. A., D. F. Johnson, D. P. Nierlich, and I. Zabin. 1985. DNA sequence of the lactose operon: the *lacA* gene and the transcriptional termination region. *Proc. Natl. Acad. Sci. USA*, **82,** 6414–6418.

Celebrating the naturalization of Agnes Ullmann (left) as a French
citizen, January 1967. The hat is a "bonnet phrygien" or Phrygian cap,
a symbol of the French Revolution (for fun). François Jacob (center),
Jacques Monod (right). (Photograph by M. Brunerie.)

Being Around

Agnes Ullmann

I arrived too late at the Pasteur Institute to participate in the great adventure: the founding of molecular biology. But I was around long enough to witness many later developments.

How did I get to Pasteur? It all began in Budapest in the late 1940s. As a young student in biochemistry—never trained in genetics—I was attending seminars and film projections on the great discoveries of a genius named Lysenko. Soon afterwards I became suspicious and started wondering why the whole thing seemed monstrous, but I didn't have the elements to judge. One day I decided to confess my doubts to a friend I could trust. Secretly he gave me a page of a French newspaper, *Combat*, with an anti-Lysenko article written by the scientist Jacques Monod. It was a fabulous discovery for somebody to whom Western information was unavailable.

A few years later, another paper by the same Jacques Monod impressed me profoundly. This time I could read it openly; it was the BBA article on the de novo synthesis of proteins. Once again I was overjoyed because it put an end to the old and very dialectic Schoenheimer myth of the "dynamic state of living matter." This time, I decided that I would do everything to try and go to the Pasteur Institute, in order to work with that man.

In 1958, I managed to go to Paris for six weeks and obtain an appointment with Professor Monod. I was nervously waiting in Madeleine Brunerie's office, when to my surprise a young man came whistling down the corridor and introduced himself as Jacques Monod. He was both friendly and distant. Since Arthur Pardee was waiting for him, the best way he found to get rid of me was to invite me to give a seminar the next day. After the seminar he politely asked me what I was doing in Paris and how long I intended to stay. (He had obviously forgotten our previous evening's talk; at that time I did not yet realize that he always forgot unimportant things.) I gathered all my courage and told him that if he would allow me to work in his lab I could arrange to stay six weeks, and if not I would have to return to Budapest soon. He then took me to his office, told me all about the work going on in the lab, and asked me what I would like to do. I said that I would prefer to collaborate on François Gros' project that he had mentioned, and very scared I added—if Monsieur Gros accepts me. He burst out laughing, "François? He is the nicest man on earth; he never says no." It was true; the next morning I started working with François and had a most exciting time during all my stay. François quickly realized that I had personal problems. I felt that I could speak to him safely, and I confided to him that I wanted to leave Hungary for good, come to France, and go on working forever in the lab. Somewhat puzzled, François advised me to discuss the problem with Jacques. I answered that I would never dare. He tried to persuade me that Jacques was the most understanding and nicest person on earth, and if I was scared to talk to him myself, he would do so for me. The next day Jacques invited me for dinner at his home and, after a few hours' discussion, told me that he was willing to do everything he could to help me. I asked him why. He answered, "It is a question of human dignity."

In 1960, Jacques arrived in Budapest with a carefully prepared plan for my escape and that of my husband from the country. Three months later I was in Paris. But he would never accept my thanks or agree to talk about all the trouble he had taken to insure the success of our adventure. For him it was a settled affair; at that time the important thing was to isolate the *lac* repressor—the one supposed to be an RNA. I tried hard first with François Gros, then alone with no more success. After a while, I began having doubts about the RNA nature of the repressor; but I did not dare admit them.

However, after hesitating a long time, I knocked at Jacques' office door. He was in discussion with François Jacob and Mel Cohn. Abruptly, I told them that on the basis of the recent data, I thought that the repressor might not be an RNA. I was expecting them to throw me out of the office. But to my greatest surprise, after asking for some experimental details Mel said quietly, "After all it may be a nucleo-protein." Then Jacques added, "Why shouldn't it be a protein?" And after a few moments of thought he continued, "Of course it is a protein, it can't be anything else." While they were discussing this new possibility, I stood watching them without understanding why they didn't share my torment. That same evening Jacques came to my lab, where I was sadly sitting at my desk, and said cheerfully,

"You know, you made an important discovery: the repressor *is* a protein."
He was right: Benno Müller-Hill and Wally Gilbert proved it.

By the end of 1961, one evening quite late Jacques walked into my lab. His tie was loose and he looked tired and worried. He stood silently at my bench and after a few long minutes he said, "I think I have discovered the second secret of life." I looked at him a bit alarmed; my first impression was that he was not well, so I suggested he should sit down and have a drink. After downing his second or third scotch he started to explain *The* discovery to which he had already given a name: *allostery*. As he realized that I was not quite ready to accept his arguments at once he got up and said, "Believe it or not, the regulatory role of allosteric proteins is absolutely fundamental; it explains everything: hormonal action, repressor function, non-Michaelian enzyme kinetics . . . " and I don't remember the end of the list. During several weeks everybody discussed allostery in the lab—in fact, we couldn't help it, as any other kind of discussion turned back in a few minutes to this topic. One evening I used a "sentimental" argument concerning his enzyme, β-galactosidase, which lacks the virtue of being allosteric. His reply was quick, as usual: "I am sure we can find experimental conditions to make even β-galactosidase allosteric." Since I was skeptical he turned to my technician, Françoise Tillier, and asked her whether she didn't want to try. Of course she did. After a great number of experiments, it turned out that β-galactosidase was completely refractory to allostery! It was the only time his enzyme disappointed him. During this period, Jacques desperately searched for a suitable experimental model, with the idea of doing some benchwork himself. He asked me one day whether it was difficult to prepare muscle phosphorylase and to assay it. I answered that it was a matter of a few days. Ten days later, with Roy Vagelos we obtained experimental evidence that 5'AMP might be an allosteric effector of phosphorylase *b*. From that time on, new French students arriving at the lab—Marie-Hélène Buc, Maxime Schwartz—were kindly persuaded to choose an allosteric enzyme as their first research project. American postdoctorals were still allowed to do microbial genetics.

In fact, Jacques felt very happy with phosphorylase. Having Eddy Fischer around telling us what not to do, Jacques set up a dye-binding technique, and spent hours counting drops from columns and doing binding stoichiometry. After a long and successful experiment I told him that I was amazed to see how good he was at the bench; he answered, "I know everybody thinks I am a theoretician able only to construct hypotheses, and not to do an experiment." In my opinion he was the fastest, the most rigorous, and the most imaginative experimentalist I ever had the opportunity to work with. Moreover, he could enjoy an elegant experiment as much as he enjoyed meeting a beautiful woman.

At that time, he was so convinced that only noncovalent interactions are important that at a Gordon Conference on proteins he even remarked that after all, peptide bonds were irrelevant. I remember that Fred Richards got up with consternation, and said, "But Jacques, if you were right, proteins should be gases."

For a few years—in spite of my respect for allostery—I went on being the devil's advocate, by insisting on the importance of covalent modifications. One day, in 1967, I told Jacques that I had decided to look for covalent modifications, triggered by cyclic AMP, in bacteria. After a while I showed him the first data demonstrating that cyclic AMP relieves the glucose effect. His first comment was that it was too bad to get involved with catabolite repression, which was too complicated, and certainly had nothing to do with covalent modifications. I was convinced that he was absolutely not interested in this problem. Some weeks later, coming back to the lab after a short leave, I saw him working with my technician, both looking very busy. When I asked them what they were doing Jacques proudly answered, "Do you know that cyclic AMP relieves diauxic growth?" Gérard Buttin entered the lab at this point and, looking at the growth curves, made the spontaneous remark to Jacques, "How nice, now you can at last finish your thesis."

Jacques was now ready to admit that cyclic AMP plays an important regulatory role, but as far as I was concerned further experiments convinced me that covalent modifications were not involved in this phenomenon. Jacques ironically noticed, "You see where the search for covalent modifications leads? Now you are just about to go on and identify a new allosteric protein: the cyclic AMP receptor. . . ."

The year 1968 was difficult for everybody. The May events stopped all scientific activities in the lab for a while. Jacques was very much involved in various aspects of the students' problems.

Besides, many of the members of the group had decided at that time to assume new directions. This gave Jacques mixed feelings of loneliness, nostalgia, and freedom. He was seriously thinking about a sabbatical leave: "My students don't need me anymore, and I too feel I need a change, a kind of self-renewal," he used to say.

At the beginning of 1970 he was asked to become Director of the Pasteur Institute. For weeks and months he discussed this matter over and over with me; on the one hand, his sense of duty and of responsibility, on the other hand, the temptation of having a year of freedom in the States where he could think and write undisturbed (the project of his second book was already born), discuss science, and give lectures.

I tried my hardest to persuade him to leave. Convincing Jacques was nearly always impossible; he listened attentively and generally said, "You may be right but let me think it over." In July 1970, while I attended a meeting in Seattle, he sent me a telegram saying that the decision had been taken: he was not to be Director of the Pasteur Institute.

When I got back to Paris, Jacques seemed relieved; he told me about his plans concerning his next book, and among other things about a future transatlantic sail (he even bought a few maps). These dreams did not last long. In the fall, his wife Odette became seriously ill.

Jacques immediately gave up the idea of a sabbatical leave and a few months later he became Director of the Pasteur Institute. By accepting this task he wanted to pay a tribute to, and help to save, a unique institution which had given him and many others the opportunity to do the kind of research which could have been done nowhere else in France at that time. A feeling of indebtedness certainly played a part in his decision, but in my opinion, the key factor was his Protestant sense of responsibility. He considered that preserving Pasteur from bankruptcy, safeguarding its independence, and insuring the freedom of its scientific activities should be given top priority.

At the beginning of his directorship, Jacques believed his administrative tasks would take him only a few hours a day and that he could spend the rest of his time discussing science, attending seminars, and could even participate personally in some experimental project. After a few months he realized to what extent he had underestimated what administration and business meant. He suddenly became conscious that he was engaged in a different "profession." And by changing profession he had to change his attitude. He became more firm—sometimes harder—and he often regretted being obliged to do so. Gradually he discovered that a gap was widening between him and some of his friends, former collaborators, and students.

One day in November 1972, Jacques arrived at my door. Almost shyly, he took out of his pocket a charming little three-week-old puppy. "May I introduce Vicky?" he said. (I could not have imagined that within a year it would become a sixty-pound Airedale!) I needed some time to understand why he had decided to have a dog: he needed a companion—faithful, loving, and never arguing, except when he wanted to go out for a walk. Jacques took him to the Institute practically everyday. Around 4 p.m. Vicky used to bark at my lab door; according to Jacques he preferred the smell of β-mercaptoethanol to the tediousness of directorial meetings. Vicky knew that Jacques would come along around 8 p.m., a fact some of his friends and former students also knew. It was his moment of relaxation. He wished to be kept informed about the latest results and he was always happy to discuss any detail of a current experiment. He listened and argued with his usual solid logic. But if he happened to discover a mistake in a sophisticated mathematical demonstration of Maxime Schwartz he was as proud as a schoolboy!

In the summer of 1975 during twenty-five days of quite hard sailing, Jacques was in an excellent mood, happy as usual on his boat; his health and physical condition were excellent and he was intellectually impatient to start working on his own. His term as a director would be over in eighteen months, after which he meant to spend part of his time in his house in Cannes working quietly. He then asked me somewhat anxiously, "Do you think I am still capable of doing something worthwhile?" Without waiting for an answer he went on, "I don't mind, I want to learn new things, to feel free to think and write, and also to invite all sorts of people for long discussions on all kinds of topics." Back in Paris he ordered a large number of books and embarked on serious preparation for his future book, *L'homme et le temps.* He discussed experimental work with increased interest and pleasure, was full of ideas and projects; he felt rejuvenated.

Once again these dreams did not last. A few months later he became ill and had to slow down his activities. He still came practically every day to the Institute and continued to assume his functions as Director. In the evenings, however, when he came down to the lab to have a short rest and a drink and to pick up Vicky, he looked frightfully tired. But he refused to stay at home and rest unless a high fever kept him in bed. As soon as he could get up he was back at the Institute.

Early in May 1976 his health improved. He told me he was glad to report that even his physician seemed satisfied and had given him the permission to attend the Lactose meeting at Cold Spring Harbor. He regained strength to such a point that he insisted that we should spend some evenings writing an article—his last one.

At the end of May he went to Cannes for a week or so. When we spoke on the phone on May 28 he told me that he was feeling very well and questioned me about several details such as his subscription for the next season's concert programs, his plane reservation for the Lactose meeting, and many other things. Finally he reminded me that I had promised to finish the article for his return on May 31. He ended the conversation by saying, "See you Monday in the lab." I saw him that Monday—not in the lab, but on his bed in Cannes—before he closed his eyes forever.

From this day on, sitting in my lab with Maxime or others, we have understood that nothing would ever be as it was before. Vicky was the only one who for months and months still desperately waited every evening at the door and could not understand why his master did not call for him anymore.

Agnes Ullmann

Agnes Ullmann obtained her doctorate (Doctor rerum naturalium) at the University of Budapest in 1958, where she was assistant professor. There she discovered (with T. Erdos) that streptomycin acts by inhibiting protein synthesis. She started to work in Jacques Monod's laboratory at the Institut Pasteur in 1960. She participated in the studies leading to the definition of the promoter (with F. Jacob and J. Monod) and elucidated the intracistronic complementation mechanism of *Escherichia coli* β-galactosidase. Beginning in 1968, she studied the role of cAMP in bacteria and showed that cAMP relieves catabolite repression. Her later research, while she headed the unit of Biochemistry of Cellular Regulations (from 1979 to 1996), was on structure-function relationships and regulation of *Bordetella pertussis* adenylate cyclase toxin, which led to the use of recombinant adenylate cyclases to induce cellular immunity and to generate protective antiviral and antitumor protective activity.

She has been affiliated both with CNRS (1962–1996) and with Institut Pasteur (1978–1996). In addition to directing a research unit, she has served as Scientific Director of Development of the Institut Pasteur. She retired as research director of CNRS and professor at Institut Pasteur and is now Emeritus, still working at Pasteur.

In the hall of the Director's office, with his young Airedale Vicky, ca. 1973. (Photograph by Madeleine Brunerie.)

Ca. 1970.

Another Route

Maxime Schwartz

O ften, after a long day, he would go downstairs and have a scotch with Agnes. Almost invariably, he would sit in the red chair, squeezed between the side of Agnes's desk and the small blackboard. This was a good location. Just at hand were the glasses, the ice cubes, and the bottle of scotch, on a small pull-out slide extracted from the desk for that purpose. At hand, too, was the blackboard, on which he could draw straight genes and roundish proteins. He was facing Agnes, sitting at her desk, and also the blue chair. From time to time, somewhat unexpected in the office of an elegant woman, a dirty piece of old bone could be seen on the ground. This meant that Vicky was around. Vicky was an Airedale—not a dog—because dogs were not allowed in the laboratories.

"Tiens, voilà le petit Maxime," Monod would say when I came in. There really was no need to call me "petit." Calling Jacob "le grand François," as some people did, was in a way necessary, because of all the other "François" around. But there was no other Maxime. Well, even if it made no sense, I did not object to being welcomed as "le petit Maxime." This "petit" was one of the very few words expressing the affection which existed between us. Vicky, who was much more outgoing than his master, would generally jump on me, to express his feelings in a noisy and damp manner. Monod would then laugh loudly, because he knew I was not so fond of dogs, sorry, Airedales. I would sit in the blue chair and Agnes would ask me, "Vous voulez un verre?" Somewhat reluctantly, because I don't really like scotch, I would say yes, and have some. And then, in spite of the Airedale, and in spite of the scotch, I would spend some very warm moments. We were friends, and it was good to be together. This was during the last years, when Monod was Director of the Institut Pasteur. From Agnes and from me he would get a feeling of what was going on at the bench level, so to speak, in what had always been his field of research. It was a deep pleasure for me to tell him about an exciting result, or a new idea, either from my own work, or from others. From him, I was gaining self-confidence, when he agreed with me, but perhaps even more when we disagreed. And this happened more than once. Scotch and Airedales, indeed, were not our sole matters of disagreement—allostery and the mode of gene control were others.

When Monod first asked me what I wanted to work on, in early 1963, I had just completed my first course in genetics, and had seen at the University a beautiful time-lapse movie on mitosis. I had marveled at the gracious ballet of chromosomes, and thus answered Monod that I wanted to understand what were the forces driving the chromosomes. Monod laughed, raised his hands to the sky, and this was the end of my first project. Having so destroyed my impetus, he was obliged to provide me with some other idea, and suggested that I work on an ill-defined concept called "allostery." I therefore started to work with Agnes Ullmann, with the aim of demonstrating the existence of an allosteric conformation change in the glycogen phosphorylase *b* from rabbit muscle. Not that I particularly enjoyed grinding rabbits to extract the enzyme, but I had no choice. Anyway, that's how I started to learn the job. Even though Agnes, David Perrin, and Alex Fritsch were my main interlocutors at this time, Monod played a great role in my development. When I look back, I am truly amazed by his patience in answering my somewhat naive questions. If he was in discussion with someone else in the lab, and I wanted to ask something, he always let me speak, answered, and never made me regret my question.

After a few months of rabbit grinding, after a few thousand phosphate assays, I had obtained some results. In particular, I had provided kinetic evidence for the existence of what was later called "homotropic" as well as "heterotropic" interactions between various ligands of the phosphorylase. In the fall of 1963, when I started to work full-time in the laboratory, I had a discussion with Monod about the orientation of my work. Obviously he expected me to repeat on phosphorylase the same type of work which Jean-Pierre Changeux was finishing on threonine deaminase. Clearly apparent from Jean-Pierre's work was the importance of working with an enzyme of bacterial origin and therefore open to genetic analysis. Monod therefore suggested that I look for a bacterial glycogen phosphorylase. Since, as everybody knows, what is true for the rabbit is true for *E. coli*, the bacterial phosphorylase had to be allosteric, and I would only have to study the enzyme produced by the wild-type strain and compare it with that produced by various mutants.

Since I had achieved the unbelievable record of working for six months in Monod's laboratory without growing a single bacterial culture, perhaps even seeing a colony on a plate, Agnes showed me how to inoculate a culture; Michael Malamy, then a postdoctoral fellow in the lab, showed me how to streak bacteria on a plate, and from then on, I was essentially on my own.

One week after drawing my first growth curve, I had found a polysaccharide phosphorylase in *E. coli*, and shown that the synthesis of this enzyme was induced by maltose. A month later I was convinced that this phosphorylase was not an allosteric enzyme. By then my decision had been made. I would not be a second Jean-Pierre Changeux, rather I would be another Gérard Buttin. Gérard had just completed his thesis on the galactose operon, and I had been very impressed by his seminar. I would work on the "maltose operon."

I often wondered about the reasons which led me to make this decision. For one thing, I saw much more clearly where I was going in a study of the "maltose operon" than in that of an allosteric enzyme, were it of bacterial or mammalian origin. It seemed to me that I could really "prove" things in a study of gene regulation. With an allosteric enzyme, on the other hand, what could I do but accumulate evidence for a somewhat mythical conformation change, which I had no hope to ever see with my eyes. Other factors must also have contributed to my decision. The explosions of hilarity from Jean-Pierre may have scared away the rather shy beginning student that I was. It could also be that the proud young man that I was felt the necessity of a weaning from Daddy Jacques and Mama Agnes. During the first weeks of my adventure in maltose, Monod made some timid attempts to drive me back to allostery. Was I really sure that this bacterial phosphorylase was not allosteric? The truth was that I simply did not want it to be allosteric, and it is really under Monod's slight, but firm, pressure that I did the minimum number of experiments required to study this point. Had I really wanted to find an allosteric polysaccharide phosphorylase in *E. coli*, I would probably have found the constitutive glycogen phosphorylase described a few years later by Chen and Segel. I never asked Monod what his feelings had been about my change of orientation. Up to a point he was perhaps disappointed that I was not displaying more interest for allostery, his passion at the time. Mainly, I believe, he was secretly pleased that I was paying a tribute to one of his first loves.

Indeed, when he suggested that I look for a bacterial polysaccharide phosphorylase, Monod was not sending me to some terra incognita. He knew that such an enzyme was almost certainly involved in maltose metabolism. This was an old story, which brings us back to 1946. Monod then thought that the so-called "adaptive enzymes" could be created by the action of substrates on nonspecific, somewhat "shapeless" protein precursors, according to a kind of "induced fit" mechanism. Enzymes involved in the degradation of substrates similar in structure seemed likely to derive from the same precursor. For this reason Monod decided to study the enzymes involved in the degradation of two similar disaccharides: lactose and maltose. As a result, two communications were presented on July 12, 1948, at the French Academy of Sciences. One, signed by J. Monod, A. M. Torriani, and J. Gribetz, described the occurrence, in *E. coli*, of lactase, only present in lactose-grown cells. In the other, signed by J. Monod and A. M. Torriani, was reported the existence of an "amylomaltase," present exclusively in maltose-grown cells. Amylomaltase was an enzyme of a new type. Instead of simply splitting maltose into its two glucose moieties, as anyone might have expected, it transferred one of the glucose moieties to a growing chain of polysaccharide of the amylose type, liberating the other glucose in the free form. One year later (1949) Doudoroff et al. demonstrated that the amylose synthesized by amylomaltase was most probably degraded by a phosphorylase. This was the enzyme which Monod sent me to look for.

Why is it that Monod, after the initial characterization of amylomaltase, essentially abandoned it to concentrate on lactase? The reasons, it would seem now, were contingent. First, because the "lactase" was in fact a β-galactosidase of broad specificity, it became possible to synthesize a chromogenic substrate for this enzyme, and therefore to render its assay both very simple and very sensitive, while the assay of amylomaltase still required the use of the cumbersome Warburg apparatus. Then, more importantly perhaps, because analogues of β-galactosides were much easier to synthesize than analogues of maltose, a study of the compared stereospecificity of substrates and inducers of β-galactosidase could be undertaken, leading to the decisive discovery of gratuitous inducers. So, deserted by Monod, amylomaltase found a first "cavalier" in H. Wiesmeyer, who, as a student of Melvin Cohn, characterized this enzyme more thoroughly and, in addition, demonstrated the existence of a maltose-inducible maltose permease (1960). By what mechanism did maltose induce the formation of maltodextrin phosphorylase, as I called the enzyme I had characterized, as well as of amylomaltase and maltose permease? This was what I wanted to find out.

To say that I was "on my own" to do so, as I wrote above, was somewhat of an exaggeration. To be sure, I was no longer working directly with Monod. Still, "Monod était mon maître, et j'étais son élève": "maître" and "élève" in the old European tradition, words which cannot be translated into American English. Jean-Pierre, David, and I would call Monod "Monsieur," even if, according to the American custom, everyone in the lab called each other by their first name. No American word can convey the load of love and respect endowed in the word "maître." To have Monod as a "maître" was a real blessing. This I was not long in realizing after meeting so many students complaining either that their boss was always on their back, or that their thesis director had abandoned them. Monod was always available, never on my back. From time to time he would come around and ask me how things were going, and we would always have an interesting discussion. Whenever I had a question, a problem, or a new result, I knew I could knock on his office door, or see him at tea time, when he would regularly come into his lab. If I was depressed, and went to see him, he would invariably cheer me up by showing me how to look at the situation in a positive way.

If Monod's intellectual and moral support were of extreme importance to me, others also had a great influence on me. Jacob was one of them, since I feel that I am most indebted to him for my development as a geneticist. I already mentioned Agnes Ullmann, Michael Malamy, and also David Perrin, who always knew before anyone else what was going on in the laboratory. But there were many others, both among the "permanent" staff of the laboratory and the crowd of first-rate postdoctoral fellows attracted by Monod and Jacob. I say Monod *and* Jacob because their two groups were almost completely mixed. As a matter of fact, the maltose saga itself soon became a "coproduction" of the two groups, with Maurice Hofnung, one of Jacob's students, starting to work on the system in 1965.

Alas, with Monod as my "maître," with Jacob as my adviser, and with all the good angels around, how could I give birth to such an ugly little duckling?

At first, after a few months of work with maltose, I thought I could foresee very well what my thesis seminar would be like. It would take place in the library of the Service de Biochimie Cellulaire, which served as a meeting and seminar room (as well as a lunch room!). This room contained a small worn-out blackboard which displayed "$R \leftrightarrows T$" when the last discussion had been about allostery, and "iozyAc" when it had been about gene control. For my thesis seminar I would choose an "iozyAc" day and, before a petrified public, I would erase the i, o, z, y, and Ac, only to replace them with the symbols representing the gene of the "maltose repressor," the operator sequence, and the structural genes of amylomaltase, phosphorylase, and maltose permease. Indeed, my thesis had started well. I had obtained mutants affected in the structural gene for amylomaltase and phosphorylase, and obtained evidence that these genes were in the same operon. Unfortunately, however, I had also obtained other mutants, which were all monsters! The most frequent class had a pleiotropic negative phenotype, i.e., lacked simultaneously amylomaltase, phosphorylase, and permease and, in addition, were resistant to phage λ. The corresponding mutations generally mapped in the vicinity of the amylomaltase-phosphorylase operon. Some of them, however, mapped in another region, a quarter of the chromosome away. Other mutations, mapping in this second region, led to the most incredible phenotypes. Some mutants were resistant to phage λ, others were sensitive. Some were pleiotropic negative; others lacked permease, but expressed the amylomaltase-phosphorylase operon in a constitutive manner. And there were still other complications, so much so that even now, to tell the truth, we still haven't quite cleared up this mystery.

At the time, the complication of the system rapidly became a serious hindrance in my communicating with others, Monod included. Every time I wanted to relate even a minor result to someone, it took me half an hour to put him or her up to date. This very complication, in fact, made a very strong impression on Monod. A few years later, when molecular biologists, with "le grand François" as one of the color-bearers, started to leave the world of *E. coli* for new horizons, he often said to me something like "How can they expect to unravel the regulation circuits involved in cell differentiation when you, with all the power of bacterial genetics, can hardly understand the intricacies of the maltose system in *E. coli?*" When I came up with all these horrible mutants, Monod's advice was clear. I should first try and isolate the "simple" mutants, i.e., those affecting the repressor and the operator, and then I might be in a position to explain the monsters. He was all the more inclined to give such advice since Germaine Cohen-Bazire and Madeleine Jolit had isolated in 1953 from *E. coli* ML a mutant synthesizing amylomaltase constitutively, and which he believed to be affected in the "maltose repressor." Since nobody knew how to perform genetic crosses with *E. coli* ML, I was unable to analyze this famous mutant. I therefore had to isolate my own, in *E. coli* K-12. I failed, in part for technical, but mainly for psychological reasons. I did not find the constitutive mutants (which existed, since I found them much later in 1976) (but they are not affected in a "maltose repressor"!), just for the same reason I did not find the allosteric glycogen phosphorylase of *E. coli*: I didn't really want to find them. The truth was that, rather soon, I started to believe that the maltose system was not regulated in the same way as the lactose system, that the "maltose repressor" simply did not exist.

It must have been in the fall of 1964, or very early in 1965, when a scientist by the name of Ellis Englesberg gave a seminar in the Service de Biochimie Cellulaire. He spoke of regulation in the L-arabinose system of *E. coli* B. He tried to say that this system was controlled by a "positive" regulator gene. This gene would code for a product, called an "activator," which, once it is itself activated by L-arabinose (the inducer), would promote the expression of the arabinose operon. Being a young student, very susceptible to influence, I was convinced. It is probably fair to say that I was the only one. Not only was I convinced but I was stricken by the many similarities between the L-arabinose system and the maltose system. It is from the day of this seminar, I think, that I must have been working under the secret hypothesis that the maltose system was positively regulated. Still, at this point, it was only an intuition and I had a long way to go before I would convince myself, not to mention Jacob and Monod. In a way the atmosphere of friendly scepticism which surrounded the idea that the maltose system was positively regulated was very stimulating. Before anyone would believe me, I had to try in every possible way to demonstrate that there was no such thing as a "maltose repressor," that what looked like positive regulation involving an activator was not in fact disguised negative regulation involving a repressor. This was not an easy thing, and Ellis Englesberg had experienced, and was still experiencing, the same problem. A positive-regulator gene would be mainly defined by mutations which are pleiotropic negative, i.e., which prevent the expression of all the structural genes controlled by its product. The difficulty is that several classes of pleiotropic-negative mutations can also be expected in a negatively controlled system. Deletions, polar mutations, "i^s type" mutations leading to super-repression are some, relatively easy to discard. But others, much more vicious, are mutations in a transport system, such that the inducer cannot penetrate and inactivate the repressor; or mutations in an enzyme which might convert the external inducer, added in the medium, into an internal inducer which interacts with the repressor. How could one rigorously exclude all these and a few other possibilities? "Proving" the existence of a regulatory circuit turned out to be more difficult than I originally thought. After all, I was not much better off than if I had chosen to demonstrate the existence of a conformation change in an allosteric enzyme! Nevertheless, I did accumulate evidence that there was a positive-regulator gene (*malT*) in the maltose system, to the point where, some days, Monod would start believing it. One such day he even proposed a name for the product of *malT*, and of positive-regulator genes in general. This name was "provocateur." He thought that "activator" was too general, and often implied the involvement of energy. It was a good word, and I suggested it in a paper for the *Annales de l'Institut Pasteur* in 1967. However, when Dolph Hatfield, Maurice Hofnung, and myself tried to use it in the next paper, which was to be published in an American journal, it was strongly rejected by the editors. I suspect that my friend Ellis was not totally innocent in this affair.

Monod was very difficult to convince about the existence of positive control. In a letter that he sent me in the States, in October 1968, he wrote, "Il faudra bien qu'on arrive à savoir un jour si les provocateurs existent réellement. J'avoue que je continue à être un peu sceptique." In view of this attitude it is not surprising that, when I presented my thesis, in 1967, I thought that Ellis Englesberg and I were more or less the only two persons in the world to believe in positive control.

Why is it that Monod was so reluctant to believe in positive control? One reason can possibly be found in the evolution of his own ideas about regulation in the lactose system. The concept of a regulatory molecule, distinct from the enzyme, and responsible for receiving the induction signal, apparently became clear to him around 1951, after the discovery of gratuitous inducers. It was only in 1958, however, that the famous "Pardee–Jacob–Monod" experiment led to the idea that the regulatory molecule was a repressor. It had therefore taken seven years for Monod and his colleagues to come up with the hypothesis that "l'inducteur agissait, non pas en provoquant la synthèse de l'enzyme, mais en 'inhibant un inhibiteur' de cette synthèse" (Monod, Nobel Prize Lecture, 1965). It took seven years to demonstrate that the product of *lac i* was a repressor and not a "provocateur"! No wonder Monod was not immediately willing to accept that the reverse was true for the regulator gene of the maltose system. In addition, Englesberg and his colleagues were soon led to the somewhat unaesthetic idea that the product of the arabinose regulator gene was not only an activator in the presence of L-arabinose, but also a repressor in its absence. In the maltose system, on the other hand, the phenotype of many mutants remained unexplained. This unwarranted complexity of the two most well-known positively regulated systems made Monod, and many others, suspicious that something basic could be wrong. Still, over the years I could not escape imagining what would have happened if Monod had focused on amylomaltase, rather than on β-galactosidase. I might have had to prove, against an atmosphere of general scepticism, that the synthesis of β-galactosidase was regulated in a strictly negative way. And I probably would not have convinced Monod.

Maxime Schwartz

Born in 1940, Maxime Schwartz prepared his doctorate under the supervision of Jacques Monod from 1963 to 1967. His thesis dealt with the regulation of protein synthesis in bacteria. Studying the maltose-utilizing system of *Escherichia coli*, he was among the first to show that activators could control genes just as well as repressors. His later studies, also performed at the Institut Pasteur, where he headed a unit from 1975 to 2002, were on the structure and functions of membrane proteins, including the phage lambda receptor, which is a component of maltose permease, and on protein secretion. He spent some time in the United States, including a two-year period in the laboratory of James Watson at Harvard University in 1967–1969 and several months, on various occasions, in that of Jon Beckwith at Harvard Medical School. He has been affiliated both with the CNRS (since 1964), where he is a research director, and with the Institut Pasteur (since 1973), where he is a professor.

From 1985 to 1987 Maxime Schwartz was Deputy Director of the Institut Pasteur; he then held a 12-year term as Director General of this institution, from 1988 to 1999. Since 2001, he has served as Scientific Director of the French Food Safety Agency. In addition to his scientific papers he has written a book entitled *How the Cows Turned Mad*, which was first published in France by Odile Jacob in 2001 and then, in 2003, in the United States by the University of California Press.

In the Cellular Biochemistry office, June 1970.

Ca. 1970.

The Lively Corridor

Marie-Hélène Buc

During the spring of 1962, a biochemistry course on "the biosynthesis of macromolecules," presenting experiments which were being done at that very time, was given at the Faculté des Sciences de Paris by Monsieur Jacques Monod.

The courses one usually attended at that time consisted of the boring recitation of formulas, like that of vitamin B_{12} or testosterone—formulas we were more or less bound to learn by heart for the exam—without any reference to the ideas and the experiments which had led to the comprehension of these structures. It was like teaching the precepts of the Bible without referring to the history of the Jewish people.

By contrast, our new teacher appeared personally concerned with the theory of the genetic code. Was the code universal, degenerate? What was the basic unit? What kind of punctuation existed? Rather than trying to simply teach us facts, he had us participate in an intellectual exercise which was his own approach: the hypotheses, the type of methods which could test them, and the actual experimental verifications. I had the impression that he was not teaching, but trying his own logic on us.

His approach had a great power of attraction and even of seduction. Postponing my own aim, medical research, and totally ignorant of the international reputation of my teacher as a scientist, I asked Monsieur Monod for a position in his laboratory.

The interview I had with him one week later was different from what I had imagined: no curriculum vitae, no test of scientific ability. He opened a debate about the potentialities of women and about their place in research. We had a general discussion as if I was not myself a woman asking for a position in his lab. He stated that he very much appreciated the intellectual qualities of women; however, he was totally against women working in research because of their necessary investments outside their work. Moreover, he told me that he was against women in science, especially if they were good, since serious conflicts would develop between their work and their family. How not to agree? I thought I knew what he meant since I already had to organize myself between work at the hospital and a baby at home. In fact, he was not thinking only in terms of time being a limiting factor; he knew that a free mind is an absolute requirement to do good research, a situation which is unhappily less common among women due to what he called "the necessary investments." At the end of the discussion, I heard that I had to come back the next Friday. When I arrived, I was greeted with, "Here is this ambitious young lady. Come and visit your new lab." I realized then that the word "ambitious" had a different meaning for Monsieur Monod and for me. For him, my main ambition was not to do research instead of being a physician, but it had been to ask for a position in his group though I was a woman.

I quickly discovered that various aspects of regulation were attacked at the same time in the lab; each student was working hard, trying to analyze "his" operon or "his" regulatory protein. Our results were daily put in the frame of the models which had been elaborated by Jacques Monod and François Jacob, and the models would be modified, after long discussions, according to the experimental data. It seemed to me that most people were working with enthusiasm and thought they were participating in important issues. It was the golden age of allostery.

Trying to recall that period, I wonder what specific conjunction of people and circumstances had made this group both creative and happy.

The topography of the lab favored communication: it was a long corridor with one pole of attraction, "le labo bleu," which opened to Jacques Monod's office. Another pole of attraction came later at the other end of the corridor: François Jacob's door. In between in one room were senior workers tutoring a few students; in another a mixture of young students from Jacob and Monod's teams were deliberately placed together with no tutor; and in the largest room, "le grand labo," American postdocs learning French from other students, who would in turn learn some science from them. The last room was the kitchen, in which Jacques Monod used to come and talk with people or listen to Lucie Barnard sing. All labs opened to the corridor and people therefore had many opportunities to meet. One could see at any time during the day a group of three or four persons in discussion. The composition of the group was constantly changing, but one of the members very often was David Perrin, who knew everything from the papers in the *Journal of Molecular Biology*, to the method for getting a specific mutant, to such things as the number of cattle in Argentina in 1936. There, in this forum, theoretical and technical problems were discussed amiably, and we learned to know and appreciate one another.

Another important and particular place was indeed "le labo bleu." If I remember so well who was sitting at which desk, it means that I probably felt, and I was not the only one, that these individuals were privileged people. They were called "mes enfants" by the boss, who many times during the day would open the door between his office and the lab and appear, his glasses in one hand, immediately to start a discussion about a specific point. When I entered the room by chance, I always had mixed feelings. Science was certainly the concern of all these people; at the same time, the atmosphere was clearly different from that in the corridor. I do not mean that everyone wanted to be thought superior by the boss, but it was clear that everyone was very sensitive to his intellectual seduction and wanted to seduce in turn.

I come now to the third important place in the lab: the seminar room. We were supposed to meet there almost every day at noon. The student who was giving the talk had to be prepared for a difficult job: to defend the qualities of the paper as if he were himself the author, and criticize the content and the methods, thereby schooling himself in self-confidence. Jacques Monod would manifest his scientific rigor by interrupting at any time to question, or to criticize, the questions of others. He was sometimes so hard on us that we hesitated to ask questions. He used to say at the end of the lecture: "It is late; time for just one, but very interesting question." As we were wondering if the question we had in mind was clever enough, he had time to formulate his own comment, which was exasperatingly good.

I have described Jacques Monod's laboratory as objectively as I can, trying to put myself in the position of an outsider: admiring, but sometimes irritated and frustrated. But I am not in that position. Day after day, I have lived in this friendly atmosphere, where everyone contributed a lot because of Monod's own generosity. He was indeed generous, giving us his time and his warm affection, particularly when we had professional or personal problems. No one knew it, but he was present and efficient. Clearly, he was interested in people; he enjoyed discovering their particular talents and helped develop them. Not only did we tolerate this father-to-child relationship, but I feel we were happy there partly because of it.

May 1968 was the turning point. A picture in *Le Nouvel Observateur* shows Jacques Monod taking care of a young girl wounded by the police. It was for me a symbol of the kind relationship Jacques Monod had with us, but young people did not want any more of it. When Jacques Monod became Director of the Institut Pasteur, he encountered objection to some of his choices for the Institute among those who never fundamentally contested his scientific options. This was certainly difficult for him, and not easy for us. In 1971, when I did not agree with one of his decisions and happened to mention it in public, he was so furious with me that he told me, "You act as an irresponsible. I know better than you what is good for you?" I remembered this sentence.

Three weeks before he died, he invited me to his home to have tea with him—again, his presence, his warm smile, his questions ("Are you happy in François' group?"), his suggestions concerning my new work ("Why not try to induce specific differentiation in teratocarcinomas with these new hormones?"). Also he told me, "When I am not too tired, this period of my life is so interesting. I shall die soon, I know it, and because of that, the intensity and the quality of the relations I have with my friends is of another nature." He then spoke about my sons and his sons; and some of his last words to me were: "One never knows what one gives to his children. Did I help them in finding their way?" I was with him for the last time and I knew it. As he accompanied me to the elevator, his eyes followed me for a long while, for he also knew.

Marie-Hélène Buc

After completing her M.D. degree, then her Ph.D. on allosteric enzymes with Jacques Monod, Marie-Hélène Buc-Caron joined François Jacob's laboratory in 1973 and switched to mouse embryology using the teratocarcinoma model. She derived immortalized cell lines having the properties of precursor cells of the three embryonic lineages, which could be induced to differentiate in vitro into neuroectodermic, mesodermic, or endodermic derivatives. These cell lines now serve as model systems to analyze early steps of differentiation and to study the modulation of receptors and transporter functions.

In 1993, Marie-Hélène Buc turned to human cell biology, isolating and propagating in vitro multipotential neural progenitors from the embryonic telencephalon. These progenitors have now been evaluated in cellular and gene therapy of neurodegenerative diseases (Parkinson's, mucopolysaccharidoses), using animal models of these pathologies.

To Fold or Not to Fold: The Way Toward Research

Michel Goldberg

I came to Monsieur Monod's laboratory in October 1962, a year or so too late to have known the great excitement of the "operon" and "allostery" periods. Yet, in spite of my total ignorance of biology, I very rapidly became aware of the importance of these two concepts: the main principles governing the regulation of polypeptide chain biosynthesis were understood and the regulation of the enzymatic activity of a protein could be interpreted in terms of subtle changes in the conformation of the polypeptide chains. At that time, these regulation models were far from being universally accepted; but Monsieur Monod was convinced that, though some "refinements" undoubtedly would have to be added, the fundamental principles had been uncovered. He therefore became more and more interested in the single link still missing, at that time, between the genetic message and its expression as a functional protein: the folding of a newly synthesized peptide chain into the specific, three-dimensional structure achieved in the native molecule. How did he involve me in that field of protein folding?

He thought that this problem, like so many others in the then new field of molecular biology, had to be approached with the aid of physical chemistry; and therefore, he tried to attract to his laboratory a few students who had a fairly solid background in physical sciences. I was among his first victims.

As soon as we met, he explained to me how he intended to organize my biological training. To start with, a complete brainwashing to erase from my consciousness the academic knowledge I had gained during my studies. Then, some elements of biochemistry would have to be learned by attending the courses in which he participated at the University of

Paris. He insisted that these courses were of utmost importance because classical biochemistry was the real, concrete basis of molecular biology. (A few weeks later, I was somewhat surprised to discover that he himself taught essentially the exciting aspects of molecular biology, relying on others to teach what he claimed to be more basic.) Last but not least, what he considered as essential was the apprenticeship in the laboratory.

He exerted extreme care in choosing the appropriate sponsor for each beginner in his group, and through a subtle combination of intuition, psychology, and reasoning he usually succeeded in reaching an excellent match. In my case, I consider that he reached perfection. When I started working at the bench, Edwin Lennox was spending an extended sabbatical stay in Paris, waiting for the Salk Institute to be completed. When Monsieur Monod asked him to train me as an experimentalist (starting from pipetting and weighing correctly!) Ed was not very happy, since he had expected to be free of such a chore while on sabbatical. Yet he accepted this burden. Because he was so patient, so profoundly generous, he developed a real interest in my evolution and rapidly became a most helpful, exquisite, and efficient adviser. When, after a few months, I told Monsieur Monod how happy I was with Ed Lennox's teaching, he explained the reason for his choice: because Ed himself had undergone a reconversion from physics to biology, nobody better than he could guess, and therefore fill, the gaps in my understanding and knowledge of biochemistry.

Monsieur Monod also considered attending seminars of predominant importance in learning biology; and he literally did not allow anyone to miss a lecture. I once continued an experiment, instead of going to the smoky, overcrowded, uncomfortable library, where what I considered as an uninteresting talk was being given. Monsieur Monod noticed my absence and, in the afternoon, upbraided me in an extremely severe way, insisting that I should never consider my own experiments more important than those of others! Indeed, the seminars were usually quite enlightening and of excellent scientific level. They were also extremely lively, mainly because of Monsieur Monod's sharp, often aggressive and merciless questions to the speaker. It was very noticeable to anybody who regularly attended these noon seminars that Monsieur Monod had a highly variable degree of causticity, depending on who was giving the talk. He obviously could not tolerate an unintelligent or self-satisfied speaker. On the contrary, a graduate student giving a moderately interesting and not too disorganized lecture would frequently be congratulated publicly with some emphasis. I do remember my intense emotion when this first happened to me. Monsieur Monod had asked me to give a talk on van der Waals forces and hydrophobic interactions. I did my best to explain the nature of these interactions without going too much into details concerning their thermodynamics, but I was not certain that I had been understood, nor that I had really captivated the audience with this esoteric subject. As usual, after the seminar, we all went to a neighboring delicatessen to get some food and, back on the veranda, close to the library, gathered together at a long table to have lunch. Just as I sat down, Monsieur Monod started congratulating me on my talk; these were, as far as I remember, the first compliments I had received from him after six months at the Pasteur Institute. Of course, I tried to hide my emotion, which was so intense that I was unconsciously and nervously squeezing the bag of Italian-style spaghetti I had brought with me. And my pride rapidly turned to extreme confusion when I noticed that all the tomato sauce had spread on my trousers. That day, I remained unusually long at the table, and nobody noticed that my clothes had also been blushing!

However, Monsieur Monod was as efficient in condemning as he was in praising. I have a vivid memory of an extremely unpleasant scolding I once received. After Ed Lennox had left to join the Salk Institute, I usually asked David Perrin to teach me any new method I had to use. He thus showed me how to perform sucrose gradient centrifugations and helped me throughout my first experiment. The centrifugation went on overnight. On Friday morning, we stopped the centrifuge, collected the fractions, and analyzed them through a fairly long procedure which kept us busy well after midnight. Because the gates of the Institute closed at 10 p.m. at that time, we had to climb over a wall to escape from the laboratory. When we were about to jump onto the pavement, a police car passed by. The men in the patrol looked at us, smiled, and passed by. A few hours later, I was back at the bench, worked all Saturday, slept about two hours on the sofa in the library, and went on pipetting, assaying, measuring, until Sunday afternoon. Having then finished my experiment, I was so tired that I decided to go home immediately and have a long sleep. When, on Monday morning, I came back to the laboratory, I found a large note, clearly written in huge black letters, hanging above my bench. It said something like: "Michel! This is the last time I tolerate such a mess on your bench. Jacques Monod."

I was shocked, vexed, and became furious; after having worked so hard, having completed such a nice experiment, all I got was that stern and public humiliation! I rushed into Monsieur Monod's office. He greeted me with a friendly smile; I was about to explode, but he told me, in his deep, quiet, and cheerful voice, "Well, my little Michel, what about your sucrose gradients? I heard you have been working like mad!" I was a bit comforted by the fact that he at least was aware of how hard I had been toiling, and I started explaining why I had left my bench somehow crowded with dirty glassware, pipettes, and tools. His answer came, as sharp as a knife-edge: "I don't care why you left your bench messy. You just should never let it become filthy like that. The type of work you are doing requires an extreme care and I'll never trust an experiment of yours as long as you are not able to get organized!"

A few days later, having completed the analysis of my experimental results, I carefully and cleanly drew the curves corresponding to the four centrifugations I had performed. To allow for a better comparison of these experiments, I represented the four curves on the same sheet of graph paper, using a different color for each diagram. Proud of my work, I waited for the first occasion to show and explain it to Monsieur Monod. When, at last, I could point out to him the meaning of each curve, I realized that something was wrong; he did not understand what I was explaining. This was quite unusual since he ordinarily would have grasped an idea before I could finish expressing it. So, I tried again, more slowly, "The red curves represent wild-type β-galactosidase, the green is U178, the blue is CZ1, and the black one corresponds to the complemented enzyme." A dull, interrogative look showed me that I was not yet clear enough. I was completely baffled, and had no idea why Monsieur Monod could not immediately see the interest of my results. But David Perrin, who just had joined us, whispered in my ear, "He is color-blind. How could he sort out your curves?" Understanding the situation was quite a relief and, by reproducing my graphs one by one on the blackboard, I could at last convey my message. While Monsieur Monod was discussing my results, I could not help being amazed by the slight coquetry which led him to conceal his sight defect. The discovery of this weak point in his otherwise apparently monolithic personality was indeed my first insight into his complex, sensitive, and fascinating character.

And that is how, within two years, and by a wise admixture of stimulating compliments, of severe exactingness, of constant scientific interest, of permanent availability to discussion, of total confidence, Monsieur Monod molded me into what he considered an honest biochemist, who could be decently sent out from his laboratory.

This period of introduction to biochemistry came to an end when, on a sunny Saturday morning, Monsieur Monod called me into his office and, playing as he often did with my sensitivity, formally asked me to sit down and warned me that he had a very serious matter to discuss with me. I was impressed by his solemn attitude and, when he started telling me that there was nothing more I could learn in his laboratory, I became extremely worried. Judging my deep distress, he then burst into cheerful laughter, and reassured me. What he really meant was that he was no longer able to direct my research, because what I needed at this stage was good training in protein physical chemistry. And this, he said, was far beyond his capabilities. He had come to the conclusion that I should leave his laboratory for some time, and spend one or two years abroad. He just wanted to know whether, in principle, I would agree to such a project. This was a somewhat unexpected question, and I tried deferring the answer by saying that it would depend on where I would go and with whom I would be working. But Monsieur Monod was not caught off guard. He warmly praised Robert Baldwin, who was just completing a sabbatical year with François Jacob, telling me how he enjoyed discussions with him, and how he admired his sharp mind and acute knowledge of macromolecular physical chemistry. He informed me that Buzz Baldwin had already accepted the offer to sponsor my Ph.D. work; that the "D. G. R. S. T." would support me during my stay in the United States and pay for my fare; that I could apply to NATO to be aided with my wife's travel expenses, that California was a great place. He had organized everything and thought over every little detail. So, a few months later, in October 1964, I left for Stanford University.

When I came back to Paris, many things had changed. The Nobel Prize had drastically modified the status of molecular biology in France. This hitherto minor branch of biology, which itself had been a neglected area when compared to the flourishing fields of mathematics and physics, was now recognized as a real science. And the "Monod–Jacob–Lwoff trio" had become very familiar to people in the street. Yet, as soon as I returned to the laboratory, I could feel that the quality of the intellectual and affective relations between Monsieur Monod and his collaborators had not been altered.

For me, however, the situation was clearly different. He no longer wanted to treat me as a student, and I rapidly discovered that I would be entrusted with some of his scientific preoccupations. Very shortly after my return to Paris, he asked me to give a seminar on the work I had been doing at Stanford. He thought that this was the best way for him to learn about my results. He was apparently happy with them and decided that, after some short experiments which I had to do to complete this work, I should start writing my Ph.D. thesis. At that time a thesis at the University of Paris consisted of two parts: the main one described the examinee's own research, and the minor one was a bibliographical essay which had to be written on a subject chosen by the jury. By deciding that my essay should bear on "prediction of the conformation of a protein from its amino acid sequence," Monsieur Monod hurled me right into the field of protein folding, a theme which has since remained the subject of my research.

When, after a decade, I try to understand why I got so deeply involved in that subject, I can find two main reasons, both related to the fact that Monsieur Monod himself had become particularly interested in protein folding. He was then preparing his first series of lectures at the Collège de France, a very important part of which was devoted to a dissertation on protein structure and evolution. This stimulated endless discussions, which would last for hours, with the help of a glass of scotch and tens of cigarettes; a huge number of butts thrown on the floor around a stool clearly indicated, the next morning, the bench at which Monsieur Monod had been sitting during the last discussion.

Also, there was an extremely efficient complementarity in our approaches to that problem: he was the biologist, asking questions of physiological importance; and I was the physical chemist, trying to answer these questions in quantitative terms. And this complementarity, because of his exacting confidence, turned me into one of the many facets of an intellectual mirror he had created with several scientists to enrich, through a permanent dialogue, his scientific creativity.

At the risk of boring nonscientists with some words incomprehensible to them, I should like to briefly illustrate the mixture of intuition, common sense, and logic which made so fruitful Monsieur Monod's approach to protein physical chemistry. He was always interested in very general questions dealing with features which could be easily generalized. For instance, one common feature of proteins is that they are very large molecules. Thus, one of the first questions which Monsieur Monod asked in his lectures at the Collège de France was "Why are proteins so large?" To answer this question, he started by emphasizing that a protein, in addition to being a molecule obeying the laws of physics, is an object reflecting the history of millions of years of evolution and that therefore its "function," on which evolution is based, should serve as a guide to the interpretation of its physical properties. Thus to account for the large size of proteins, he tried to analyze the constraints which the functional features of a protein exert on its structure: the existence of a catalytic site, requiring the exact relative positioning of half a dozen amino acids; the stability of the protein conformation, requiring, according to the "oil drop model," a hydrophobic core; the solubility of the molecule, requiring the presence on its surface of polar residues able to interact with the solvent. By using molecular models, and rather rough energy calculations, we came to the conclusion that a minimum of about 70 to 100 residues were necessary to fulfill these requirements. And surprisingly enough, this is indeed the size of the smallest naturally occurring globular proteins.

Such reasoning was made in the beginning of 1967, when so little was known about protein conformation. Since then, the unraveling of many protein structures by X-ray crystallographers has abundantly confirmed the validity of the then daring assumptions underlying Monsieur Monod's approach to protein conformation.

And today, it may even seem strange that Monsieur Monod could have been so deeply involved in such rough and elementary questions as "Why are proteins so big?" or "Why are proteins so often oligomeric?" or "Are oligomers symmetrical?" or "What is the basis of protein folding and stability?"

In fact, these questions were at the heart of his philosophical preoccupations. It is not by mere coincidence that his lectures at the Collège de France on proteins and evolution were given while he was writing *Chance and Necessity*.

Indeed, he wrote in his book, "By these standards proteins must be deemed the essential molecular agents of teleonomic performance in living beings." Hence, no wonder that he was so much interested in trying to reproduce, in vitro, some of the steps of protein evolution. First, as a model for the origin of evolution, he imagined "the precambrian protein" the amino acid sequence of which would have been determined at random. And for years he tried to convince people to construct and study such a protein. Second, as a model for the most elaborate achievements of molecular evolution (i.e., the acquisition of allosteric properties), he got involved in trying to "allosterize" a nonregulatory enzyme. And I still remember his excitement when I reported some observations I had made on the ion activation of the β-galactosidase produced by a cold-sensitive mutant. When, after a closer examination, it turned out that the observed activation was simply due to an ion-dependent subunit association, Monsieur Monod told me, "You now have the ancestral allosteric transition in your test tube. Try to progress by one more step in evolution."

His interest in protein conformation survived Monsieur Monod's involvement in the direction of the Pasteur Institute. Whenever I entered his directorial office and told him that I had some interesting results, he would sit up in his chair, light up with his deep, affectionate smile, and say, "It is so kind of you, Michel, to bring some science into this office." And he would easily spend a whole hour discussing my experiments, while the upper limit for an administrative discussion was about ten minutes.

About two weeks before he died, he felt physically much better than during the preceding months and he told me, "When I retire, I want to write my second book. So many ideas matured during these years. I have so much to say. But do you know, mon petit Michel, what will happen when I am sitting in front of a white page? Shall I ever be able to start again?"

He was not given the chance to face this new challenge. A pity for him. And for science.

Michel Goldberg

Most of Michel Goldberg's research dealt with experimental studies on protein folding. By identifying and characterizing folding intermediates, his group provided an experimental basis for recent models developed by theoreticians. Goldberg's studies on protein aggregation led to original procedures for protein renaturation that are widely used in the biotechnology industry. These studies are also often quoted as seminal in understanding the molecular origin of "misfolding diseases" like spongiform encephalopathies and Alzheimer's disease. Michel Goldberg's group also developed original strategies to sort mouse chromosomes by flow cytometry, thus obtaining unique chromosome-specific gene libraries for the murine chromosomes X, Y, and 21.

Dr. Goldberg was Deputy Director of the Institut Pasteur from 1976 to 1979 and Professor at the Paris University from 1968 to 1998. Since 1982 he has served as Professor at the Institut Pasteur, where he chaired (1988–2001) the Cellular Biochemistry Unit founded by Jacques Monod. He currently chairs the Protein Folding and Modeling Unit and is active in prion research.

A Ph.D. with Jacques Monod: Prehistory of Allosteric Proteins

Jean-Pierre Changeux

A beautiful model or theory may not be right; but an ugly one must be wrong.
JACQUES MONOD, IN "SYMMETRY AND FUNCTION OF
BIOLOGICAL SYSTEMS AT THE MACROMOLECULAR LEVEL."
NOBEL SYMPOSIUM 11, 1969

October 1958. After three years at the École Normale Supérieure and the preparation of the Agrégation des Sciences Naturelles, I was still unacquainted with Jacques Monod's scientific work. That year, Jacques Dauta, a colleague who was also preparing the Agrégation, asked me to be present during the defense of his Diplome d'Études Supérieures. His subject, dealing with the biochemistry of bacteria, was quite remote from the concerns of the zoologist and would-be embryologist that I was. Dauta had spent a few months at the Pasteur Institute, in the newly created Service de Biochimie Cellulaire, where his sojourn appeared to have been rather stormy. On the examining board sat a man with severe and regular features, who used a rich and accurate vocabulary, and expressions that were sometimes cold and cutting—nothing which could attract a young biologist. This man, nevertheless, was to become my master.

Following several prolonged summer stays at the Laboratory of Marine Biology in Banyuls-sur-Mer, I became convinced that the future lay in embryology. I spent a few weeks in Jean Brachet's laboratory in Brussels and came back to Paris with a subject: the study of the molecular mechanisms of the egg's activation by the spermatozoon. It had been shown that this activation is accompanied by an increase in the activity of several enzymes: different phosphatases and proteases. In Louvain, De Duve had just discovered the lysosome. The hypothesis was that this increase in activity results from the breakdown of lysosomes. Working in a small laboratory at the Institut de Biologie Physico-Chimique, I was vainly trying to measure the phosphatases of the sea urchin's egg. These failures leaving me hopeless, I went and talked to Jeanine Yon, who was working in a neighboring room. According to her, the only laboratory in Paris where people were both interested in fundamental biological problems and had a profound knowledge of enzymology was Jacques Monod's, at the Pasteur Institute. I had to meet Jacques Monod.

Chance had it that a few days later Jacques Monod called on René Wurmser at the Institut de Biologie. He was a candidate for a vacant professorship at the Faculté des Sciences and, therefore, had to have meetings with each of the current professors. I jumped at the opportunity. Recognizing the person I saw stepping out of René Wurmser's office as the man I had seen at Dauta's thesis defense, I approached him boldly. He looked surprised at first, but his face lit up when I asked him my first question: "Could you teach me how to assay phosphatases?" He asked me why and I started to explain the scientific problem I was facing and the difficulties I was encountering. After the first amusement, interest came. We could not discuss the matter in a hall; I should come and give a seminar at the Pasteur Institute.

A few days later, on a Saturday morning, I found myself in a dusty room on the first floor of the Pasteur Institute. A few books on shelves decorated the room, but most of it was crammed with heaps of jars filled with silkworm cocoons and obsolete machinery. What was to become one of the high places of molecular biology looked more like a thrift shop than a lecture room. Nevertheless, I presented my project enthusiastically, unaware of (as I heard later) the amused but benevolent responses among my small audience. Following the discussion, Jacques Monod, holding my arm, took me to his office, a small and severe place dominated by a beautiful engraving by Piranese. Monod spoke briefly and firmly. Nobody in France was working on chemical embryology. If I wanted to continue my ongoing project, I had to emigrate for a while to the United States. Otherwise I had to "... choose another path, switch to simpler systems, learn enzymology, genetics, immunology..." and then, after a few years, I would be able to come back to my early love.

If I were to choose this latter possibility, Monod would give me the opportunity to come and work at the Pasteur Institute. François Jacob, who had no students at that time, could offer space for me in his laboratory. Jacques Monod gave me two months to think about his proposal. At the moment I did not fully realize what an incredible opportunity had come my way; Jacques Monod's offer did not appeal to me immediately. I read

François Jacob's thesis. His theory of episomes, the experiments on chromosome transfer between bacteria and phages, reminded me of Briggs and King's embryology experiments. I inquired about the biological material. Gérard Buttin, then one of Monod's students, took me to the 37°C incubator where the coli-bacillus cultures were regularly agitating. The sight of the milklike suspension and its nauseating smell disgusted me to the point that I questioned the whole thing. Reason was stronger, however, and I called Jacques Monod to tell him I had chosen the Pasteur Institute.

From Bacterial Genetics to Regulatory Enzymes

A room in the attic, the heat of a Turkish bath without the charms of Oriental perfumes, the rhythmic noise of a waterbath, piles of petri dishes, and, sitting at his desk in the faraway corner of the room in front of the window, François Jacob, his head bent like a car racer's. The first dominant constitutive mutants had just been isolated and identified. The "operon" was in gestation. There was no time to lose. I was, obviously, not needed here. My first experiment consisted in following the infection of a culture of bacteria by a bacteriophage. The experiment succeeded: the culture lysed within the expected time. In the second one I had to establish the growth curve of a bacterial population. The only culture medium available on the third floor was broth: the experiment was carried out in broth. It so happened that the very evening the curve was drawn, Jacques Monod walked into the laboratory and asked for the results. His eyes turned suddenly cold. None of the parameters he was interested in—growth rate, yield, etc.—could be seen on my diagrams. The main part of the analysis he had developed in his thesis could not be applied to this curve, which had been established in too complex a medium.

His irritation at the nonquantitative aspect of one of my first experiments in the laboratory made me go down two floors. I found myself in the Service de Biochimie Cellulaire. For the next four months I worked under the direct supervision of Jacques Monod. The proposed subject, which had come out of a discussion between François Jacob and Jacques Monod, consisted of a comparison between the expressions of a gene—that of β-galactosidase—in different cytoplasms, that of coli-bacillus and of Salmonella. Every week a protocol was established, remarkably detailed, precise enough to be given to a technical assistant. Week after week I explored each of the main properties of the enzyme synthesized by the strain of Salmonella: dissociation constant, thermal stability, immunological properties, etc. Within a few months I had gone through nearly all of enzymology, and even tackled the regulation of the synthesis of β-galactosidase as well as the active transport of β-galactosides across the bacterial membrane. All this was done very concretely on the experimental level, while resting on a sound theoretical basis. Jacques Monod had prepared and organized for me a research training program, the exceptional quality of which I still appreciate today.

Results proved the identity between the coli-bacillus enzyme and the enzyme synthesized by the hybrid *Salmonella*. They had to be written up. The format of a note to the *Comptes Rendus* was adopted. I did not know it would be such a painful ordeal. My first text, written in absolute quietude, was violently rejected because of its lack of organization, the looseness and vagueness of style, its excessive length, in short its lack of scientific rigor. For Jacques Monod a note to the *Comptes Rendus*, precisely because of its conciseness, had to be written up with great care. I should use as a model the "style of a sonnet rather than that of a novel by Proust. As in classical tragedy, everything has to contribute to the action." Rich with these criticisms and comments, I wrote up a second version, which, of course, did not receive the imprimatur: I had made but little use of the remarks my first paper had provoked. A third version followed, then a fourth. By the fifth one I could no longer construct a sentence; by the sixth one I was losing my vocabulary.

The ninth version was transmitted to Jacques Trefouel, who presented it to the Academy. I started to realize that being a reliable experimenter, or even having inventive genius, was not enough to make a research worker. One also had to master one's writing style. Jacques Monod attached as much importance to the manner in which ideas were expressed as to the ideas themselves. I have even heard him say that an idea existed only insofar as it had been written up. Once I had finished this training, Jacques Monod decided the time had come for me to choose a subject for a Ph.D. A meeting with François Jacob took place in Jacques Monod's office in the beginning of September 1959. Jacques Monod gave me a choice of three subjects: (1) the analysis of the fine genetics of double mutations β-galactosidase permease, a problem the solution of which was to contribute to the concept of operator; (2) the genetic and biological analysis of the enzymes of maltose metabolism, a system from which positive regulation of enzyme synthesis was to be discovered; and (3) the follow-up of the enzymology and biochemistry of β-galactosidase and its different mutants. None of these subjects appealed to me. François Jacob informed us, then, of Umbarger's recent results. He had proven that the synthesis of L-isoleucine in the coli-bacillus was regulated by a negative feedback mechanism. L-Isoleucine, the end product of the biosynthetic pathway, inhibited selectively the activity of L-threonine deaminase, the first enzyme of the chain, and this effect was preserved in vitro after the enzyme's extraction. Moreover, the L-threonine deaminase showed high-order kinetics as a function of substrate concentration. No one at the Pasteur Institute had worked on this system. Jacques Monod mentioned a recent conversation with Arthur Pardee, who had obtained similar results with another enzyme, the aspartate transcarbamylase, the first enzyme of pyrimidine biosynthesis. It seemed advisable that somebody work on these regulatory enzymes. I was interested in the problem, which reminded me of the activation of the enzymes of the egg during fertilization. The choice was made, but we still had to decide on a specific enzymatic system. We chose, at first, the acetolactate synthetase, the first specific enzyme of the biosynthetic pathway of L-valine, also identified by Umbarger. François Jacob had L-valine-requiring mutants; moreover, the given coli-bacillus was sensitive to this amino acid, and L-valine-resistant mutants had already been selected. According to François Jacob, the best way to tackle the problem was to isolate mutants with an acetolactate synthetase modified on the feedback regulation by valine. Several valine-resistant mutants did happen to possess an acetolactate synthetase insensitive to valine. But it was a difficult enzyme, with a still unidentified cofactor apparently necessary for its activity. We decided to abandon this system in favor of L-threonine deaminase, a very stable protein, simple to assay. My first task was to repeat Umbarger's experiments. The study of regulatory enzymes was becoming one of the research topics of Jacques Monod's laboratory.

The Concept of Allosteric Interaction (1960–1962)

My subject had been defined. The task of proceeding was difficult. Isolated in a laboratory crowded with American postdoctoral fellows, it was out of the question for me to discuss the protocol of my experiments each week with Jacques Monod. I assume this weaning was done on purpose. Jacques Monod may also have considered my topic of minor importance. He was giving much attention, at that time, to the regulation of protein synthesis, and spent considerable time with François Jacob. Every, once in a while he would visit me. During a conversation with Gérard Buttin, who was sharing my bench, he made a remark underlining the secondary character of the regulation of enzyme activity compared to the regulation of protein biosynthesis. I had to resist.

Several results helped me regain confidence. First, the fact that Umbarger's observations were confirmed: the inhibitory effect of L-isoleucine was specific. Moreover, in spite of an obvious difference in structure with the substrate, L-isoleucine inhibited the activity of the enzyme in a strictly competitive manner. Was this effect due to a steric hindrance following the classical rules of competitive inhibition by a structural analogue of the substrate? An important observation went against this idea. Various chemical treatments of the enzyme, such as heating, exposure to SH reagents, and purification, modified its properties in a remarkable way. The threonine deaminase was losing its sensitivity to isoleucine, while keeping its enzymatic activity: the enzyme was becoming "desensitized." Even more important was the fact that along with the loss of the inhibitory effect of isoleucine, the nonlinear relationship between activity and substrate concentration disappeared. A unique structural property of the regulatory enzyme was to account for this phenomenon.

In June 1961, a scientific event took place that was to be of importance in the history of regulatory proteins: the 26th Cold Spring Harbor Symposium on Cellular Regulatory Mechanisms. The moment had come to present the data collected on L-threonine deaminase. Several weeks before leaving for the United States I wrote up a first paper, which I submitted to Jacques Monod. He read it with an interest that surprised me. One conclusion had to be drawn: the classical scheme by steric hindrance could not be maintained. We had to think of a new type of interaction between regulatory ligand and substrate: an interaction between nonoverlapping sites. A discussion was then started on the relationship between the nonhyperbolic shape of the substrate saturation curve and the inhibition by the regulatory ligand. The hypothesis (later verified) of several existing sets of sites for the substrate and for the regulatory ligand was mentioned. Jacques Monod found it needlessly complex. As a rule, he used to tell me, one should always limit oneself to the smallest number of hypotheses. Two sites should be enough, one for the substrate and another for the effector. The few structural analogies between threonine and isoleucine (both are amino acids) would allow the substrate to bind nonspecifically to the site of the inhibitor, and vice versa. This economy of hypotheses actually led to some questioning—although minor—of the fundamental postulate of the existence of two distinct and specific sites. I did not feel satisfied with this model but there was no mention of presenting another one. When the final text was written up I asked Jacques Monod whether he would cosign it. Why did he refuse? Did he consider the results too light for him to engage his scientific reputation, or did he wish to spare my feelings?

The puzzling results on threonine deaminase drew the attention of the participants of the Cold Spring Harbor Symposium. Immediately after the presentation, Bernard Davis stood up to make an important remark. According to him, there was a fundamental analogy between the properties of hemoglobin and of threonine deaminase. Both of them exhibited "cooperative" effects for substrate binding. This property had to be related to their regulatory function, and could result from the presence on the protein molecule of several sites for each specific ligand. Hemoglobin was becoming the prototype of regulatory proteins.

One of the highlights of the symposium was, undoubtedly, Jacques Monod's presentation of the concluding remarks. Combining ease with rigor, Monod gave, with his exceptional synthesizing mind, a general review of nearly all of molecular biology in 1961. Dealing with the regulation of enzymatic activity by negative feedback, he went back to the paradox of the apparently competitive inhibition by a compound very different in structure from the substrate. Converging results achieved independently by Gerhart and Pardee on aspartate transcarbamylase (results that were known to him although they had not yet been published) and on L-threonine deaminase led him to draw the general conclusion that the interaction between regulatory signal and substrate was taking place between distinct sites. In his oral report, Jacques Monod used the expression "Novick–Szilard–Umbarger" effect as a tribute to those who had, according to him, first observed negative feedback inhibition. (We actually owe the first description of this phenomenon, in 1941, to Zacharias Dische.) It is in the final text of these concluding remarks, which was written in the course of July and August, that the word "allosteric" appears for the first time—its two Greek roots express the difference in structure between regulatory effector and substrate. In Jacques Monod's writings the word "allosteric" qualifies nouns such as "inhibition", "effect", "enzyme", or, "protein"; never does the substantive "allostery" appear. To account for the interaction between distinct sites, Jacques Monod mentioned already the possibility of an induced fit of the enzyme molecule, according to the scheme proposed by Koshland.

These ideas were further elaborated in the *Journal of Molecular Biology*, in a review article entitled "Allosteric proteins and cellular control systems." This text was written by Jacques Monod during 1962. Models of steric hindrance and allosteric effect were being completed by a less plausible model by direct interaction. Actually, the introduction of this last model helps better to define the word "allosteric" and to identify this type of interaction as *indirect*, a general definition still entirely valid today. In Chapter 2,

Jacques Monod tries to describe more precisely the mechanism of these interactions. A conformational transition—or "allosteric transition"—is postulated. However, in almost all the given examples (except hemoglobin) a change in the state of protein aggregation takes place. Indeed, it is now known that these molecular-weight effects are usually secondary to the allosteric transition triggered by the regulatory ligand. This idea was actually suggested in a footnote about the preliminary results of an ongoing work on phosphorylase *b* by Agnes Ullmann and Roy Vagelos (who was then on a sabbatical year at the Service de Biochimie Cellulaire). Contrary to expectations, the activation of phosphorylase *b* by AMP does not depend directly on the dimerization of the molecule: the allosteric transition is more discrete than a change in the state of aggregation.

Enough progress had been made by Perutz and his associates in their work so as to be able to imagine the extent of the changes accompanying oxygen binding to hemoglobin. First, the oxygen binding sites, the hemes, are about 30 Å apart. Affinity interactions between hemes are therefore allosteric interactions. Secondly, a comparison between the three-dimensional structures of oxygenated and reduced hemoglobin discloses a relative displacement between subunits, minor (about 19% of the distance between subunits), but still significant. Jacques Monod's report of these preliminary results in the 1963 review shows his early interest in the relation between subunits in regulatory proteins. This key idea was to be further developed during the following years, leading to the theory of 1965.

The general discussion in the 1963 review finishes with the generalization of the concept of allosteric protein. These molecules would be the key component of any system of biological control, from the regulation of enzyme activity to enzymatic adaptation, passing by hormone action. The hypothesis is put forward that gene repressors are also allosteric proteins. In 1961, Jacob and Monod thought that the repressor was a polyribonucleotide; as a result of further development of the work on regulatory enzymes, this idea was abandoned.

Interpretation of Allosteric Transition in Terms of Quaternary Structure (1963–1965)

As soon as it appeared, the 1963 paper gave rise to much controversy. Voices rose, denouncing its lack of originality; the idea was old, only the word "allosteric" was new. Others criticized the idea: "Its ability to explain was so great that it excluded nothing or nearly nothing" (Monod, Nobel Prize Lecture), which led Boris Magasanik to quote it as "decadent." Finally, the most serious criticism dealt with the difficulty of distinguishing between a strictly allosteric mechanism and other possible types of direct interactions at the experimental level. Perutz's structural studies on hemoglobin and those of Gerhart and Pardee and of Gerhart and Schachman on aspartate transcarbamylase brought irrefutable proofs of the existence of interactions at a distance between topographically distinct sites.

A wave of publications on regulatory enzymes, nonetheless, began flooding the biochemical literature. Allosteric proteins were fashionable. In the laboratory, Jacques Monod himself was passing phosphorylase *b* through columns of Sephadex equilibrated with bromthymol blue, while Agnes Ullmann followed each of his gestures with warm and apprehensive attention. I was finishing up my thesis work. Kinetic data on threonine deaminase were piling up. Several of its properties were shared by other regulatory enzymes, for example, the existence of "cooperative" phenomena for the binding of substrate and regulatory ligands. I was beginning to feel the temptation to propose a mathematical model accounting for the kinetics observed.

I consulted Adam Kepes, who proposed a scheme derived from Michaelis' classical equations. I was put off by its extreme complexity and abstraction. In the course of the year 1963, I presented to Jacques Monod

Agnes Ullmann and Jean-Pierre Changeux, 1965.

a first draft of my thesis work. The meeting took place in his new and much larger office, which had become vacant following Gabriel Bertrand's death. This room consisted of a desk, a bookcase, and walls of light oak, a comfortable armchair, in a very "rétro" style, although this was not yet fashionable; above the desk was a beautiful watercolor, a Provencal landscape, reminding one that Jacques Monod's father had been a painter. We started talking. It soon became essential for us to propose a simple molecular mechanism which could account for all the kinetic data on threonine deaminase.

Jacques Monod then launched into a series of very general reflections about the structure of globular proteins. Actually, this was a continuation of old ideas that were close to his heart. As early as 1949 he was already discussing the origin of the specificity of enzymes in a paper entitled "Genetic and specific chemical factors in the synthesis of bacterial enzymes," using terms that were surprisingly presaging the 1965 theory. The matter was quite different though, since the problem was to explain the enzymatic adaptation which Jacques Monod himself later proved to be due to protein synthesis. At that time, however, he considered, among others, a mechanism where the substrate acted as an "enzyme selector." He wrote that

> ...every hypothesis on the mechanism of enzymatic adaptation implies a choice between a "true adaptative" mechanism and a "selective" one at the intracellular level. If we accept that the structure of the enzyme is determined by a gene, instead of the substrate, then we admit that the role of the substrate is to *select* a certain type of molecule, at least virtually preexisting. How does this selection operate? This is the problem. One assumes naturally that this effect results from the combination of the substrate with the enzyme. This is what Yudkin was the first to suggest, adding the hypothesis—probably hazardous—of an equilibrium between the enzyme and a hypothetical precursor: Precursor $\leftrightharpoons$ Enzyme. According to Yudkin, the role of the substrate was to shift the equilibrium toward an increased formation of enzyme....

This excerpt is a good illustration of the basic thought of Jacques Monod to look for the elements of protein structure responsible for the specificity of a biological reaction. It is at the heart of Monod's work on enzymatic adaptation, of his ideas on antibody synthesis (1959), and, finally, of his interest for regulatory proteins.

Why did Jacques Monod, in the course of 1963, turn toward the functional organization of proteins? It is quite likely that he was influenced by David Perrin's work on intracistronic complementation, which was done in the laboratory, as well as Crick and Orgel's theoretical paper (1964) on the same topic. Actually, his interest was naturally drawn to homotropic cooperative effects, which he considered the fundamental property of allosteric proteins. Did the idea of a correlation between molecular symmetry and cooperative binding develop while he was reading the writings of Jeffries Wyman on hemoglobin (1948–1960)? Did it come out of a conversation with Jeffries Wyman himself? In any case, Wyman had expressed the idea several times, and it had been confirmed by Max Perutz's X-ray work unraveling the symmetry properties of the hemoglobin molecule. Be that as it may, Jacques Monod started studying the various modes of assembly of subunits in a protein molecule, and the relevant properties. He built models—out of cardboard at first, then with balls of clay and then with dice—which helped him think like a geometer.

First consideration: the three-dimensional folding of a single polypeptide chain gives a fundamentally asymmetrical object on which cannot exist several identical sites for a given ligand. The assembly of several of these objects into a finite structure, or "oligomer," offers the possibility to build a molecule possessing several sites for a given ligand, an essential condition for cooperative binding to appear. This assembly can be done in different ways depending upon whether the association between these asymmetrical objects, or "protomers," involves the same area (isologous association) or different areas (heterologous association) of the protomer surface. For either aesthetic or (more likely) practical reasons, Jacques Monod always preferred the isologous association. It should be noted in this respect that the latter automatically confers a twofold axis of symmetry and, thus, oligomers built by isologous association possess more symmetry properties than those of heterologous association. They always have even numbers of subunits, which explains Jacques Monod's quasi-mystical opposition to trimers and pentamers.

We then had to find a mechanism which would create interactions between identical sites within the oligomer. A purely formal and abstract idea occurred to Jacques Monod: that the conformational transition of the oligomer conserves the symmetry both of the mode of association between subunits and of the affinity of the binding sites; the cooperative binding then results from a shift in the conformational equilibrium in favor of this state. Without any ambiguity, for Jacques Monod, the postulate of the conservation of symmetry is responsible for the homotropic cooperative effects.

If the allosteric transition takes place between symmetrical states, then these states can only be present in small number. As a consequence, they must be independent from the type of ligand bound, and preexist to ligand binding. Therefore, the fundamental postulate of a preexisting conformational equilibrium was the result of a logical reflection required by the model's coherence. As already mentioned, the concept of a "selection" by the ligand can be found in Jacques Monod's writings as early as 1949. In our conversations it had also been mentioned as a simple way of explaining the effect of heterotropic ligands on the cooperativity of substrate binding. Considering the minimum hypothesis of a protein existing under two different states, the allosteric activators would then stabilize the same state as the substrate; conversely, inhibitors would favor the other state. The hypothesis of preexisting states allows a remarkable economy of means and simply explains a large number of kinetic properties. Obviously this concept goes against Koshland's hypothesis of a conformational change induced by the interaction with the ligand, leading to a multiplicity of structural states. We had adopted Koshland's hypothesis in 1961, but after three years it could no longer be considered appropriate.

The writing of the final text took a long time. The first manuscript was written by Jacques Monod during his summer vacation in 1963. If my memory is good, it followed the "historical" approach and began with considerations regarding the structure of oligomeric proteins.

The mathematical model followed, then the fitting of the experimental data by the equation of the model. The manuscript was sent to Jeffries Wyman, and also submitted to Buzz Baldwin, then spending his sabbatical year working with François Jacob. Buzz Baldwin noticed mistakes in the formulation of the equations of the model, and proposed the correct derivation of the "state" and "binding" functions. Version followed version, but Jacques Monod remained unsatisfied. A presentation beginning with the mathematical model was finally adopted, which could not fail to please Jacques Monod by its highly theoretical and formal aspect. In conclusion, speculations on the thermodynamics of allosteric transitions were added. This section did not seem essential to me, but what could I say?

More than a year and half after its conception, the text of the theory was delivered personally to John Kendrew for publication in the *Journal of Molecular Biology*. Jacques Monod made a special trip to Cambridge, where he gave a seminar to explain himself. The manuscript was immediately accepted and appeared a few months later.

Reactions were much stronger and often more critical than those which followed the publication of the 1963 text. The Anglo-Saxon audience, always very pragmatic, did not see the logical necessity, and also the experimental basis, of the postulate of symmetry conservation. Rather, they saw in it the delirium of a Frenchman intoxicated by the spirit of geometry. Several months after the publication of the theory, Koshland, Nemethy, and Filmer submitted to the journal *Biochemistry* a paper which clearly expressed this reaction. The authors initially agreed with us upon considering the hypothesis of the "induced fit" as a particular case of a general scheme of two preexisting states; but then they postulated that the conformational transition concerns only the protomer and that various artificial modes of interaction between subunits are proposed without serious attempt to relate these phenomenological schemes to any concrete protein structure. They wrote:

> ... The "concerted" scheme is represented by a square array merely for convenience, but this model does not depend on the geometry of the interactions since all subunits change simultaneously. The allosteric model of Monod *et al.* (1965) utilizes such a "concerted" change, but it also assumes various symmetry requirements *which are not an essential part* of the "concerted" model shown here....

Biochemists found the symmetry conditions shocking; yet they seduced the solid-state physicist Charles Kittel and the crystallographer Max Perutz. The proposed scheme does appear to be an extreme case. Experimental observations may lead to moderate it. It remains true, though, that experimental data on hemoglobin—still the best-known example of allosteric protein—fit both quantitatively and qualitatively with the theory's main postulates.

Epilogue

After my Ph.D., I had to leave the laboratory. Military service and a growing interest for pharmacology were distracting me from regulatory enzymes. Daniel Blangy and the Bucs had taken the relay. Allosteric proteins were in good hands at the Pasteur Institute.

During the years I spent at Jacques Monod's laboratory we established a deep relationship. My departure for the United States was truly heartrending. In the dedication of his book *Chance and Necessity*, Jacques Monod wrote, "To my true spiritual and of course slightly parricidal son..." This was after I had come back from the United States.

Why would the desire to be independent make me a parricide? Jacques Monod was an exceptional master. He has been much more than that—in many respects, a true father.

Jean-Pierre Changeux

After his Ph.D. in the laboratory of Jacques Monod, Jean-Pierre Changeux continued his research as a postdoctoral fellow in the laboratory of Howard Schachman at the University of California, Berkeley (1965–1966), then in the laboratory of David Nachmansohn at Columbia University College of Physicians and Surgeons (1967). Returning to the Pasteur Institute in 1967, he established his own research group and tackled the issue of the molecular nature of the receptors for neurotransmitters and their possible allosteric properties.

He identified the acetylcholine nicotinic receptor from fish electric organ (1970–1974), demonstrated its allosteric properties (1974–1980), and investigated its biosynthesis and the regulation of its transcription during development of the neuromuscular junction.

Together with his group, Dr. Changeux extended his research to the brain nicotinic receptors. They deleted in mice four defined nicotinic receptor subunit genes and demonstrated their differential contribution to nicotine addiction, to nicotine anti-nociception, and to the regulation of cognitive functions. The allosteric properties of the nicotinic receptors have important implications in medicine because their alteration is at the origin of severe human pathologies like the congenital myasthenic syndromes and frontal lobe autosomal dominant syndromes such as nocturnal epilepsy.

Discussions about Proteins

Robert L. Baldwin

W hen I went to the Institut Pasteur in 1963–1964, I was to work with François Jacob. My plan was to learn phage genetics for use in later physicochemical studies of genetic problems. Two surprises were awaiting me. One was a student (Alex Fritsch) who wanted to work with me; he was particularly interested in the uses of the analytical ultracentrifuge. I had wanted to leave both students and the ultracentrifuge behind at Stanford while I learned a new field, but working with Alex proved to be both a pleasure and a big help scientifically.

The second surprise was a manuscript by Jacques Monod on a model for allosteric enzymes. I was taken aback by its contents. It proposed a general model for allosteric enzymes. The particular protein which was discussed in detail was hemoglobin, which was not then known to be an allosteric protein. There were no allosteric ligands for hemoglobin; the Benesches had not yet found 2,3-diphosphoglycerate, and the role of the Bohr protons in affecting O_2 binding was unclear.

The cooperative binding of O_2 by hemoglobin had intrigued chemists for decades. Such eminent physical chemists as Linus Pauling, G. S. Adair, F. J. W. Roughton, and Jeffries Wyman had studied the "heme-heme interaction." Was it likely that they had missed the answer which would now be supplied by a biologist without formal training in physical chemistry, who had never worked on hemoglobin? The manuscript had no names on it, and I always assumed that Jacques was the sole author of that first draft.

My skepticism grew as I studied the manuscript. It stressed the importance of symmetry: the allosteric transition from an inactive to an active conformation was postulated to be a change from one symmetrical conformation of an oligomeric protein, the R state, to a second symmetrical conformation, the T state. But individual protein chains were notoriously asymmetric objects, and almost nothing was known about the symmetry of their packing arrangements in oligomers. The model was extremely simple in view of the complexity of the problem. There was an equilibrium between molecules in the R and T states; mixed-state intermediates, molecules containing protomers both in the R and T states, were forbidden. All protomers in the R state were assumed to be identical and in equivalent environments, and also in the T state.

In that first draft, the statistical factor for ligand binding, which was well known from the work of Klotz and of Scatchard, was missing. It describes the increase in the probability of binding when a molecule with n binding sites is only partly liganded, caused by the several possible arrangements of i ligands among n binding sites. When I pointed out its omission, Jacques studied the changes that it would produce in the equations. His quick eye caught the fact that the statistical factor rises to a maximum for the half-liganded species and falls off symmetrically for species on either side. He looked down at the set of four dice, glued together, which he used as a model in thinking about the symmetry properties of oligomeric proteins. Then he looked up and smiled, "Well, that makes the equations more symmetrical, doesn't it?"

And so the year began. It progressed through six drafts of the allosteric manuscript. Everyone at the Pasteur enjoyed debate, and I found it easy to take part. But in order to turn it off and sleep nights, I had to resort to strenuous exercise on weekends. I began to climb the tall rocks in the forest of Fontainebleau, a sport in which Jacques once excelled. When I asked him if he had trouble shutting off his thoughts about science when he wanted to sleep, he gave me a surprised look and said, "Why no, do you?"

I was intrigued by Jacques' approach to the problem. First of all, I was surprised that he expected to solve the problem of cooperative binding just by thinking about it and studying the clues in the literature. He didn't consider it necessary to have training in physical chemistry. Instead, he studied evidence from genetics on the properties of mutants and crystallographic data (then scanty) on protein structure. A precedent had been set by Crick and Orgel, who had developed a model for intracistronic complementation based on the probable symmetries of protein oligomers. Then too it surprised me that Jacques was convinced that a simple model would provide a general explanation for allosteric enzymes.

At that time the crystal structure of myoglobin was known and the structure of lysozyme was just being completed. The results could be represented by brass rod models with standard bond angles and distances and with every atom locked in place, as in a crystal of NaCl. Most chemists viewed proteins as massive, rigid objects. But the logic of the allosteric model seemed to require that proteins be flexible. A ligand binding at a distant regulatory site had to send a signal through the polypeptide chain to turn on or off the catalytic site. This was the way that transmission of the signal was envisaged in Jacques' manuscript.

I had a slightly different idea. Since the allosteric transition from an active to an inactive conformation was a concerted transition in his model, perhaps the transition was simply a change at the quaternary level between two different packing arrangements of the subunits. The protomers could remain as rigid objects with very little change in tertiary conformation. Also, in a model of this kind, I felt that one could discard the requirement for a transition between two symmetrical conformations, which pleased me. My idea that allosteric transitions might take place at the quaternary rather than at the tertiary level undoubtedly came from the finding in Perutz's lab that the β chains of hemoglobin move substantially closer together on oxygenation. I remember Jean-Pierre Changeux saying, "But the tertiary conformation of the globin chain *must* change on binding O_2." Jacques was intrigued with my idea, and two years later he told me he wished he had adopted it. Today, I'm sure he would be glad he didn't.

Paris weather in winter looked even more inhospitable through the streaked glass walls of the old dining room next to the small library of Biochimie Cellulaire. But the atmosphere at lunch was extraordinarily cheerful. To begin with, one could bring delicious hot food from a nearby épicerie, close to the corner of rue du Dr. Roux and rue des Volontaires. Lunch had its proper place in the day, as a time for discussion and reflection. No lunch seminars! François Jacob, Pierre Schaeffer, André Lwoff, and Mme. Lwoff regularly came down from the genetics section on the third floor and joined Sarah Rapkine, Agnes Ullmann, Jacques Monod, and other regulars for a leisurely lunch at the long table. Dr. Lwoff encouraged American visitors to try out their French.

The seminars were lively occasions with a highly responsive audience. A speaker had to choose his words with care. Twice in the year a visiting speaker paused as his next slide came on and mused, "Now what was I going to say about this?" whereupon Jacques, sitting in the front row, promptly answered the question for him.

Jacques became interested in physical chemistry that year and ordered books for the library. He visited Manfred Eigen's laboratory in Göttingen and was deeply impressed by the possibilities of studying allosteric mechanisms by fast-reaction techniques. I was surprised that, until then, molecular biology had developed in Paris with so little contribution from physical chemistry. I asked David Perrin why this had happened. His grandfather (Jean Perrin) and father (Francis Perrin) had both made notable contributions to physical chemistry. David smiled and replied, "Probably because my grandfather was a poor teacher."

Jacques had earlier discussed his ideas with Jeffries Wyman, and a collaboration between them now sometimes brought Wyman to Paris. The physicochemical behavior of the allosteric model was being studied in earnest. Jean-Pierre Changeux arranged for computer simulation studies. At the end of that year Jean-Pierre went off to Berkeley to join John Gerhart and Howard Schachman in their study of the allosteric transition in aspartate transcarbamylase. They had just made the remarkable discovery that ATCase had separate catalytic and regulatory subunits. Jacques admitted then that the answers to allosteric mechanisms would probably come from physicochemical studies of purified proteins, together with their crystal structures.

In the spring of 1964, Jacques gave a few seminars on the allosteric model and sparks began to fly. In the autumn, when I had returned to Stanford, he gave a seminar there. Afterward, Dave Hogness turned to me and said, "Well, Jacques has done it again: he has a model which will make a lot of people mad, and they will work on his problem to try to prove him wrong." That certainly is what happened. Hemoglobin, in particular, became one of the most intensively studied molecules known to chemists. One would suppose that today the dust would have settled and one could say definitely whether the change from the oxy to deoxy conformation is a concerted transition without detectable intermediates, as the allosteric model postulates. But no, the controversy goes on. I was told that when Jacques learned in 1970 of the first experiments demonstrating that certain enzymes show negative cooperativity or half-site reactivity, he was pleased and excited although these allosteric enzymes didn't fit his model.

The last time I remember discussing allosteric proteins with Jacques was in the spring of 1968 at Schloss Elmau in the Bavarian Alps. Manfred Eigen had organized a long meeting on relaxation kinetics and molecular biology. Francis Crick was there. Had he thought of doing so, he might easily have applied the concepts of the Crick–Orgel model for intracistronic complementation to the problem of allosteric proteins, and he was keenly interested in Jacques' model from the beginning. At Schloss Elmau he was still thinking about alternative models. But Jacques was beginning to lose interest. He felt that his role in opening up the field was nearing an end. He gave me the set of bonded dice which he had used in his talk and said,

"Now, Buzz, if you will admit the importance of symmetry in allosteric enzymes, I will give you these dice. And if you will still not admit it, I will give them to you anyway." Later, in Paris, he inscribed his initials, symmetrically, on each die.

When I was at the Institut Pasteur, I was strongly conscious of the difference between Pasteurian science and what I was used to at home—especially the Pasteur emphasis on logical analysis and on the development of alternative models. I was not conscious then of any particular influence on me of Jacques Monod, outside of my sharing the general appreciation of his brilliance in analysis and debate. But later I realized that when I thought about research problems I was attempting to carry on an inner dialogue in a manner inspired by discussions with Jacques, and I am grateful to him for his example.

Jacques once told me his method for choosing a research problem: "The way to do interesting research, Buzz, is to find a paradox; but you must be sure that it is a true paradox." I never asked him which paradox led him to the study of allosteric enzymes, and I can only guess. Certainly he enjoyed the clash of ideas that his model provoked, and it pleased him that physical chemists as well as molecular geneticists and enzyme chemists entered the debate.

Robert L. Baldwin

Jacques Monod was very interested in the mechanism of protein folding, and the last discussion I had with him was on this subject in 1972, at his apartment. He was ill, but he asked to see me in order to discuss this.

Since 1979, the date of the publication of the last edition of this tribute, I remained a Professor of Biochemistry (in Arthur Kornberg's department) at Stanford University until I retired in 1998. I was chairman from 1989 to 1994 and have been Emeritus since 1998. I've been a member of the U.S. National Academy of Sciences since 1980. Awards (received for research into the mechanism of protein folding) have included the Stein and Moore Award of the Protein Society (1992); the Wheland Award in Chemistry, University of Chicago (1995); the Merck Award in Biochemistry, American Society of Biochemistry and Molecular Biology (1999); and the Founder's Award in Biophysics (formerly called the Cole Award), American Biophysical Society (1999). I have served on a number of advisory panels: 1984–88, NIH Study Section on Molecular and Cellular Biophysics (1984–1988); Searle Scholars Advisory Committee (1993–1996); and Advisory Committee, Burroughs-Wellcome Interface Program (1996–2001).

My research interests have centered on the mechanism of protein folding, specifically, characterization of folding intermediates and the mechanism of alpha-helix formation by peptides. Currently my specific interest lies in the role of peptide backbone solvation in the energetics of protein folding.

Anne and I have two sons: David, a biochemist, who studied in Michel Goldberg's lab for a few months before beginning Ph.D. work, and Eric, an architect. Eric is married to an architect and he has two sons. Anne now composes music; she finished her second string quartet recently. I am still active in research, through collaborations with younger scientists.

Some Memories of Jacques Monod

D. E. Koshland, Jr.

B y his theories and experiments, Jacques Monod permanently altered the thinking of scientists in genetics and molecular biology. He did this largely by cleverly designed and beautifully written papers in the scientific literature. But perhaps as important was his impact as a person in private interactions with friends, visitors to his laboratory, scientists he met on travels, and anyone who enjoyed logical discourse at a high intellectual level. I was one of those fortunate enough to have a number of contacts with him over the years, and each was unusually rewarding on both the scientific and personal level.

One of Jacques' most impressive characteristics was his deep joy in discussing scientific theories. There were many occasions on which I witnessed this deep commitment to scientific puzzle-solving, but one in Berkeley stands out as typical. Jacques was visiting the university to give a seminar at a time when I was home convalescing from an operation. I reluctantly had assumed I would miss him and had instructed my students to take careful notes which they could recount to me later. To my surprise, because I knew of his tight schedule, Jacques phoned the day before and said he'd like to visit with me, a particularly thoughtful gesture. I was, of course, delighted and my wife cancelled her university appointments and devoted the morning to making a meal worthy of a distinguished French visitor. Jacques arrived, exchanged pleasantries, and then we plunged into a discussion of regulation, protein structure, gene expression, and transport. Suddenly I realized it was very late. He had already missed several appointments and he might even be late for his speech, but the time and our lunch had swept by unnoticed. It was one of Jacques' abiding attractions that he became totally intense in the pursuit of scientific problems and as a result swept his listeners along with him in his enthusiasm. Scientific discussions with him always seemed to me like a deep-sea diving expedition in which one plunges into a new world with exciting terrain and a colorful fish flashing by at breathtaking speed. You were never bored and frequently had to sit quietly after he left to assimilate the many ideas that had streamed by in your conversation.

One discussion is particularly memorable to me as illustrating the strong philosophical and Descartian logic which permeated Jacques' thinking. We were discussing our alternate theories of cooperativity which had recently appeared in the literature and in particular the role of symmetry. I was arguing that nature undoubtedly used symmetry but only as a tool to minimize energy, and cooperativity resulted from ligand-induced distortions which altered the energetics between subunits. Jacques argued that symmetry was a ubiquitous characteristic of nature, and conservation laws were pervasive in science. Hence, conservation of symmetry would be an almost inevitable fundamental principle guiding the construction and behavior of protein molecules. At one point Jacques said, "I have no formal training in physical chemistry. I'm a geneticist who views the world from principles of evolution." I was impressed not only by the candor of his statement but also because his basic insight and sharp mind had led me to assume he was an expert in the area. He had approached the subject from cosmic generalizations and collected his thermodynamics as he went along. The more we talked the more it became apparent that the two theories were internally consistent but each proceeded from different premises. No logical argument can prove or disprove internally consistent theories any more than they can prove or disprove Euclidean versus non-Euclidean geometry. Yet in the process of discussion we clarified our thoughts and the discussion, like a fencing match, was a pleasure in itself. I was impressed at the extent to which Jacques utilized philosophical and historical precepts to guide his thinking. And in the end I believe that any theory which attracted him would have to be aesthetically appealing as well as logically impeccable.

In New Hampshire in the relaxed science plus leisure atmosphere of a Gordon Conference I remember spending most of one week talking to Jacques on subjects ranging from politics and lifestyles to science and mysticism. One discussion on sailing particularly intrigued me. I enjoy sailing but consider it largely a device to converse with congenial friends in a relaxed and mildly athletic atmosphere. The occasional crises of wind and waves add a pleasant piquance, particularly if it is clear that the danger is surmountable and a happy landing is inevitable. To Jacques, sailing offered these virtues, but he liked his dangers more intense. He said a crisis that was potentially calamitous excited and stimulated him enormously and sailing in too tepid waters essentially bored him. We discussed at length the analogy between scientific dangers and physical dangers. We both agreed that scientists must venture into uncharted seas to make progress and fear of the turbulence of controversy and the disaster of being wrong inevitably leads to repetitive science. To me, the dangerous adventures of science were enough. To Jacques, adventure was part of his blood, not excluded from any facet of his life. He described to me some of his dangers in the French resistance and his attitude was the same. He was a man not seeking danger in a foolhardy way but nevertheless exhilarated and excited by physical as well as intellectual danger.

Jacques enjoyed controversy. He did not seek controversy per se, at least not consciously. However, he had strong convictions both in science and in applying science to philosophical and social phenomena. He discussed his forthcoming book, *Chance and Necessity*, with me, at one point saying he knew it would create a storm. The expected attack from those whose religious convictions were threatened he viewed with equanimity. The expected displeasure of scientists worried him more, as those were his intimate peers, whom he respected and with whom he broke his daily bread. I asked, "Why do you do it if you are so concerned?" He replied, "I must. I have a conviction of the importance of extending scientific reasoning and one cannot turn back because of the expectation of criticism." There was a twinkle in his eye and no shadow of gloom, and I had the distinct impression that he partially welcomed the furor he knew he'd create. He liked the approbation of his peers, but he also liked to be in the center of a rousing intellectual controversy. His assumption of the directorship of the Pasteur Institute was not surprising to me, even though many scientists didn't understand his willingness to accept an administrative position. He felt an obligation to help the Pasteur, but perhaps even more he wanted a challenge in a new sphere in which his abilities had not yet been tested.

Jacques' sense of humor added greatly to his charm and to his scientific effectiveness. Humor is always attractive, but it is sometimes absent from individuals of extraordinary scientific dedication. The kind of mind that desires to pursue ideas to their logical extremity is frequently impatient with the nonproductive side paths or illogical juxtapositions that are at the heart of humor. And frequently the most successful scientists become enamored of themselves to the point where the gentle prod of humor becomes an unacceptable irritant. Jacques was more than willing to poke fun at himself. Once when I was talking with him and a number of other scientists during a visit to the Pasteur, I was impressed by his artful use of humor to provide restful interludes during a long and arduous scientific discussion. It was not calculated, but he periodically relieved the tension by a momentary humorous diversion and then quickly brought us back to the pursuit of the elusive scientific problem. And he enjoyed jesting with colleagues and peers. At one session of the 50th anniversary of the Biochemical Society in Paris, he gave me a carefully wrapped present which he told me to deliver to my wife. When she opened it, there was a French cigarette lighter with the admonition "Please help Dan throw some light on his theories."

An interesting sociology develops between scientists who compete in a common area and yet have a deep respect for each other. My relations with Jacques were always tinged with a slight tension based on our interests in similar problems and perhaps they were made special because of it. Whatever the cause, I always found his company unusually exciting and stimulating. He was invariably challenging and penetrating, and I remember our meetings with a special sense that they were rare events which should be prolonged like a good book. His enjoyment of science bubbled over into a world of personal charm and eloquent advocacy. He left contributions of great brilliance in the annals of science and memories of enchanted interludes in the hearts and minds of friends over the whole world.

Daniel E. Koshland, Jr.

Since 1979 I have been up to no good, wasting my time while writing one book, 195 journal articles, being editor of *Science* from 1985–1995, collecting a few awards (Lasker, National Medal of Science, Merck, etc.) and a few honorary degrees (Univ. of Chicago, Brandeis, etc.), and giving lectures at various institutions. A building was named after me at the University of California-Berkeley (they probably hoped I would die). My research interests are mainly devoted to enzymology and allostery and Alzheimer's disease. Briefly, our research is focused on the mechanisms of enzyme action to explain the enormous catalytic power of enzymes and to use that knowledge to design better catalysts. A second focus is on the biochemistry of Alzheimer's disease and discovery of agents for its prevention.

I am very well and enjoying myself on the faculty at Berkeley, where I am a professor in the graduate school. I just published two papers, one on allostery and one on "The Seven Pillars of Life" (*Science*, 2002), so I am still carrying on in Jacques Monod's tradition. I still miss his enthusiastic and original spirit very much.

In New York, January 1973.

Mother Nature and the Design of a Regulatory Enzyme

Henri Buc

If we were to follow Plato we would consider such perfect figures as endowed with more significance and "reality" than any actual object. Very often indeed, a scientist cannot help feeling a much closer affinity to Plato, the radical idealist, than to some of the supposedly "realist" or "materialist" thinkers. A beautiful model or theory may not be right; but an ugly one must be wrong.

JACQUES MONOD

At the Institut de Biologie Physico-chimique, scientists used to meet in the library for the Friday afternoon seminar. In 1958, when I had just entered research, an American scientist was presenting her data on the biochemistry of bacterial cell walls. Suddenly, the course of the seminar was totally diverted. Somebody whom I could not see (because the tiny place was overcrowded) had started asking questions which, up to then, I had never considered. He was not arguing about the experimental facts, but questioning the logical conduct of the work. Stressing the implicit assumptions made by the speaker, he counterproposed the set of hypotheses he would have formulated; then he showed that no decisive experiment had yet been performed which could have ruled out either of the hypotheses. It was obvious that he thought *his* approach to the subject was more controversial and of a more general interest than the speaker's view. In fact, as soon as the discussion resumed, it focused on the questions this man had posed. I left the seminar, quite puzzled. For the first time I realized that it was possible to openly question the very reasons for which others were doing research.

It was roughly at that time that molecular biology started to be recognized in France as a new attempt to understand the structural basis of biological functions. Yet most French physicochemists, though deeply interested in biochemistry, had an instinctive scorn for the oversimplifying

255

darings of molecular biologists. Most of them refused to accept the idea that the organization of at least some living systems could be progressively reduced to the complex interplay of macromolecules. They were telling me: "Are you interested in biology? Well, spend ten years to become a good physicochemist, and then find a problem of biological interest. You will become a good biophysicist." The trouble was that I did not understand for what reason they had themselves decided to be physical chemists rather than, for example, embryologists or mathematicians.

In the early 1960s, the arrogant character who had so early ruined my obedient confidence in the work of the scientific establishment was giving a series of lectures on advanced biochemistry. I could not attend the course, but my wife, Marie-Hélène, gave me her notes each week. The excitement around the genetic code was at its peak. Monod's lectures were so logically arranged that students could understand what the specialists in the field were currently trying to demonstrate or to rule out at this very moment. Probably as many others, I was reading these notes as a scientific "roman feuilleton." Looking back at them recently, I realized that part of the excitement came neither from their novelty nor from their logical content. Monod was looking at a nascent science from his peculiar point of view, molecular evolution. Each new result was presented in the lecture as one tiny piece in the puzzle that Mother Nature had solved. We were contemplating "Mother Nature herself in the process of studying new developments for better and finer control of cellular metabolism." Between such a fruitful attitude and the endless studies on "structural features of macromolecules of biological interest," there was no reason to balance.

I had the chance to better understand Jacques Monod's quest when he invited Marie-Hélène and me to his home. He spent the evening explaining the content of the manuscript he had written with Jean-Pierre Changeux and François Jacob. He stressed the type of performances which had to be expected for certain proteins "acting at critical metabolic steps which appeared as electively endowed with specific functions of regulation and coordination." Probably because I had not spent too much time studying the intricacies of classical enzymology, it sounded reasonable to me that when a ligand was binding at a specific place on a protein subunit, this event could be transmitted across a distance to the substrate site, allowing the control to take place.

But Jacques Monod was looking for something else. As he wrote later on: "The disadvantage of this concept is precisely that its ability to explain is so great that it excludes nothing, or nearly nothing. There is no physiological phenomenon so complex and mysterious that it cannot be disposed of, at least on paper, by means of a few allosteric transitions." He agreed with Boris Magasanik to consider the present state of the theory as "decadent" because "there was no a priori reason to suppose that allosteric transitions for different proteins need be of the same nature and obey the same rules." As he was actively looking for some unifying principles, he pointed out to me that in some cases the binding of the regulatory signal was affecting the molecular weight of the protein while in other instances no change of molecular weight could be noticed. We therefore discussed possible means of showing that, in this latter case, the conformational change corresponded to a weakening of the noncovalent bonds between the subunits.

Three years later, when I came back from Harvard University, the "model" was written down in its final form. It came to me as a great surprise. If anybody takes care to read in historical order the article by Monod, Changeux, and Jacob and then the reports at the Cold Spring Harbor Symposium of June 1963 devoted to the same topics, he would be struck by the convergence of theoretical reasonings between the main groups engaged in this field. In 1963, everyone was stressing the role of flexibility in enzyme action, the most precise structural analysis being certainly given by D. Koshland. Viewed from outside, the first quantitative theory on allosteric control was expected to be a general induced-fit theory. However, between Pasteur and Berkeley, the prospects were already different. D. Koshland, refuting the argument given above that "the flexibility theory itself is too flexible, i.e., that it can explain anything and hence explains nothing," noted that "the very ease with which a wide variety of phenomena can be explained by this theory is a cogent argument for it," a statement which radically differed from Jacques Monod's objectives. In the same article, Dan Koshland went on to stress in what respect his theory was going to be fruitful: "The detailed arguments lead to predictions of correlation between protein structure and specificity patterns which are verifiable by experiment." Therefore it has always seemed evident to me that the two schools differed, almost from the very beginning, in the very content of what they were trying to explain. Most of the scientists wanted to describe how local, tertiary, and quaternary changes were geared together on a particular protein when a regulatory molecule was binding to it. No general principle was expected to emerge immediately, for the rules of the game (i.e., the correlation between protein tertiary structure and specificity patterns) were not even understood. To express this feeling with an image, it seems to me that, for a "realistic" biochemist, Mother Nature was a rather easygoing creature, trying to do her best with the "molecular tinkering"

she has at her disposal. This permissive divinity had devoted all her inge-
nuity in finding out and in maintaining in good order various mechanical
devices which allow a regulatory protein to sense the instructions given
by the external medium. To understand Mother Nature will take very long
because we shall have to become acquainted with a great number of tricks
of which we have no idea, for the time being.

Many scientists appreciate the Monod–Wyman–Changeux model because it is a simple, elegant, and imaginative proposal. (Monod liked to call
this model a theory since each statement could be independently proved
or ruled out on specific examples.) But few people seem to realize that this
apparent simplicity had been gained at the cost of painstaking meditations
on how evolution could have proceeded at the molecular level. These reflections have extended long after the original manuscript. They have been
best expressed in the Nobel symposium on symmetry and functions in biological systems. It is because evolution had been constantly working on
allosteric proteins that it was plausible to offer, as early as 1965, a simple
theory accounting for their behavior. Mother Nature wanted to have specific devices able to open or to close metabolic pathways with a maximal
efficiency; she had thought of proteins and had arranged them into closed
structures. The imperious need of obtaining both the maximal sharpness
of the regulatory response and the maximal stability of the regulatory proteins imposed the present solution: a concerted transition between two
symmetrical states. When Monod was speaking of evolution, Mother Nature looked like a Greek divinity gathering in her hand anarchic monomers
and fusing them into platonic figures.

I feel very sad when I see that a very significant confrontation between
two different approaches of molecular biology is reduced to a formal conflict between two hypothetical kinetic pathways. What is really at stake is
a triple issue. First, it is a good example of classical opposition between
informative and selective theories in biology (in the M–W–C model, preexisting states are selected by small metabolites, the concentrations of which
reflect the various physiological needs of the cell; according to Koshland
and his collaborators, the ligand informs the protein structure and directs
its conformational change). Second, they diverge on the basic unity or on
the diversity of the structural solutions historically retained by evolution
to solve a problem of regulation. Third, Monod's theory is falsifiable: to
refute it does not simply mean to show how significantly the real solutions
differ from a model. It will demand to gradually put forward a unifying
view of the various structural solutions invented by evolution.

I wish to emphasize again that it was this evolutionary point of view which oriented most of the discussions in the laboratory. Monod used to ask: "Suppose you synthesize, according to Merrifield's method, a polypeptide chain, the primary sequence of which is taken completely at random. What is the chance that this Precambrian protein will fold with a hydrophobic core inside and with most of its hydrophilic residues outside?" With respect to protein assembly, a constant debate was going on: Was it reasonable to assume that, in a first step, evolution had disfavored those cells where proteins had a large tendency to form aggregates? In such a situation, was it not plausible to assume that only similar monomers would have had the chance to assemble together? More precisely, let us assume that we knew exactly the three-dimensional structure of several monomers and their local flexibility around their equilibrium position: Would it be possible to find a physical criterion which would predict which proteins would polymerize into a symmetrical oligomer, and which ones would lead to closed asymmetrical structures? At what point of a metabolic pathway, enzymes exhibiting negative cooperativity (or functioning according to the "flip-flop" mechanism proposed by Michel Lazdunski) would they be selectively advantageous?

Most of the time, those speculations were initiated according to a well-defined ritual. At around ten or eleven o'clock in the morning, the door between Monod's office and the "labo bleu" opened. Usually Monod started by spending ten to twenty minutes with Madeleine Jolit, looking at the colonies which had grown overnight. By a smile or a long glance, he made it evident that some new speculations had been going on in his mind overnight. The discussion started on the blackboard in his office. The precise function of this particular type of colloquium was to check with some of us how far one could go from general speculations to logical proposals and eventually to experimental verifications. It is of course impossible for me to describe in English his typical blend of seduction and of demanding logic. In a very personal tempo, rather slow, each logical step was carefully marked and the voice used to adopt a particular impatient vibrato whenever a plausible objection was expected to be raised ("Mais enfin, cela crève les yeux!"). If the whole laboratory worked together for so long, with such an enthusiasm and so few conflicts, it is certainly because he led us to basically adopt his very personal approach to molecular biology, and he managed to reduce the slightly different subjective appreciations of the problem to a few precise alternatives which sometimes could be experimentally solved.

Thus the scientific questions he was raising appealed to us in those days as deep intellectual pleasure. But, as coined by opponents, this attitude would have been at best a dazzling game of fashionable models, at worst "allohystery," a dogmatic oversimplification, first, if Monod had not applied to his own dreams his demanding logic; and second, if the whole permanent staff of the laboratory as well as the numerous postdoctoral fellows and the foreign friends or visitors had not permanently challenged his reductive view with the wealth of their experience. For example, during a long contest with Monod on the proposal that oligomeric proteins are able to react to lower concentrations of ligand than the "equivalent monomer," Francis Crick begins one of his letters with an exasperated call: "Jacques, be reasonable." Monod was sensitive to similar warnings whenever they were justified by strong experimental evidence. Conversely, I fear, he often looked rigid and deaf to logical arguments on extrascientific matters.

It is at this period, as we were all anxious to confront theory and experiment, that another facet of Monod's talent became evident to me: his ability to delineate those dilemmas which were at a given time amenable to solution. Jean-Pierre Changeux had already left for H. Schachman's laboratory to work on the aspartyl-transcarbamylase system. Before leaving, he lent me a very precious gift: "la bibliothèque allostérique," an impressive collection of preprints and reprints on regulatory enzymes. He was sending from Berkeley epic letters on his struggle with the ultracentrifuge and on his verbal fights with the misbelievers. We were trying to find out with D. Blangy how to check quantitatively on *E. coli* phosphofructokinase each theoretical prediction of the model. Passionate attempts to allosterize β-galactosidase were discussed in the long corridor of the laboratory. Monod insisted on two objectives: to find out a reliable method that could directly reveal in solution the point groups of symmetry to which an oligomeric protein could belong, and to be able to determine the elementary steps of the reaction between an allosteric protein and at least one of its regulatory effectors. The first problem is still not really solved; the second one had been previously discussed with Manfred Eigen before I came to the lab. During the spring of 1966, K. Kirschner, a gentleman having the inimitable port of an officer of the British Raj, brought to Paris exciting kinetic data. They allowed for the first time the visualization of a concerted transition of an oligomeric enzyme as the coenzyme was colliding with its stereospecific site. Jacques Monod told me, "Enrico, we must go to Göttingen." So we went there, a few months later. I was expecting a rather exotic meeting, something like Haroun-al-Raschid paying a visit to Charlemagne, so distant were the cultures of Manfred Eigen and Jacques Monod at that time. I was really baffled to see, on the contrary, how Manfred Eigen's ability to reduce complex physical problems to a set of binary conjectures was in basic harmony with the scientific approach of the geneticist.

Thus in 1967, a large fraction of the allosteric problems, those related directly to structure, were in a fair way to be reduced to specific physico-chemical questions, while most of the evolutionary aspects still remain unsolved and challenging.

This period corresponds also to the foundation of the European Organization of Molecular Biology, in which M. Eigen, J. Monod, and F. Jacob had taken a very active part, and to Monod's election at the Collège de France, an institution that he liked almost as much as the Pasteur Institute. He opposed these types of institutions to the medieval organization of the French university that he bitterly attacked in several public interviews. In Jacques Monod, as a public man, we recognized our basic need for creative freedom. May 1968 was not very far, another struggle against the stubborn vanity of the establishment, the rise of another creative fancy, the physical solidarity with the students; May 1968 perhaps also led him to the tough discovery that the younger generation seemed to reject almost any intellectual leadership. Along the night wanderings through the Quartier Latin, amidst the noise of explosions and hand grenades, I realized that this exceptional relationship was going to take a new course.

Henri Buc

When he became head of his own research unit at the Institut Pasteur, Henri Buc continued to determine changes in the ternary and quaternary structures of several allosteric enzymes by kinetic approaches. At the end of the 1970s, he switched to the study of the interactions occurring between proteins and nucleic acids in bacteria. In particular, he has investigated how *Escherichia coli* RNA polymerase and the cyclic AMP receptor protein recognize their DNA targets and cooperate to increase the rate of initiation of transcription. In order to do so, he developed several strategies that permit following the establishment of these processes in real time. He used similar approaches to characterize how HIV reverse transcriptase replicates its RNA template and is able to generate in vitro a large spectrum of point mutations and of recombination events.

Henri Buc is now Emeritus Professor at the Institut Pasteur, and looks back at the scientific exchanges that followed the publication of theories of allosteric regulations.

Recollections of Jacques Monod

Jeffries Wyman

My first encounter with Jacques Monod was over forty years ago when he gave the Dunham lectures at the Harvard Medical School. But at that time our paths were widely separated. He was largely concerned with genetic adaptations in microorganisms, I with dialectic studies on amino acids and proteins and the physicochemical phenomena involved in ligand binding by respiratory proteins, especially hemoglobin. Little inkling did I then have of the extent to which our interests were to converge in the years to come.

It was some time after this, during the 1950s, that my real friendship with Jacques began. That was when I went to the U.S. Embassy in Paris as science attaché to initiate the program of science advising which had just been established by our State Department. In that role it was incumbent on me (and a pleasant duty it was!) to get to know the community of French scientists and to promote relations between them and their U.S. counterparts in a postwar era of relative isolation. It was the period when the paranoia of McCarthy and his followers was approaching a crescendo and creating a serious barrier between the two countries. Naturally for me, as a biologist, the Pasteur Institute was an obvious place at which to begin my activities. Moreover, it so happened that at the time of my arrival in Paris, Alvin Pappenheimer, who was married to a cousin of mine, was spending a sabbatical year at the Institute. As a result of all this I used to walk over to the Institute for an impromptu lunch with a group of which Jacques was a member (this was before the days of the present cafeteria). Others included André Lwoff and François Jacob, as well as another American, the then youthful Mel Cohn. These lunches, with their lively conversation about current scientific developments, were a pleasant change from the somewhat bureaucratic air of the Embassy. Sometimes during that period, at Jacques' suggestion, I would join another and larger group that used to gather for supper before the seminars held in the library of the Institut de Biologie Physico-chimique in the rue Pierre Curie. These contacts marked the beginning of a friendship with Jacques which was to extend well beyond the bounds of pure science.

In those years Jacques was an enthusiastic rock climber, and I remember his taking me out to a place in the forest of Fontainebleau where we could practice. I don't think I could find the site today, but the image of those sheer and challenging rocks stands out clear in my memory. This was in Jacques' mountaineering period, before he had developed his passion for sailing, although I do remember his glowing accounts of an early voyage to the east coast of Greenland in an old-fashioned square-rigger, of which he showed me some striking photographs.

A much closer relationship with Jacques, however, came more than a decade later, after I had left the Embassy in Paris as well as a subsequent post in the Middle East, and had returned to scientific work as a guest at the University of Rome. It was in 1965 that Monod and I published our joint paper with J.- P. Changeux on what has come to be known as the M–W–C allosteric model (or sometimes "the concerted two-state model"). This paper, which appeared in the *Journal of Molecular Biology*, grew out of a seminar I gave in Paris in the autumn of 1964 (a successor of that series of seminars in the rue Pierre Curie which I had frequented earlier in my Embassy days); and this seminar in turn was the result of a talk given the preceding summer at Cold Spring Harbor, which is where I first met Changeux. (Changeux was at that time responsible for the program of the Paris seminars.) Well before this time it had become clear that the interactions, both the homotropic (cooperative) and heterotropic (control) ones, characteristic of many enzymes and respiratory proteins, notably hemoglobin, could be in a large number of cases explained in terms of ligand-linked conformational changes. In particular, in the case of human hemoglobin, there was evidence, drawn partly from kinetic studies involving the uptake of dye but even more from the X-ray work of Max Perutz in Cambridge, that the tetrameric molecule existed in two different quaternary conformations: one characteristic of the deoxy, the other of the oxy form, the transition between them being controlled by the activities of oxygen and proton. The underlying linkage principles, in accordance with which conformational changes can give rise to both homotropic and heterotropic control, had already been laid down, and it was this subject that formed the basis of the talk I gave in Paris. The paper with Monod and Changeux was really the child of this talk, and its prompt publication after the talk bears witness to Jacques' quickness of response to an idea, which was one of his great qualities. The paper as it stands was written almost wholly by him and was presented to me more or less as fait accompli for discussion and criticism. It should be pointed out here that the introductory part dealing with the possible role of symmetry in the assembly of oligomeric proteins was largely due to Jacques. Although it is true that I had introduced the concept of symmetry in the case of such systems a good many years before, pointing out that insofar as site interactions were the reflection of the spacing of the sites, symmetry of function as represented by symmetry of the binding curve (fractional saturation vs. chemical potential of the ligand) implied a certain geometric symmetry in a macromolecule. Yet that was before the realization of the role of conformational change and the introduction of the allosteric concept, which greatly changed the picture. It is this part of the paper which is at once perhaps the most intuitively appealing from a physical point of view and at the same time the most open to criticism (and criticism was not wanting). The rest of the paper, based as it is largely on the sure principles of classical thermodynamics, stood on firmer ground.

I vividly remember a meeting Jacques and I had with several of our friends and colleagues at the MRC Laboratory in Cambridge to discuss the manuscript and its suitability for publication in the *Journal of Molecular Biology*. As I recall it, these included Max Perutz, John Kendrew, Francis Crick, and Sydney Brenner, a rather formidable group of critics. The attack mainly centered on the symmetry ideas, and Jacques bore the brunt of it. But afterward, in the calmer atmosphere at tea, we all agreed that the paper, symmetry and all, should be published as it stood, a decision which in retrospect I thought no one could possibly question. That evening John Kendrew and I went to a dinner party at Francis's house, where although there was little further discussion of such a technical matter, some doubts and reservations were further expressed.

Shortly after the paper came out there was an International Congress of Biophysics held in Naples. By that time the paper had been widely read and attracted much attention. Both Jacques and I gave talks based on it at one of the sessions. Jacques stressed the symmetry concept; I the more abstract thermodynamic ideas, taking the occasion also to introduce the closely related concept of the binding potential, which had been developed subsequently to the appearance of the paper. As I look back, that meeting stands out as one of the most exciting and pleasant I have attended. The setting was perfect: in front of us, as we looked westward, the blue expanse of the Bay of Naples with the magic profile of Capri in the distance; behind and to the east, the tangled city rising to the hills crowned by San Martino and Capo di Monte; to the left and south, the smoking bulk of Vesuvius; to the right and north, the jagged silhouettes of the islands of Procida and Ischia. At that time Francis Crick was joint owner of a fine cutter, which was lying in the port and on which he was living. One afternoon he invited several of us, including Jacques, John Kendrew, and myself, to go for a sail. The day was unexceptionable—a warm sun, a cloudless sky, and a gentle breeze. Soon after we cleared the mooring, Jacques, by that time an accomplished sailor, took the tiller. However, at the end of the afternoon, he gave it back to Francis, who was somewhat less sure of himself and, as we approached the port, showed signs of nervousness—small wonder considering the complications of entering a crowded Italian port. At this point Jacques, evidently enjoying the situation, turned to me with the remark, "It is clear to me that when Jim wrote the opening sentence of *The Double Helix* ["I have never seen Francis Crick in a modest mood."] he had never seen Francis on a boat"—a fine example of Jacques' wit.

Jacques prided himself, and with reason, on his abilities as a sailor—so far as I know, he did not take up sailing seriously until fairly late in life, though his experiences on the voyage to Greenland must have laid a foundation. I remember, after he had got his first seagoing boat, his speculating about the prudence of a sail from Cannes to Corsica. But later on, in a stouter ship, he had no hesitation in undertaking a voyage from northern France to the Greek islands, though it meant crossing the stormy Bay of Biscay and facing the sometimes furious and capricious winds of the Mediterranean. Jacques' love of the sea and its challenges, like his love of climbing, provided a fine complement, on the human side, to his passions and achievements on the intellectual one.

Jeffries Wyman

Jeffries Wyman was born in 1901 in West Newton, Massachusetts, and died in Paris in 1995. He received an A.B. degree in 1923 from Harvard and a Ph.D. degree in 1927 from University College, London. Wyman was a member of the Biology Department at Harvard from 1928 to 1951, where he studied the dielectric properties of amino acids, proteins, and related compounds. In 1951, Wyman ended his career at Harvard to become the first Scientific Advisor to the U.S. Embassy in Paris.

He then spent three years (1955–1958) working in the Middle East for the United Nations as a director of the regional science cooperation offices of UNESCO.

In 1960, Wyman returned to scientific research as a guest scientist at the University of Rome's Biochemical Institute and the Instituto Regina Elena, where his research focused on the structure and function of hemoglobin. His collaboration with Jacques Monod goes back to this period.

Wyman was the author of numerous papers on the biophysical properties of proteins. He was a member of the National Academy of Sciences (U.S.) and the Lincei Society of Science.

Francis Crick and Jacques Monod at the Salk Institute annual Fellows
meeting, January 26, 1973. (Photograph by D. K. Miller, Salk Institute.)

Sailing with Jacques

Francis Crick

I cannot remember for certain when I first met Jacques Monod. My recollection is that on my first trip to Paris after the war, during the period when I was working with Arthur Hughes at the Strangeways Laboratory, he and I visited the Pasteur Institute, together, but to see Dervichian, not Jacques. Nor, in 1949, when we were on our honeymoon (which was spent in Italy) did Odile and I pause for any significant time in Paris, although we passed through it. I think my first meeting with Jacques was some years later in the mid-1950s, probably after Jim Watson and I had put forward the DNA structure. I clearly recall giving a seminar at the Pasteur in which I suggested the quite erroneous theory that during protein synthesis the inducer was needed to fold up the enzyme correctly. Without inducer, protein synthesis would come to a complete stop. Even at that time I didn't think much of this notion and I came to think even less of it, so it was never published. In those days things moved more slowly and there was less pressure to rush into print.

I believe it was on this occasion (because I recall that there was a certain constraint between us) that I was with Jacques in his laboratory on a Saturday morning. "You'll never guess what I'm trying, my dear," he said to me. (Jacques' English was practically perfect. This use of "my dear" as the equivalent of "mon cher" was the only slip I ever remember hearing him make.) He was attempting, in a way that I cannot exactly recall, to see if he could get some sort of an immune response from *E. coli*. I thought it was a silly idea and clearly so did he, but I was impressed that he was prepared to give it a try. It seemed to show a commendable spirit of adventure.

I must, over the years, have had many scientific conversations with Jacques, and I am surprised to discover that I remember hardly any of them. But then I've had many more with Sydney Brenner and almost all those have also been swallowed up by time. I think that, as science advances, one recalls only the results, tending to forget the process of discovery unless it was unexpectedly memorable. Certainly I shall not easily forget the Good Friday morning, in the Gibbs Building in Kings, when François Jacob was talking to us. Suddenly the scales fell from our eyes and we saw that the ribosomal RNA was not the messenger RNA, but that the "Volkin and Astrachan" RNA was. But Jacques was not there on that occasion, although he was present at a lunch Odile prepared for us, some months earlier, in the basement of Portugal Place when Sydney and I were first told about the PaJaMa experiment. I recall how puzzled we were by it and how reluctant to believe it. It was this meeting which sowed the seeds for the eventual conceptual breakthrough. Once it was realized that the ribosome was basically a reading head, the world never looked the same again.

There was at least one occasion when we did have a long discussion together. This was in the house in Cannes, where his parents had lived and which the family had kept on as a holiday home. A lovely, solid house, high up on the hillside surrounded by trees, it had a period feeling, partly due to the many paintings by his father on the walls. Jacques had given the Pomona Lectures under the title "Chance and Necessity" and had written, in English, a first draft of his book. He had sent me a copy to read and I stayed with him for a few days, on my way from Greece to the Ile du Levant, to go sailing with him and to discuss the manuscript. During a long morning's conversation together, I remember expressing a number of criticisms and reservations, but again time has wiped from my memory exactly what we said. Perhaps my notes on the manuscript still exist somewhere. Jacques eventually rewrote the book in French and I never read it again until the English version (translated by someone else) made its appearance. My impression was that he hadn't changed it very much.

There were a number of topics about which we seldom spoke, though I knew they were important in his life: his experience with the French resistance, during the war; his friendship with Camus, or any literary topic for that matter; his love of music, though he had told me that at one point he had had to choose between playing the cello professionally and a career in science. Perhaps he sensed that my taste in music was, compared to his, very unsophisticated. He told me once that he greatly admired George Orwell, an author who is not very much to my taste. I nearly said I preferred Proust but thought better of it, which, looking back, was a pity. He did tell me that he was color-blind and produced a striking description of what it felt like—"The word 'green' is a meaningless word to me"—but in spite of this he must have had some interest in the visual arts because I remember once running into him, quite by accident, at an art exhibition in Paris. But most of the time we talked science or gossiped or fussed about the running of The Salk Institute, of which we were both nonresident fellows. In fact, I saw as much of him in La Jolla as anywhere else. He and I were both appointed in 1962. I served two six-year terms, while he was in the middle of his third when he died. We must have met there in the winter (the California winter!) for most of the years from 1962 to 1973, though one or other of us missed a year here and there. I remember writing him a letter from La Jolla with a P.S. "We have just discovered that your head is full of actin." Unfortunately the typist, confronted with an unknown word, wrote "action," and I didn't spot the error before it went off. I can't imagine what Jacques thought when he received it.

The other times we met were connected, in one way or another, with sailing. Jacques was an excellent sailor who could manage his 37-foot sloop by himself, whereas I was always a rather bumbling amateur. One year there was a scientific meeting at Naples. I recall criticizing Jacques' overenthusiasm for symmetry in allosteric reactions, referring to myself, somewhat to the surprise of some of my colleagues, as an "old Jesuit" in these matters and comparing him to a more recent convert. At that time I owned a half share in a splendid sailing boat, a 47-foot racing cruiser, which was based on Naples. It was crewed by an elderly Italian and, as he could speak hardly any English and I spoke less Italian, communication was not always easy. A party of us persuaded Jacques to come out for a sail (not that he needed any real persuasion) and he soon assumed command. The sailor, who had very early spotted my limitations, was soon respectfully taking Jacques' orders. Fortunately the wind was reasonably brisk and we had a pleasant turn outside the harbor and back.

My most rewarding sailing with Jacques was in his own boat. While I was staying with him in Cannes he decided we should run over to Corsica. On the trip out one of his sons and daughter-in-law were with us, so my assistance was hardly required. The trip was uneventful apart from coming close to three whales who frolicked around the boat for a while. On the way back there were just the two of us. We set out in daylight, bound for St. Tropez, intending to sail through the early part of the night. "We'll be there in time for the nightclubs," Jacques told me. It was pleasant enough when we set out, but we ran into a considerable storm. As the waves got higher and the wind blew more strongly I became mildly apprehensive, although trusting in Jacques to get us through, even with the handicap of having me to help him. He was clipping himself on, as he moved about the boat, lit dramatically from above, with darkness all around us. Finally I said to him, "Jacques, exactly what do I do if you fall overboard?" (I didn't think that he would, but I felt I'd better know.) He explained to me what maneuvers to make and I felt a little more relaxed.

Our next problem was to find out where we were. We had a fairly good idea of what our bearing had been but our reckoning of the distance we had covered was far less accurate. This didn't bother Jacques, because he had radio direction-finding equipment on board. Unfortunately, the electric storm was so severe that as we neared the coast, Jacques found that he couldn't get any proper bearings, so we had to look out for the lighthouses and hope we were roughly where we thought we were. Eventually we located the channel and slipped into St. Tropez a little after dawn. The nightclubs were closed, but even if they'd been open we would probably have been too tired to go to one.

It was the sequel, however, which was more revealing. The storm was still blowing strongly. Jacques' motor—he always hated having one—had given up completely so I was confident that we were stuck there till the wind abated. But I had not allowed for Jacques' determination. The following morning it was still blowing strongly from the north. Yachts from St. Tropez are never in a hurry to leave the harbor and on that morning not a single vessel, either motor or sail, made any attempt to leave. We had picked up a passenger and Jacques was determined to get us back to Cannes for the weekend. So he arranged for us to be towed out of the harbor by a local boatman and off we went. Since the wind came from off the shore, the waves, though high, had not had time to grow enormous, although the wind was blowing half a gale. Under very light sail we sped along at high speed (or what passes for high speed on a sailboat—that is, rather slower than a bicycle), making the journey, if I recall correctly, in about three hours. It was the best sail I have ever experienced. There was less wind at Cannes and, even with no motor, we managed to berth without too much trouble.

During most of the time Jacques was head of the Institut Pasteur I saw very little of him. He told me, initially, that the job was turning out to be not too difficult, but that was before the real problems started. I was no longer going to La Jolla for winter meetings, looking for sun instead at Marrakesh. In the spring of 1976 Odile and I passed through Paris on our way to Iran. We'd heard that Jacques was seriously ill although now back at work. When we met him he seemed fitter than we expected although sobered somewhat by the threat hanging over him. He was having blood transfusions twice a week. We had a long lunch at his apartment. I broached the question of writing a book about his work and, hopefully, we made plans to meet at

Jacques Monod, Salvador Luria, Jacob Bronowski, and Francis Crick at the Salk Institute.

The Salk, after Jacques had retired from his Pasteur job. I think we both realized how precarious such plans were. Even so, his death was a surprise; certainly it came much sooner than I had expected. It produced in me the numb shock one suffers when somebody dies who one feels is quite irreplaceable.

I have written elsewhere (in an obituary for *Nature*) a studied estimate of Jacques' career and character. Here I have tried to convey something of him in a less formal way. He was a few years my senior but his youthful manner and his close friendship made me think of him as a contemporary. He was small in stature, but I never consciously felt of him as being much shorter than I am, though once sitting in the bar of the old Del Charro in La Jolla I was suddenly surprised to notice how small his feet were. Jacques had a natural charm, together with courteous good manners, though less formal than one might expect for a Frenchman in his position. He could be formidable, due to his force and his clarity, and sometimes aggressive, but he never seemed that way to me. In fact, he was very companionable— one always enjoyed an evening when he was there—and it's not everyone that one can go sailing with. Our friendship was not the friendship of those who were young together, nor were we intimate in the sense that we discussed our personal problems with each other. Rather it was based, I think, on a steady admiration, seasoned with an affectionate recognition of each other's failings. Our general attitude to most scientific matters was very similar, yet our backgrounds were sufficiently different to make both of us eager to hear what the other thought. And this is where I feel most strongly a sense of loss. As each new thing comes up, one regrets so much not being able to talk it over with Jacques. He would understand so quickly; he would appreciate the importance of the point; he would say something illuminating that hadn't occurred to one; one could reach agreement together and a deeper understanding. It is for this reason that I find, now that he is gone, that I have no stomach for writing a book about his work even though I still feel a small nagging sense of duty. It would have been such a rewarding experience to write it with his cooperation, his clarifications, and his criticisms—and of how many people could one say *that*?

Francis Crick

Born in 1916, Francis Harry Compton Crick majored in physics at University College, London. He worked with the British Admiralty for seven years during and after World War II. He then turned to biological research and in 1949 joined the Cavendish Laboratory at Cambridge University. He received his Ph.D. in 1954 with emphasis in X-ray diffraction.

In 1953, Crick, together with James Watson and with Maurice Wilkins and Rosalind Franklin at King's College, London, discovered the molecular structure of DNA, for which Crick, Watson, and Wilkins received the 1962 Nobel Prize in Physiology or Medicine. After winning the Nobel Prize, Crick worked on problems connected with protein synthesis and the genetic code. Francis Crick is currently J.W. Kieckhefer Distinguished Research Professor at the Salk Institute for Biological Studies, where he also served as President in 1994 and 1995.

Crick has been an integral member of the Salk Institute since the 1960s. For the past 20 years, Crick's research has focused on understanding the neurobiological basis of consciousness, joining the molecular and cellular aspects of neurons, observations in neuroanatomy, and psychological studies of behavior in an attempt to arrive at a theory of what makes us aware and conscious.

Francis Crick is Fellow of The Royal Society and foreign member of the National Academy of Sciences. In addition to the Nobel Prize, Crick was a recipient of an Albert Lasker Award and received in 1992 the Order of Merit from the Queen of England.

At the Salk Institute, February 8, 1969. Jacob Bronowski, Jacques Monod, Edwin Lennox, and Salvador Luria. (Photograph by D. K. Miller, Salk Institute.)

The Ode to Objectivity

Gunther S. Stent

Einmal dem Fehlläuten der Nachtglocke gefolgt—es ist niemals gutzumachen.
FRANZ KAFKA, *Ein Landarzt*

[Once having responded to the false alarm of the night bell—it cannot ever be made good.
FRANZ KAFKA, *A Country Doctor]*

To my everlasting regret, the final years of my relations with Jacques Monod were under a cloud because of my critical review in 1971 of his *Chance and Necessity.* During the two years from 1948 to 1950 that Elie Wollman and I collaborated as young postdoctoral fellows in Max Delbrück's laboratory in Pasadena, Elie's description of Jacques (of whom I, an only recently liberated high-polymer chemist, had never heard) caused me to form a mental portrait of Jacques as Superman, straight out of Hollywood Central Casting: handsome, tough, courageous, artistic, brilliant. And when, on my first visit to the now celebrated microbial-physiologic attic, I eventually met Jacques, he did not fall short of the romantic image that I had constructed: he was everything that Elie had claimed on his behalf. I could hardly wait until it was time to move down from Copenhagen to Paris, to begin my final postdoctoral fellowship year and work down the corridor from that fabulous combination of Darwin and Prince Charming. And again my expectations were not disappointed. Having chosen him for a role model, the year I spent in Jacques' ambience had a profound effect on my formation as a scientist. At the time of leaving Paris for Berkeley to take up my first (and so far, only) job in Wendell Stanley's Virus Laboratory, I considered it as one of the main accomplishments of my sojourn at the Pasteur Institute to have been able to win Jacques as a friend.

The sparkling lunchtime conversations at Jacques' table (to which I was graciously admitted, even though as a member of the *équipe* André

Lwoff I was supposed to eat in François Jacob's lab) covered a broad range of scientific, political, and cultural subjects. But I don't remember the subject of philosophy being discussed much. None of us, Jacques no more than such regulars of his table as Annamaria Torriani, Germaine Cohen-Bazire, Roger Stanier, Mel Cohn, or Alvin Pappenheimer, appeared to be interested in discussing epistemology or the foundations of morality. In 1951–1952 there seemed to be more pressing subjects to think about than the metaphysical nature of man. The only topic that brought us even close to philosophy was the then still festering Lysenko affair. Naturally, everybody was on the side of the angels, and all there was to talk about was how it could be possible for otherwise seemingly intelligent "Free World" scientists to stand up, as some then still did, in defense of the hereditary transmission of acquired characteristics. Accordingly, the rich intellectual Pasteurian dowry, to which Jacques had so heavily contributed and with which I set up business on my own in Berkeley, did not include much philosophical baggage. And so, when a decade later both Jacques and I began to write in the philosophical domain, our ideas turned out to be rather divergent.

I had chosen philosophy as the required arts "minor" for my undergraduate B.A. degree in chemistry at the University of Illinois. Among my teachers there had been the outstanding philosopher Max Black. But it was only in the early 1960s that two, at first sight wholly unconnected, developments of great affective personal significance caused me to revisit the books of my old "minor." One of these developments was the culmination of molecular genetics in the breaking of the genetic code and in the Jacob–Monod operon model. I had made up my mind to join the search for the physical nature of heredity after learning, while still a chemistry student at Illinois, from Schrödinger's book *What is Life?* that if Delbrück's quantum mechanical model should fail, we would have to give up all attempts to fathom the gene. At the age of twenty-two, the project of proving Delbrück's model true seemed to me to offer scope for a lifelong romantic quest. But now, just barely turned forty, I saw that my Holy Grail had been found. The other development was the eruption on the Berkeley campus of the free speech movement, whose trauma caused me and many of my colleagues to perform a critical self-examination of our own motivational infrastructure. As one of the strategies to defuse the explosive situation, the university administration instituted a special category of Professor of Arts and Science, whose only charge was to think big for a year and try to heal the evident cultural breach between students and faculty. For reasons never fully explained, I was given one of the professorships (now long defunct).

So, nolens volens, I was cast into the role of hometown, cracker-barrel philosopher. (My fellow appointees were a mathematician and an art historian, all three of us of Central European provenance. Curiously, none of the local distinguished bona fide philosophers or sociologists were called on for this service.) In discharging my duty as a philosophical orator before my campus constituency (later I was gratified to meet a younger colleague who had taken his wife-to-be to one of my speeches on their first date), I managed to connect the denouement of molecular genetics with the free speech movement, showing that both are symptomatic of the inevitable End of Progress.

Thanks to an invitation arranged by François Jacob as Visiting Professor at the Collège de France, I presented a synoptic version of my grandiose apocalyptic, macrohistorical insights in Paris in the summer of 1967. I was rather disappointed that Jacques, to whom I was most eager to expose my views, did not come to hear me, although he did attend the reception that the Collège's Director, Etienne Wolff, gave for me at the conclusion of my course. Word later reached me indirectly that Jacques, just as most of my friends, thought my whole thesis nonsense. His lack of appreciation of my insights should not have been unexpected, though. Just two or three years earlier, on one of Jacques' occasional visits to Berkeley, I had taken him to the Haight-Ashbury district of San Francisco, in order to show him the then brand-new hippie scene. (The word "hippie" had only just been coined, as the diminutive of the earlier San Francisco North Beach "hipster," or "beatnik"; it was then still largely unknown outside of Northern California.) I tried to impress on Jacques my notion that all these unwashed, drugged youths from Squaresville, Middle America, loitering on the street were an omen of a coming profound transformation of the human condition. But Jacques was bored, and even slightly annoyed, that I had dragged him all the way across the Bay to show him what he saw merely as the California homolog of the inebriated *clochards* that have lain under the arches of the Seine bridges since the days of Vercingétorix.

When I was giving my lectures at the Collège, I did not realize that at that very time Jacques himself was getting ready to work the philosophical vineyard, laying the groundwork for his *Inaugural Lesson* as Professor of the Collège. I received the manuscript of the *Lesson* about a year later and was surprised to find Jacques holding views that I had never known to be his. At the very outset of the *Lesson*, Jacques announced his finding that modern man is in a state of alienation. Why? Because, according to Jacques, whereas the present human condition is, in the main, a product of science, "the very source of science, in objective knowledge and in the ethic which grounds it, remains obscure for the majority of mankind." And how is this global "schizophrenia" to be cured? Jacques answers, "by the deepening of knowledge itself, by constantly extending the objective method into new domains." Thus, although I had on previous occasions heard Jacques make disparaging remarks about "scientism" (the ideology holding that the presuppositions and methods of science are valid for the entire sphere of human activity), I suddenly realized that he himself in fact still embraced the scientistic world view, passed down from Francis Bacon via the Encyclopedists and Marx and Engels. Before setting forth the details of his recommended cure, however, Jacques presented an overview of the contributions of molecular biology to our understanding of the evolution of life, celebrating *en passant* the definitive interment of vitalism. Now with our understanding of the role of DNA as the transmitter of hereditary information, evolution can be seen to be nothing other than a series of accidents. In this connection, however, Jacques exaggerated the claims that can be made on behalf of molecular-biological neo-Darwinism. The stochastic nature of the DNA mutation process notwithstanding, it is by no means certain that the aspect of life to which Jacques referred as "emergence," or the evolutionary creation of increasingly complex structures, is, as he alleges, "pure chance." On the contrary, it is formally possible that there does exist some physicochemical principle that made "emergence" a necessary, rather than contingent, feature of the history of the earth. For instance, as Ilya Prigogine once proposed, nonequilibrium thermodynamics might embody a principle to the effect that the flow of solar negative entropy through the earth not merely allows but actually "drives" the emergence of life and the evolution of ever more complex biological structures. Perhaps there is no such principle, but it hardly seemed fair to stigmatize, as did Jacques, persons who are searching for it as "vitalists."

According to Jacques, the most recent of these evolutionary accidents was responsible for the emergence within the biosphere of a new realm, the noosphere, or realm of knowledge. Once the noosphere had come into being there began within it an evolutionary process based on the natural selection, not of genes, but of ideas. Of all the ideas in the noosphere, the most powerful to have emerged is that of objective knowledge. And now Jacques, for the first time, attempts to explicate the concept of objectivity that was to be central to *Chance and Necessity:* for him objective knowledge is that which has no source but the systematic confrontation of logic and experience. Here the *Lesson* already presages a disturbing feature of the later book, namely, the somewhat cavalier usage of its key terms. In either ordinary speech or philosophical parlance "objectivity," or its adjectival form "objective," does not denote anything resembling the meaning that Jacques assigned to it. Moreover, in *Chance and Necessity* he gave it a variety of other, equally unconventional meanings. For instance, sometimes he claimed that objectivity is a property of Nature itself, rather than, as in the *Lesson*, a method of studying Nature. At other times he returned to a methodological meaning, but implied that objectivity is a "principle" that forbids use of "final causes" in the interpretation of natural phenomena. However, as ordinarily understood, the term "objective" has nothing to do with the use of logic or the rejection of final causes, but as the antonym of "subjective," denotes something like freedom from personal bias in the consideration of the external world. Other key terms that Jacques employed with meanings other than those normally understood are "animism," and indeed "science" (to which he explicitly assigned a meaning that is normally covered by the term "metaphysics").

According to the *Lesson*, the practical use of objective knowledge is as old as man himself. But the conscious awareness of the idea of objective knowledge evolved only in Western Europe with the pre-Socratics: "The Chinese never managed to evolve this idea." (Apparently, Jacques was unfamiliar with the history of Chinese philosophy. Socrates' Chinese contemporary Mo Tzu, and the so-called "Logicians" he inspired, knew all about and advocated objective knowledge. But their more successful Taoist adversaries weighed the idea of objectivity and found it wanting.) But why, if objective knowledge is such a powerful idea, has the modern science that it spawned caused the contemporary condition of schizophrenic alienation? One reason, according to Jacques, is that scientific technique is now beyond the understanding of most men and for them represents a cause of permanent humiliation. A more profound reason, however, is that science has made "man a stranger in the cosmos, without his appointed or necessary place." (In being cited in support of this idea, Kant, whose contributions to the subject under discussion are at least as important as those of the ubiquitous Darwin, makes one of his rare appearances in Jacques' writing.) Moreover, science "has succeeded in destroying the traditional foundations of the various religious ethics [while] it cannot, by its very nature, provide any other." But, fortunately, science embodies within it its own value system, namely, an "ethic of knowledge." And it is that ethic of knowledge which Jacques recommends for general adoption, now that the traditional value systems lie in ruins. What is the supreme good, the summum bonum, of the ethic of knowledge? Not the happiness of utilitarian ethics, nor the self-knowledge of deontological ethics. No, it is objective knowledge itself. And immanent in that quest for objective knowledge are such values as the (formerly Christian) scorn for violence and temporal domination, as well as the (formerly Protestant) love of personal and political liberty.

Since Jacques did not solicit my opinion, I did not communicate to him my feeling that his *Lesson* fell short of the stratospheric intellectual standards for which I had always admired him. Moreover, I had no idea that he was planning to expand his lecture into a full-fledged book. No sooner had *Chance and Necessity* appeared than it became an ideal, sitting-duck target for critics of every philosophical stripe (and not merely for those beholden to Jacques' favorite twin targets of Marxism and Judeo-Christianity). Jacques' stridently aggressive tone, his practice of handing out, ex cathedra, merits and demerits to a vast cast of characters, from Heraclitus through Hegel, Marx, and Nietzsche, to Chomsky, coupled with the infuriating fact that a recent Nobel laureate in molecular biology had managed to write a best-seller on epistemology and moral philosophy, must have caused itching in the trigger finger of every philosophical hired gun. And so, when the Boston monthly *The Atlantic* (in which Jim Watson's *The Double Helix* had not long before made its debut in serial form) asked me to review the American edition of Jacques' book, I thought I would try to offset some of the more hysterical attacks by writing a critical, though friendly and respectful, analysis from within molecular biology. In my review, entitled "An Ode to Objectivity," I said that I thought *Chance and Necessity* was an important book. And for this very reason "this philosophical statement by one of the major scientific figures of our time... must be subjected to critical analysis, as no doubt (in line with the beliefs he offers here) Monod would be the first to wish it to be."

Unfortunately, I was mistaken. Jacques did not seem to have the wish I attributed to him. In response to the manuscript of my review that I sent him prior to publication, I received from him a five-page letter, the longest he had ever written to me. "My first advice to you," he began, "would be to read the book again, and this time read it carefully." But apparently I had read the book sufficiently carefully for Jacques to have found it worthwhile to refute my critique in thirteen specific points. And he closed his letter as follows: "Of course, my dear Gunther, I leave it to you to decide what you wish to make of these remarks. Let me add this: We have had many friendly disagreements in the past. Almost always I have seen you ready to attack my position before you had made any genuine attempt to really understand what I meant. I very strongly believe that the only useful and constructive type of discussion is one when the discussants both feel that what the other party says, even though one may not grasp it right away, contains valid elements. Since you have apparently written your paper even before really reading my book, please, if you have enough time and it is not too much of a bore, read it once more in that spirit. Then perhaps we may be able to talk about it in a significant way."

Naturally, I was crestfallen upon receiving Jacques' letter. It especially pained me to learn that, in all these years, Jacques had interpreted my bent for probing his ideas critically (or, possibly, *hyper*critically), not as I had thought of it, namely, as the best way of showing the authenticity of my esteem, but as perverse eagerness to attack him. Although it certainly transpired from my review that, much to my disappointment, I had perceived some conceptual defects in Jacques' global picture, my essay did in no way intimate that his book was devoid of valid elements, an intimation which would have been totally contrary to what I actually thought. Alas, his letter made me realize too late that, appearances to the contrary notwithstanding, Jacques' scientific lifestyle was not that of my earlier mentor Max Delbrück. In Max's orbit unrelenting criticism really *was* the genuine currency of sincere friendship, and, as the autobiographical essays in the Delbrück Festschrift "Phage and the Origins of Molecular Biology" show, it was hard on the psyche climbing up Mt. Olympus on Max's trail. I responded to Jacques' letter with a ten-page letter of my own, mainly to convince him that, if nothing else, at least I had read his book carefully. I did not hear from Jacques again, and there never presented itself an opportunity to discuss with him his book in some significant way. But two years later, I happened to see him on television being interviewed by Edwin Newman in the *Speaking Frankly* series. It relieved me greatly to hear that, in response to Newman's enquiry about criticism of *Chance and Necessity*, Jacques spoke without rancor of his "friend Gunther Stent" as one of his more substantive critics. Also, while preparing this article I was gratified to discover that in a reply to his critics published posthumously in 1977 under the title "Notes de bas de page" in *Prospective et Santé*, Jacques, though not mentioning me by name, did try to deal with many of the very issues I had raised in my review.

I met Jacques one more time, at the 1974 Colloquium "Biology and the Future of Man" held at the Sorbonne. He was drinking coffee at a bar in the foyer of the Grand Amphitheatre, a lone and seemingly melancholy figure. I joined him to say hello, and we exchanged a few amicable civilities, without finding words for a substantive conversation. Nevertheless, I could feel that our friendship had survived the unintended offense I had given him, and the souvenir of that friendship will always remain for me a source of pride and inner strength.

Gunther S. Stent

Gunther Stent was born in 1924 in Berlin, Germany. He received his education in Berlin and in Chicago, obtaining his Ph.D. in 1948 at the University of Illinois. His professional experience began as Research Assistant in the War Production Board Synthetic Rubber Research Program (1944–1948), then Document Analyst, U.S. Field Information Agency, Technical, Occupied Germany (1946–1947). His research career started at California Institute of Technology (1948–1950), and after having spent two years at the University of Copenhagen and the Institut Pasteur, Paris, he became affiliated in 1952 with the University of California-Berkeley, where he has been successively professor and chairman of different departments. Since 1994 he has held the position of Professor of Molecular Biology Emeritus and Professor in the Graduate School.

From 1948 to 1968 Stent's professional interests were centered in molecular biology (structure and replication of the genetic material; regulation and mechanism of expression of genetic information). After 1969 he became interested in neurobiology, particularly the structure, movement control, and embryological development of simple nervous systems.

He has published numerous books: *Molecular Biology of Bacterial Viruses*, 1963; *Phage and the Origins of Molecular Biology* (with J. Cairns and J. D. Watson), 1966, 1992; *The Coming of the Golden Age*, 1969; *Molecular Genetics*, 1970, 1978; *Function and Formation of Neural Systems*, 1977; *Paradoxes of Progress*, 1978; *Morality as a Biological Phenomenon*, 1978, 1981; a critical edition of *The Double Helix* by J. D. Watson, 1980; *Neurobiology of the Leech* (coeditor with K. J. Muller and J. G. Nicholls), 1981; *Shinri to Satori: Kagaku no Keiji Jogaku to Toyo Tetsugaku* [*Truth and Spiritual Awakening: Metaphysics of Science and Oriental Philosophy*], 1981; Max Delbrück's *Mind from Matter*, 1986; *Nazis, Women, and Molecular Biology: the Memoirs of a Lucky Self-Hater*, 1998; and *Paradoxes of Free Will*, 2003.

Dr. Stent is a Member of the National Academy of Sciences (U.S.) and several other Academies. He was awarded the Runström Medal (Stockholm, 1986), the Urania Medal (Berlin, 1990), and the John Frederick Lewis Award from the American Philosophical Society (2002).

Jacques Monod: Scientist, Humanist, and Friend

S. E. Luria

He had to choose. But it was not a choice
Between excluding things. It was not a choice

Between, but of. He chose to include the things
That in each other are included, the whole,
The complicate, the amassing harmony.

WALLACE STEVENS

It may seem strange that I should have thought of Jacques Monod as my closest friend even though we lived thousands of miles apart, met only sporadically, and never actually worked together. Ours was a friendship of choice rather than of familiarity. If we had lived in the eighteenth century instead of the twentieth, as I am sure we would both have liked, we might have produced a body of written correspondence, which would now be left for me to reminisce with and perhaps for someone to turn into a postdoctoral thesis. But we did live in our time, the time of telephone and airplane, a time that has made letter writing obsolete. Ours were exciting but tormented times, times when deep friendships were rare and precious.

I first met Jacques in 1946 at Cold Spring Harbor. Immediately I admired his mind, his delight in elegant science. Also I was struck by his unhesitating humanitarian instincts. I recall that there was in Cold Spring Harbor the family of a German-American scientist, desperately worried about the fate of some relatives in French-occupied Germany. When Jacques was apprised of it he immediately undertook (I believe successfully) to help reestablish connections between the members of this family. The former resistance fighter would readily ignore the past enmity when there was a chance of helping human beings.

Jacques and I met again briefly a number of times in the following years. I recall a visit of Jacques to Urbana, Illinois, in the early 1950s, in the course of which we talked of enzyme induction and of host-induced phage modification. A visit of Jacques to MIT in 1958 had a more significant impact on me, as we then planned my visit to the Institut Pasteur in the spring of 1959, a visit that finally brought us closer together in friendship.

It was a brief visit that allowed little time for the experiments I wanted to do. To gain time, I flew to Orly carrying in my pocket some partially grown bacterial cultures and I deposited them in the shaker bath in Jacques' laboratory before checking in at the hotel. After a brief nap I did the first experiment that same afternoon; Monod's Service was extremely conducive to work. It was then that I first met many brilliant scientists, residents, and guests, whose dedication to research made that laboratory a joy to work in.

Not necessarily pure joy, of course. My experiments at that time dealt with the expression of phage-transduced lactose genes. It was a well-guarded secret (well-guarded from me too) that similar experiments on lactose-transferring episomes were underway in the laboratory, the experiments which later led to the formulation of the operon theory. I suspect that Jacques enjoyed playing a game of cat-and-mouse with me, appearing to predict what my results would be. Only later I found out what had been going on and was a bit angry.

But science was not the most important accomplishment in that brief visit. The windfall was that my real friendship with Jacques started at that time, cemented by the delightful discovery of the close affinity of our intellectual tastes and beliefs. It was a surprising affinity between two persons so different in background and experiences, except in science. It extended beyond science, and was in fact not centered on science as content, but on science as a chosen commitment.

One of our main affinities was our common existentialist persuasion. We shared an intense distrust and dislike of abstract loyalties. We wished our beliefs to be commitments as freely chosen as possible. We viewed loyalties as blank checks, commitments as rationally explored endorsements— endorsements that are acts of will and therefore imply active participation.

Jacques Monod used to assert that an existential philosophy is the only philosophy appropriate to scientists. Later, during a year I spent at the Institut Pasteur, and still later, when I was lecturing at the Collège de France, the question of the relation between science and existential philosophy was often the topic of our Saturday morning conversations. We insisted that our commitment to science—and more generally to the advancement of rational knowledge—was a choice to be vigorously affirmed but not an absolute value of the human mind. When, in *Chance and Necessity*, Jacques translated this idea of commitment into the formula of an "ethics of knowledge," some confusion arose. The affirmation of commitment to rational knowledge as ethical choice was interpreted by some critics as an affirmation of rational knowledge as an absolute ethical principle. This issue of ethical theory is one that neither this essay nor its author is particularly suited to clarify.

I recall a meeting in La Jolla with Jacques when he was preparing the draft of his Pomona College lectures, whose text later became the substance of *Chance and Necessity*. At that time I had become intrigued by Noam Chomsky's ideas about language, and had timidly been "pushing" on Chomsky a genetic and evolutionistic interpretation of the evidence for innate linguistic structures. I was delighted, therefore, when one day at breakfast, as we talked of genetics and anthropology, Jacques blurted out (I paraphrase), "I am absolutely convinced that the evolution of the human language structure was the central driving force in human evolution." ("Je suis absolument convaincu," incidentally, was a favorite way of Jacques to preface a controversial statement.)

For me it was delightful to discover that our lines of thought, in fact our interest in a new subject, had once more converged by independent and separate ways. Such moments of intellectual convergence are infinitely precious. They are the seals of friendship, just as the kiss of Paolo and Francesca in Dante's *Comedy* was the seal of love. What is friendship but a mutual attraction and affinity of two minds, a mutual valuing and being valued, just as love is made of wanting and being wanted.

In his last years Jacques became concerned with the impact of biology on human society, and, characteristically, tried actively to promote the field of bioanthropology. He actually involved me and others, through the Royaumont Center, in a number of ventures in the border area between hard and soft science—without, I fear, generating a great deal of solid accomplishment. I believe Jacques was inclined to put more confidence in the ethological and sociobiological approaches than I thought was warranted. Even here, however, he was not dogmatically "biologizing" the human predicament, but simply trying to choose and develop a feasible approach. I suspect that for Jacques these activities actually had an additional function, providing him with an intellectual and social diversion from the concerns of administration and from his separation from day-by-day experimental work.

There was in Jacques' personality a slightly perverse streak. He consciously cultivated a certain ambivalence about issues and about people.

He had his vanities, although his sense of humor helped him discriminate attitude from substance, in himself as well as in others. His ambivalence never prevented him, however, from *se prendre au sérieux* when the task at hand warranted it, whether it was wartime resistance, or science, or concern for colleagues and students, or for the Institut Pasteur.

I miss him as a man and a friend. Mature age, even after a full life, is a sad time because one's friends start to depart. One loses one's partners in discourse, who provide excitement and validation. More important, one loses the actuality of long-lasting friendships. Yet the memory of those friendships is the evidence of existence fulfilled.

Salvador Edward Luria

Salvador Luria, one of the founders of microbial genetics, was born in 1912 in Torino, Italy. After receiving his medical degree from the University of Torino, he worked from 1938 to 1940 at the Radium Institute in Paris. When Italy entered World War II, Luria went to the United States, working first at Columbia in 1940, then between 1950 and 1964 as Professor of Microbiology at Indiana University and University of Illinois. From 1964 until his death in 1991, he was Professor of Biology at MIT. Luria joined Max Delbrück in the early 1940s, and the "phage group" was born. Their famous "fluctuation" experiment (1943) was the first demonstration of the spontaneous nature of bacterial mutations and the basis of measurements of mutation rates.

With Alfred Hershey, Dr. Luria studied genetic recombination of bacteriophages and bacteria. Their early work led to the modern knowledge of virus reproduction, gene replication, and gene function and its regulation. In 1969, Luria received the Nobel Prize in Physiology or Medicine, shared with Delbrück and Hershey. Luria's later interest in the functional organization of bacterial membranes led him for many years into research on colicins. Luria obtained a number of prizes (Lepetit, Lenghi, Academia dei Lincei, Louisa Gross Horwitz) and was a member of the National Academy of Sciences.

Conjectures and Refutations

Antoine Danchin

C ontrary to most of the authors who have been asked to contribute to this book, I have been neither a colleague nor a pupil of Jacques Monod. Our encounter occurred after many accidental events which led me from the most formal mathematics to experimental biology. Curiously, it is not in Henri Buc's laboratory, where I had many opportunities to see Jacques Monod, that I met him, but at the meetings organized by the Centre de Royaumont pour une Science de l'Homme; and after the first one in 1973 we regularly had numerous conversations. Rather than present here a review article on the various aspects of regulation at the molecular level (this has been done often by Jacques Monod himself, and developed at length in *Chance and Necessity*) or report anecdotes on the personality of Jacques Monod, I would like to continue in the spirit of the conversations I was lucky to have with Jacques Monod a short time before his death on a subject we began to discuss in detail.

It seems to me—but, clearly, it is my interpretation—that two major themes were present in Jacques Monod's thoughts at that time. The first one was concerned with the methodology necessary for investigating phenomena; it is developed at length in K. Popper's books, and, indeed, Monod has contributed to their diffusion in France: the best way to contribute to building science is to progress by making conjectures and trying to refute them. A self-consistent model, even when it is wrong, is always useful if it is "falsifiable." Oddly enough, the most frequently prevailing attitude, even in the most famous international journals, is a tendency, on the one hand, to produce "irrefutable models" and, on the other, to try to "verify" theories rather than refute them. In this respect the operon as well as the allostery theories are exemplary. In the latter, the

Monod–Wyman–Changeux model requires only *two* phenomenological constants in order to describe allosteric regulation, whereas the "equivalent" Koshland–Nemethy–Filmer model asks for at least (in the case of a dimer) *four* such constants in order to describe the same phenomenon. Thus the K–N–F model is practically unfalsifiable (whence its surprising success in the scientific community).

The second theme—as I believe I see it—is clear in the allostery theory. One can explain phenomena using two quite different approaches. Either one assumes the existence of a driving force able to "instruct" (or direct) the evolving of the considered systems, or one attributes to chance alone the ability to display sufficient variety of objects so that interactions with a changing environment are sorted out, and the most stable ones selected (those which last the longest time). Again the M–W–C model is a selective theory whereas the K–N–F model is typically instructive. It is extremely interesting to learn from the published papers that Monod himself began by an instructive view of allosteric regulation (still found in his reflections on Maxwell's devil), to end with the selective approach. In this respect I think that the word *teleonomy* used in *Chance and Necessity* carries typical instructive (and even finalist) stereotypes, although it serves to design a concept, still ill-defined in this book, which is of the selective type. The so-called teleonomy law is a kind of contingent necessity (if I dare say so) which expresses necessity a posteriori and not a priori. And after the last passionate and exciting conversation we have had on this theme, I think that *Chance and Necessity* represented a step toward a thought completely devoid of instructivism, and that the word teleonomy was an inappropriate choice rather than a mistaken concept.

Since this book is mainly devoted to the works of Jacques Monod, it seems unnecessary to consider again the fundamental regulations he has helped elucidate: allosteric regulation and regulations of transcription. I would like only to emphasize the hierarchical order of these regulations: allostery deals with one catalytic activity, the operon, with the coordinate expression of several related activities; and catabolite repression with the coordinate expression (or repression) of several operons. As we had once agreed upon, one may then ask: Is a still more general system, allowing, for instance, the coupling between the cellular metabolism and macromolecular syntheses, possible? In what follows I would like to describe the main lines of a conjecture which would help to state precisely this question, keeping in mind the methodology alluded to above. And if I dare write so, it is in order to follow one of the aspects of the research work Jacques Monod was emphasizing: its exploratory power. We must be bold and daring; we have nothing to lose!

Within the framework of this book I can only give the outlines of the conjecture and questions raised by it. I think, however, that it is noteworthy to observe that, at the time of Jacques Monod's death, when he had so many important administrative tasks to perform, he was still willing to discuss with a newcomer, as I was, the theoretical lines he was fond of.

The problem was raised by the fact that the models proposed for explaining cell multiplication all involve a *small* number of macromolecular events (e.g., the "replicon" theory). This should result in a Poissonian stochastic distribution of the initiation time which would completely prevent synchronization of cell division. Since synchronization *is* observed, it becomes reasonable to assume that a signal involving a *large* number of molecules takes part in the initiating event. A metabolic signal—which would also couple this initiation to the internal state of the cell—seems quite suitable for such a purpose.

Assuming the existence of such a metabolic signal, several aspects can be described, keeping within the Popperian doctrine often quoted by Jacques Monod:

> We hold that the ideal [of science] can be realized, very simply, by recognizing that the rationality of sciences lies not in its habit of appealing to empirical evidence in support of its dogma—astrologers do so too—but solely in the *critical approach:* in an attitude which, of course, involves the critical use, among other arguments, of empirical evidence (especially in refutations). For us, therefore, science has nothing to do with the quest for certainty or probability or reliability. We are not interested in establishing scientific theories as secure, or certain, or probable. Conscious of our fallibility we are only interested in criticizing them and testing them, hoping to find out where we are mistaken; of learning from our mistakes; and, if we are lucky, of proceeding to better theories.

This means that one does make conjectures and designs experiments capable of refuting them. For instance: How is metabolism organized: acentered or hierarchical? A hierarchy seems most stable and proper, thus the metabolite which controls cell multiplication will derive from metabolic pathways, independent of each other and organized as a hierarchy. Will the control be positive or negative? A teleological line of reasoning will suggest that a negative control is genetically stable when the system under control is rarely employed, whereas a positive control is stable if the system is often used. Thus one will conjecture that the metabolic signal behaves like cyclic AMP. What will be the systems under control? Most probably it is the general cell machinery, which suggests that stable RNA synthesis would reflect the behavior of this control metabolite. What experiments can be designed to detect the metabolic pathways involved in this control? At least two different pathways are involved, corresponding to two different pools of metabolites: an abrupt shift of one of these should both increase the synthesis of the control metabolite and induce starvation in the other pool(s). Then one is led to conclude that the cell must be somehow protected against such an unbalanced effect. The stringent control which couples RNA synthesis to protein synthesis seems a good candidate for such a protecting device. Therefore it appears that strains mutated in this control (the so-called "relaxed" mutants) would be most suitable for this study.

One could multiply the type of questions which can be asked, and the way they are answered. These were the questions I was trying to pose and answer, submitting all the problems for comment to Jacques Monod, just a short time before his death. As he always did, he offered new conjectures and then proposed experiments to refute them.

Although exposed in a brief and awkward fashion, I would like this text to be a tribute to his memory. Up to his last moments, and despite his numerous occupations, he remained available for discussion, and when he did not have free time, he asked me to write the problems in a letter to him, so that he could think about them. This is most remarkable in view of the usual attitude of many scientists. Monod very much liked *ideas*, and I would like, once again, to use the very words of K. Popper, whom Monod liked so much.

> My answer to the questions, "How do you know? What is the source of the basis for your assertion? What observations have led you to it?" would be: I do *not* know: my assertion was merely a guess. Never mind the source, or the sources, from which it may spring. There are many possible sources, and I may not be aware of half of them; and origins and pedigrees have in any case little bearing upon truth. But if you are interested in the problem which I tried to solve by my tentative assertion, you may help me by criticizing it as severely as you can; and if you can design some experimental test which you think might refute my assertion, I shall gladly, and to the best of my powers, help you to refute it.

Antoine Danchin

Antoine Danchin created a laboratory at the Institut Pasteur in 1983 (within the unit headed by Agnes Ullmann) and then created the Unit of Regulation of Gene Expression (1986–2000). In 2000 he went to Hong Kong to create the HKU-Pasteur Research Centre, which he left in 2003 to resume his activity as the head of the Unit of Genetics of Bacterial Genomes at the Institut Pasteur in Paris. After a few years of fascinating collaborations that started with the cloning of adenylate cyclase toxins from *Bordetella pertussis* and *Bacillus anthracis*, he fully embarked on the initiation of the *Bacillus subtilis* genome sequencing project, with strong emphasis on collaboration between in vivo experiments and experiments involving computers ("in silico" genome analysis). This project led to the discovery that a very large number of genes in all organisms do not have recognized functions, and this led Dr. Danchin to go back to study the metabolism of small molecules, sulfur-containing molecules in particular, as involved in the link between the architecture of the cell and that of the genome. This work is summarized in his book on the rationale for genome studies, *The Delphic Boat: What Genomes Tell Us* (translated by Alison Quayle; Harvard University Press, 2003).

Cannes, May 1976. (Photograph by J. Kosinski.)

From Enzymatic Adaptation to Allosteric Transitions*

Nobel Lecture, December 11, 1965

Jacques Monod

O ne day, almost exactly 25 years ago—it was at the beginning of the
bleak winter of 1940—I entered André Lwoff's office at the Pasteur
Institute. I wanted to discuss with him some of the rather surprising observations I had recently made.

I was working then at the old Sorbonne, in an ancient laboratory that
opened on a gallery full of stuffed monkeys. Demobilized in August in the
Free Zone after the disaster of 1940, I had succeeded in locating my family
living in the Northern Zone and had resumed my work with desperate
eagerness. I interrupted work from time to time only to help circulate
the first clandestine tracts. I wanted to complete as quickly as possible
my doctoral dissertation, which, under the strongly biometric influence
of Georges Teissier, I had devoted to the study of the kinetics of bacterial
growth. Having determined the constants of growth in the presence of
different carbohydrates, it occurred to me that it would be interesting to
determine the same constants in paired mixtures of carbohydrates. From
the first experiment on, I noticed that, whereas the growth was kinetically
normal in the presence of certain mixtures (that is, it exhibited a single
exponential phase), two complete growth cycles could be observed in other
carbohydrate mixtures, these cycles consisting of two exponential phases
separated by a complete cessation of growth (Fig. l).

Lwoff, after considering this strange result for a moment, said to me,
"That could have something to do with enzyme adaptation."

"Enzyme adaptation? Never heard of it!" I said.

* ©The Nobel Foundation 1965.

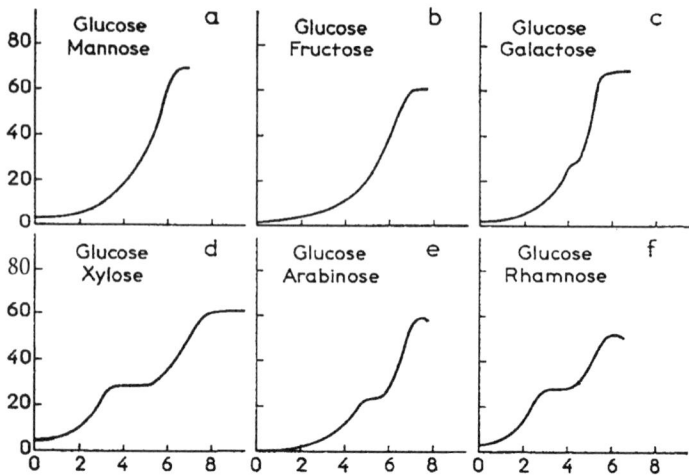

FIG. 1. Growth of *Esherichia coli* in the presence of different carbohydrate pairs serving as the only source of carbon in a synthetic medium (50).

Lwoff's only reply was to give me a copy of the then recent work of Marjorie Stephenson, in which a chapter summarized with great insight the still few studies concerning this phenomenon, which had been discovered by Duclaux at the end of the last century. Studied by Dienert and by Went as early as 1901 and then by Euler and Josephson, it was more or less rediscovered by Karström, who should be credited with giving it a name and attracting attention to its existence. Marjorie Stephenson and her students Yudkin and Gale had published several papers on this subject before 1940. (See ref. 1 for a bibliography of papers published prior to 1940.)

Lwoff's intuition was correct. The phenomenon of "diauxy" that I had discovered was indeed closely related to enzyme adaptation, as my experiments, included in the second part of my doctoral dissertation, soon convinced me. It was actually a case of the "glucose effect" discovered by Dienert as early as 1900, today better known as "catabolic repression" from the studies of Magasanik (2).

The die was cast. Since that day in December 1940, all my scientific activity has been devoted to the study of this phenomenon. During the Occupation, working, at times secretly, in Lwoff's laboratory, where I was warmly received, I succeeded in carrying out some experiments that were very significant for me. I proved, for example, that agents that uncouple oxidative phosphorylation, such as 2,4-dinitrophenol, completely inhibit adaptation to lactose or other carbohydrates (3). This suggested that "adaptation" implied an expenditure of chemical potential and therefore probably involved the true synthesis of an enzyme. With Alice Audureau, I sought to discover the still quite obscure relations between this phenomenon and the one Massini, Lewis, and others had discovered: the appearance and selection of "spontaneous" mutants (see ref. 1). Using a strain of *Escherichia coli mutabile* (to which we had given the initials ML because it had been isolated from André Lwoff's intestinal tract), we showed that an apparently spontaneous mutation was allowing these originally "lactose-negative" bacteria to become "lactose-positive." However, we proved that the original strain (*Lac⁻*) and the mutant strain (*Lac⁺*) did not differ from each other by the presence of a specific enzyme system, but rather by the ability to produce this system in the presence of lactose. In other words, the mutation affected a truly genetic property that became evident only in the presence of lactose (4).

There was nothing new about this; geneticists had known for a long time that certain genotypes are not always expressed. However, this mutation involved the selective control of an enzyme by a gene, and the conditions necessary for its expression seemed directly linked to the chemical activity of the system. This relation fascinated me. Influenced by my friendship with and admiration for Louis Rapkine, whom I visited frequently and at length in his laboratory, I had been tempted, even though I was poorly prepared, to study elementary biochemical mechanisms, that is, enzymology. But under the influence of another friend whom I admired, Boris Ephrussi, I was equally tempted by genetics. Thanks to him and to the Rockefeller Foundation, I had had an opportunity some years previously to visit Morgan's laboratory at the California Institute of Technology. This was a revelation for me—a revelation of genetics, at that time practically unknown in France; a revelation of what a group of scientists could be like when engaged in creative activity and sharing in a constant exchange of ideas, bold speculations, and strong criticisms. It was a revelation of personalities of great stature, such as George Beadle, Sterling Emerson, Bridges, Sturtevant, Jack Schultz, and Ephrussi, all of whom were then working in Morgan's department. Upon my return to France, I had again taken up the study of bacterial growth. But my mind remained full of the concepts of genetics and I was confident of its ability to analyze and convinced that one day these ideas would be applied to bacteria.

"Discovery" of Bacterial Genetics

Toward the end of the war, while still in the army, I discovered in an American army bookmobile several miscellaneous issues of *Genetics*, one containing the beautiful paper in which Luria and Delbrück (5) demonstrated for the first time rigorously, the spontaneous nature of certain bacterial mutants. I think I have never read a scientific article with such enthusiasm; for me, bacterial genetics was established. Several months later, I also "discovered" the paper by Avery, MacLeod, and McCarty (6)—another fundamental revelation. From then on I read avidly the first publications by the "phage-church," and when I entered Lwoff's department at the Pasteur Institute in 1945, I was tempted to abandon enzyme adaptation in order to join the church myself and work with bacteriophage. In 1946 I attended the memorable symposium at Cold Spring Harbor where Delbrück and Bailey, and Hershey, revealed their discovery of virus recombination at the same time that Lederberg and Tatum announced their discovery of bacterial sexuality (7). In 1947 I was invited to the Growth Symposium to present a report (1) on enzyme adaptation, which had begun to arouse the interest of embryologists as well as of geneticists. Preparation of this report was to be decisive for me. In reviewing all the literature, including my own, it became clear to me that this remarkable phenomenon was almost entirely shrouded in mystery. On the other hand, by its regularity, its specificity, and by the molecular-level interaction it exhibited between a genetic determinant and a chemical determinant, it seemed of such interest and of a significance so profound that there was no longer any question as to whether I should pursue its study. But I also saw that it would be necessary to make a clean sweep and start all over again from the beginning.

The central problem posed was that of the respective roles of the inducing substrate and of the specific gene (or genes) in the formation and the structure of the enzyme. In order to understand how this problem was considered in 1946, it would be well to remember that at that time the structure of DNA was not known, little was known about the structure of proteins, and nothing was known of their biosynthesis. It was necessary to resolve the following question: Does the inducer effect total synthesis of a new protein molecule from its precursors, or is it rather a matter of the activation, conversion, or "remodeling" of one or more precursors?

This required first of all that the systems to be studied be carefully chosen and defined. With Madeleine Jolit and Anne-Marie Torriani, we isolated β-galactosidase, then the amylomaltase of Escherichia coli (8). Our work was advanced greatly by the valuable collaboration of Melvin Cohn, an excellent immunologist, who knew better than I the chemistry of proteins. He knew, for example, how to operate that marvelous apparatus

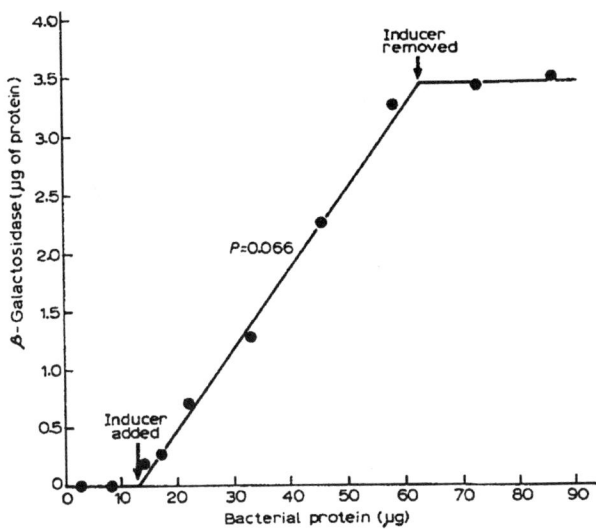

FIG. 2. Induced biosynthesis of β-galactosidase in *Escherichia coli*. The increase in enzyme activity is expressed not as a function of time but as a function of the concomitant growth of bacterial proteins. The slope of the resulting curve (P) indicates the differential rate of synthesis (11).

that had intimidated me, the "Tiselius" (9). With Anne-Marie Torriani, he characterized β-galactosidase as an antigen (10). Being familiar with the system, we could now study with precision the kinetics of its formation. A detailed study of the kinetics carried out in collaboration with Alvin Pappenheimer and Germaine Cohen-Bazire (11) strongly suggested that the inducing effect of the substrate entailed total biosynthesis of the protein from amino acids (Fig. 2). This interpretation seemed surprising enough at that time, but from the first, I must say, it won my firm belief. There is in science, however, quite a gap between belief and certainty. But would one ever have the patience to wait and to establish the certainty if the inner conviction were not already there?

We were to establish certainty a little later, thanks to some experiments with isotopic tracers done by Hogness, Cohn, and myself (12). To tell the truth, the results of these labeling experiments were even more surprising in view of the ideas then current on the biosynthesis of proteins and their state within the cell. The work of Schoenheimer (13) had actually persuaded most biochemists that in an organism proteins are inherently in a "dynamic state," each molecule being perpetually destroyed and reconstructed by exchange of amino acid residues. Our experiments, however, showed that

β-galactosidase is entirely stable in vivo, as are other bacterial proteins, under conditions of normal growth. They did not, of course, contradict the results of Schoenheimer, but very seriously questioned their interpretation and the dogma of the "dynamic state." Be that as it may, these conclusions were invaluable to us. We knew, thenceforth, that "enzyme adaptation" actually corresponds to the total biosynthesis of a stable molecule and that, consequently, the increase of enzyme activity in the course of induction is an authentic measure of the synthesis of the specific protein.

These results took on even more significance as our system became more accessible to experiment. With Germaine Cohen-Bazire and Melvin Cohn (14, 15), I was able to continue the systematic examination of a question I had repeatedly encountered: the correlations between the specificity of action of an inducible enzyme and the specificity of its induction. Pollock's pertinent observations on the induction of penicillinase by penicillin (16) made it necessary to consider this problem in a new way. We conducted a study of a large number of galactosides or their derivatives, comparing their properties as inducers, substrates, or as antagonists of the substrates of the enzyme, once more reaching a quite surprising conclusion, namely, that inductive ability is by no means a prerogative of the substrates of the enzyme, or even of the substances capable of forming the most stable complexes with it. For example, certain thiogalactosides, not hydrolyzed by the enzyme or used metabolically, appeared to be very powerful inducers. Certain substrates, on the other hand, were not inducers. The conclusion became obvious that the inducer did not act (as frequently assumed) either as a substrate or through combination with preformed active enzyme, but rather at the level of another specific cellular constituent that would one day have to be identified (Fig. 3).

Generalized Induction

In the course of this work, we observed a fact that seemed very significant. A certain compound, phenyl-β-D-thiogalactoside, devoid of inductive capacity, proved capable of counteracting the action of an effective inducer, such as methyl-β-D-thiogalactoside. This suggested the possibility of utilizing such "anti-induction" effects to prove a theory that we called, somewhat ambitiously, "generalized induction." From the very beginning of my research, I had been preoccupied with the problem posed by the existence, together with inducible enzymes, of "constitutive" systems; in other words (according to the then current definition) systems synthesized in the absence of any substrate or exogenous inducer, as is the case, of course,

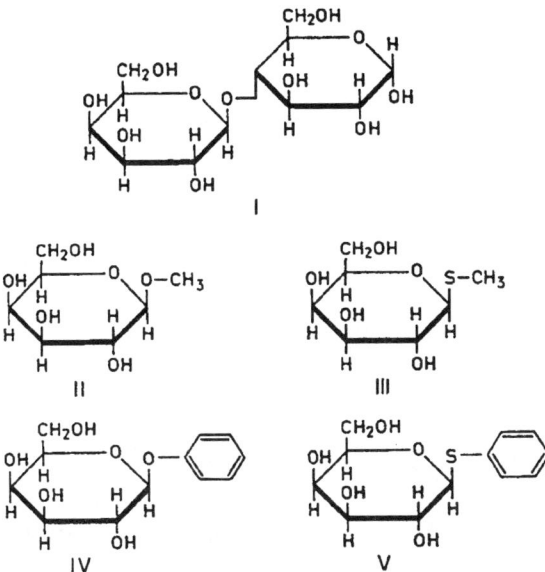

FIG. 3. Comparison of various β-galactosides as substrates and as inducers of β-galactosidase. I, Lactose: substrate of the enzyme, but deprived of inductive activity. II, Methyl-β-D-galactoside: low-affinity substrate effective inducer. III, Methyl-β-D-thiogalactoside: not hydrolyzable by the enzyme, but a powerful inducer. IV, Phenyl-β-D-galactoside: excellent enzyme substrate, high affinity, no inductive ability. V, Phenyl-β-D-thiogalactoside: no activity either as a substrate or as an inducer, but capable of acting as an antagonist of the inducer.

with all the enzymes of intermediate and biosynthetic metabolism. It did not seem unreasonable to suppose that the synthesis of these enzymes was controlled by their endogenous substrate, which would imply that the mechanism of induction is in reality universal. We were encouraged in this hypothesis by the work of Roger Stanier on the supposedly sequential induction of systems attacking phenolic compounds in *Pseudomonas*.

I sought, therefore, along with Germaine Cohen-Bazire, to prove that the biosynthesis of a typically "constitutive" enzyme (according to the ideas of the time), tryptophan synthetase, could be inhibited by an analogue of the presumed substrate. The reaction product seemed a good candidate for an analogue of the substrate, and we were soon able to prove that tryptophan and 5-methyltryptophan are powerful inhibitors of the biosynthesis of the enzyme. This was the first known example of a "repressible" system—discovered, it turned out, as proof of a false hypothesis (17).

I did not have, I must say, complete confidence in the ambitious theory of "generalized" induction, which soon encountered various difficulties. I was, however, encouraged by an interesting observation made by Vogel and Davis (18) concerning another enzyme, acetylornithinase, involved in the formation of arginine. Using a mutant requiring arginine or N-acetylornithine, Vogel and Davis found that, when the bacteria are cultivated in the presence of arginine, they do not produce acetylornithinase, whereas when they are cultivated in the presence of N-acetylornithine, acetylornithinase is synthesized. Hence these authors concluded that this enzyme must be induced by its substrate, N-acetylornithine. When Henry Vogel was passing through Paris, I drew his attention to the fact that their very interesting observations could just as well be explained as resulting from an inhibitory effect of arginine as from an inductive effect of acetylornithine. In order to resolve this problem, it was necessary to study the biosynthesis of the enzyme in a mixture of the two metabolites. The experiment proved that it is indeed a question of an inhibiting effect rather than an inductive effect. Vogel, quite rightly, proposed the term "repression" to designate this effect and thus established "repressible" systems alongside of "inducible" systems. Later on, thanks especially to the studies of Maas, Gorini, Pardee, Magasanik, Cohen, Ames, and many others (see ref. 19 for literature), the field of repressible systems was considerably extended; it is now generally accepted that practically all bacterial biosynthetic systems are controlled by such mechanisms.

Nevertheless, I remained faithful to the study of the β-galactosidase of *Escherichia coli*, knowing well that we were far from having exhausted the resources of this system. During the years spent in establishing the biochemical nature of the phenomenon, I had been able only partially to approach the question of its genetic control—enough, however, to convince me that it was extremely specific and that it justified the idea that Beadle and Tatum's postulate, "one gene-one enzyme," was applicable to inducible and degradative enzymes as well as to the enzymes of biosynthesis, which the Stanford school had principally studied. These conclusions led me to abandon an idea I had adopted as a working hypothesis—that is, that many different inducible enzymes may result from the "conversion" of a single precursor whose synthesis is controlled by a single gene; this hypothesis was also contradicted by the results of our experiments with tracers.

But genetic analysis once more encountered grave difficulties. First, the low frequency of recombination, in the systems of conjugation known at that time, did not permit fine genetic analysis. Another difficulty holding us back was the existence of some mysterious phenotypes; certain mutants ("cryptic"), incapable of metabolizing the galactosides, nevertheless appeared capable of synthesizing β-galactosidase. The solution to this problem came to us by accident while we were looking for something entirely different. In 1954, when the chairmanship of the new Department of Cellular Biochemistry had just been bestowed upon me, Georges Cohen joined us, and I suggested to him, and simultaneously to Howard Rickenberg, to make use of the properties of thiogalactosides as gratuitous inducers in attempting to study their fate in inducible bacteria, employing a thiogalactoside labeled with carbon-14. We noted that the radioactivity associated with the galactoside accumulated rapidly in wild-type induced bacteria, but not in the so-called cryptic mutants. Neither did the radioactivity accumulate in wild-type bacteria not previously induced. The capacity for accumulation depended, therefore, on an inducible factor. Study of the kinetics, of the specificity of action, and of the specificity of induction of this system, as well as the comparison of various mutants, led us to the conclusion that the element responsible for this accumulation could only be a specific protein whose synthesis, governed by a gene (y) distinct from that of galactosidase (z), was induced by the galactosides at the same time as the synthesis of the enzyme. To this protein we gave the name "galactoside permease" (20, 21) (Fig. 4).

The very existence of a specific protein responsible for the permeation and accumulation of galactosides was occasionally put in doubt because the evidence for it was based entirely on observations *in vivo*. Some of the researchers who did not really doubt its existence still reproached me from time to time for giving a name to a protein when it had not been isolated. This attitude reminded me of that of two traditional English gentlemen who, even if they know each other well by name and by reputation, will not speak to each other before having been formally introduced. On my part, I never for a moment doubted the existence of this protein, for our results could be interpreted in no other way. Nevertheless, I was only too happy to learn, recently, that by a recent series of experiments, Kennedy has identified *in vitro* and isolated the specific inducible protein, galactoside permease (22). Kennedy was brilliantly successful where we had failed, for we had repeatedly sought to isolate galactoside permease *in vitro*. These efforts of ours, however, were not in vain, since they led Irving Zabin, Adam Kepes, and myself to isolate still another protein, galactoside transacetylase (23). For several weeks we believed that this enzyme was none other than the permease itself. This was an erroneous assumption, and the physiological function of this protein is still totally unknown. It was a profitable discovery, nevertheless, because the transacetylase, determined by a gene

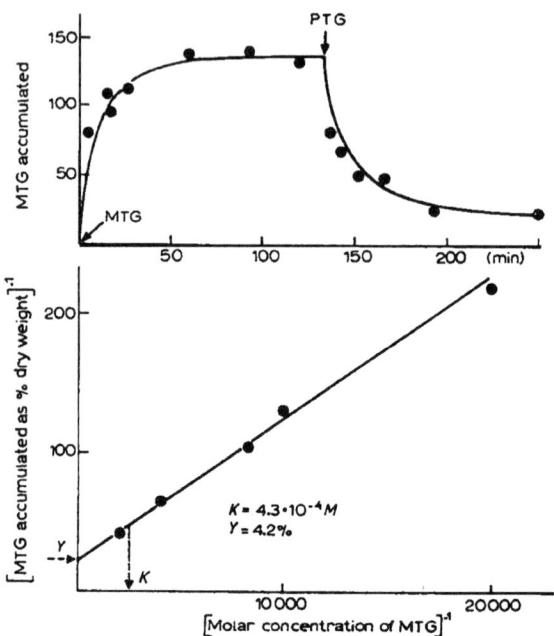

FIG. 4. Evidence for the existence of galactoside permease. (Top) Accumulation of labeled methyl-β-D-thiogalactoside (MTG) by a suspension of previously induced bacteria. Displacement of accumulated galactoside (phenyl-β-D-thiogalactoside, PTG). (Bottom) Accumulation of a galactoside in previously induced bacteria as a function of the concentration of the external galactoside. Inverse coordinates: The constants K and Y define, respectively, the constant of apparent dissociation and the constant of apparent activity of the system of accumulation (21).

belonging to the lactose operon, has been very useful to experimenters, if not to the bacterium itself.

The study of galactoside permease was to reveal another fact of great significance. Several years earlier, following Lederberg's work, we had isolated some "constitutive" mutants of β-galactosidase, that is, strains in which the enzyme was synthesized in the absence of any galactoside. But we now proved that the constitutive mutation has a pleiotropic effect. In these mutants, galactoside permease as well as galactosidase (and the transacetylase) were indeed simultaneously constitutive, whereas we knew on the other hand that each of the three proteins is controlled by a distinct gene. We then had to admit that a constitutive mutation, although very strongly linked to the loci governing galactosidase, galactoside permease, and transacetylase, had taken place in a gene (*i*) distinct from the other three (z, y, and Ac), and that the relationship of this gene to the three proteins violated the postulate of Beadle and Tatum.

New Perspectives

These investigations were given new meaning by the perspectives opened to biology around 1955. It was in 1953 that Watson and Crick, on the basis of observations made by Chargaff and Wilkins, proposed their model of the structure of DNA. From the first, in this complementary double sequence, one could see a mechanism for exact replication of the genetic material. Meanwhile, one year earlier, Sanger had described the peptide sequence of insulin, and it was also already known, from the work of Pauling and Itano (24) in particular, that a genetic mutation can cause a limited modification in the structure of a protein. In 1954, Crick and Watson (25) and Gamow (26) proposed the genetic code theory: The primary structure of proteins is determined and defined by the linear sequence of the nucleotides in DNA. Thus the profound logical intuition of Watson and Crick had allowed them to discover a structure that immediately explained, at least in principle, the two essential functions long assigned by geneticists to hereditary factors: to control its own synthesis and to control that of the nongenetic constituents. Molecular biology had been born, and I realized that, like Monsieur Jourdain, I had been doing molecular biology for a long time without knowing it.

More than ten years have elapsed since then, and the ideas whose hatching I recall here were then far from finding a uniformly enthusiastic audience. My conviction, however, had been established long before absolute certainty could be acquired. This certainty exists today, thanks to a succession of discoveries, some of them almost unhoped for, that have enriched our discipline since that time

Once the physiological relations of galactosidase and galactoside permease were understood, and once it was proved that they depend on two distinct genetic elements while remaining subject to the same induction determinism and to the same constitutive mutations, it became imperative to analyze the corresponding genetic structures. In particular, the expression of these genes and the relations of dominance between their alleles had to be studied in detail.

Precisely at this time, the work of Jacob and Wollman (27) had clarified the mechanism of bacterial conjugation; we knew that this conjugation consists of the injection, without cytoplasmic fusion, of the chromosome of a male bacterium into a female. It was even possible to follow the kinetic of penetration of a given gene. I decided, along with Arthur Pardee and François Jacob, to use these new experimental tools to follow the "expression" of the z^+ and i^+ genes injected into a female carrying mutant alleles of these genes.

This difficult undertaking, carried out successfully thanks to the experimental talent of Arthur Pardee, brought about two remarkable and at least partially unexpected results. First, the z gene (which we knew to be the determinant of the structure) is expressed (by the synthesis of β-galactosidase) very fast and at maximum rate from the beginning. I will pass over the development and the consequences of this observation, which was one of the sources of the messenger theory. Second, the inducible allele of the i gene is dominant with respect to the constitutive allele, but this dominance is expressed very slowly. Everything seemed to indicate that this gene is responsible for the synthesis of a product that inhibits, or represses, the biosynthesis of the enzyme. This was the reason for designating the product of the gene as a "repressor" and hypothesizing that the inducer acts not by provoking the synthesis of the enzyme but by "inhibiting an inhibitor" of this synthesis (28).

Theory of Double Bluff

Of course I had learned, like any schoolboy, that two negatives are equivalent to a positive statement, and Melvin Cohn and I, without taking it too seriously, debated this logical possibility that we called the "theory of double bluff", recalling the subtle analysis of poker by Edgar Allan Poe.

I see today, however, more clearly than ever, how blind I was in not taking this hypothesis seriously sooner, since several years earlier we had discovered that tryptophan inhibits the synthesis of tryptophan synthetase; also, the subsequent work of Vogel, Gorini, Maas, and others (cited in ref. 15) showed that repression is not due, as we had thought, to an anti-induction effect. I had always hoped that the regulation of "constitutive" and inducible systems would be explained one day by a similar mechanism. Why not suppose, then, since the existence of repressible systems and their extreme generality were now proven, that induction could be effected by an anti-repressor rather than by repression by an anti-inducer? This is precisely the thesis that Leo Szilard, while passing through Paris, happened to propose to us during a seminar. We had only recently obtained the first results of the injection experiment, and we were still not sure about its interpretation. I saw that our preliminary observations confirmed

Szilard's penetrating intuition, and when he had finished his presentation, my doubts about the "theory of double bluff" had been removed and my faith established—once again a long time before I would be able to achieve certainty.

Some of the more important developments of this study, such as the discovery of operator mutants and of the operon, considered as a single coordinated expression of the genetic material, and the bases and demonstration of the messenger theory, have been presented by François Jacob in his lecture (27), and I will not pause over these, in order to return to that constituent whose existence and role had so long escaped me, the repressor. To tell the truth, I find some excuses for myself even now. It was not easy to get away completely from the quite natural idea that a structural relation, inherent in the mechanism of the phenomenon of induction, must exist between the inducer of an enzyme and the enzyme itself. And I must admit that, up until 1957, I tried to "rescue" this hypothesis, even at the price of reducing almost to nothing the "didactic" role (as Lederberg would say) of the inducer.

From now on it was necessary to reject it completely. An experiment carried out in collaboration with David Perrin and Francois Jacob proved, moreover, that the mechanism of induction functioned perfectly in certain mutants, producing a modified galactosidase totally lacking in affinity for galactoside (29).

What now had to be analyzed and understood were the interactions of the repressor with the inducer on the one hand, with the operator on the other. Otto Warburg said once, about cytochrome oxidase, that this protein—or presumed protein—was as inaccessible as the matter of the stars. What is to be said, then, of the repressor, which is known only by the results of its interactions? In this respect we are in a position somewhat similar to that of the police inspector who, finding a corpse with a dagger in its back, deduces that somewhere there is an assassin; but as for knowing who the assassin is, what his name is, whether he is tall or short, dark or fair, that is another matter. The police in this case, it seems, sometimes get results by sketching a composite portrait of the culprit from several clues. This is what I am going to try to do now with regard to the repressor.

First, it is necessary to assign to the assassin—I mean the repressor— two properties: the ability to recognize the inducer and the ability to recognize the operator. These recognitions are necessarily steric functions and are thus susceptible to being modified or abolished by mutation. Loss of the ability to recognize the operator would result in total derepression of the system. Every mutation that causes a shift in the structure of the repressor or the abolition of its synthesis must therefore appear "constitutive," and this is without doubt the reason for the relatively high frequency of this type of mutation.

However, if the composite portrait is correct, it can be seen that certain mutations might abolish the repressor's ability to recognize the inducer but leave unaffected its ability to recognize the operator. Such mutations should exhibit a very special phenotype. They would be noninducible (that is, lactose-negative), and in diploids they would be dominant in *cis* as well as in *trans*. Clyde Willson, David Perrin, Melvin Cohn, and I (30) were able to isolate two mutants that possessed precisely these properties, and Suzanne Bourgeois (ref. 31) has recently isolated a score of others.

In tracing this first sketch of the composite portrait, I implicitly supposed that there was only one assassin; that is, the characteristics of the system were explained by the action of a single molecular species, the repressor, produced from gene i. This hypothesis is not necessary *a priori*. It could be supposed, for example, that the recognition of the inducer is due to another constituent distinct from that which recognizes the operator. Then we would have to assume that these two constituents could recognize each other. Today this latter hypothesis seems to be practically ruled out by the experiments of Bourgeois, Cohn, and Orgel (31), which show, among other important results, that the mutation of type i^- (unable to recognize the operator) and the mutations of the type i^s (unable to recognize the inducer) occur in the same cistron and, from all appearances, involve the same molecule, a unique product of the regulator gene i.

An essential question is the chemical nature of the repressor. Inasmuch as it seems to act directly at the level of the DNA, it seemed logical to assume that it could be a polyribonucleotide whose association with a DNA sequence would take place by means of specific pairing. Although such an assumption could explain the recognition of the operator, it could not explain the recognition of the inducer, because probably only proteins are able to form a stereospecific complex with a small molecule. This indicates that the repressor, that is, the active product of the gene i, must be a protein. This theory, based until now on purely logical considerations, has just received indirect but decisive confirmation.

It should be remembered that, thanks to the work of Benzer (32), Brenner (33), and Garen (34), a quite remarkable type of mutation has been recognized, called "nonsense" mutation. This mutation, as is well known, interrupts the reading of the messenger in the polypeptide chain. But on the other hand, certain "suppressors," today well identified, are able to restore the reading of the triplets (UAG and UAA) corresponding to the nonsense mutations. The fact that a given mutation may be restored by one of the carefully catalogued suppressors provides proof that the phenotype of the corresponding mutant is due to the interruption of the synthesis of a protein. Using this principle, Bourgeois, Cohn, and Orgel (31) showed that certain constitutive mutants of the gene i are nonsense mutants and that consequently the active product of this gene is a protein.

This result, which illustrates the surprising analytical ability of modern biochemical genetics, is of utmost importance. It must be emphasized that, with respect to the suppression of a constitutive mutant (i^-), it shows that the recognition of the operator (as well as recognition of the inducer) is linked to the structure of the protein produced by the gene i.

The problem of the molecular mechanism that permits this protein to play the role of relay between the inducer and the operator still remains. Until now this problem has been inaccessible to direct experimentation, in that the repressor itself remains to be isolated and studied *in vitro*. However, in conclusion, I would like to explain why and how this inaccessibility was itself the source of new preoccupations that we hope will be fruitful.

First of all, it should be recalled that we had tried repeatedly, even before the existence of the repressor was demonstrated, to learn something of the mode of action of the inducer by following its tracks *in vivo* with radioactive markers. One after the other, Georges Cohen, François Gros, and Agnes Ullmann engaged in this approach, using different fractionation techniques. Some of these experiments led to some unexpected and important discoveries, such as that of galactoside permease and galactoside transacetylase. But concerning the way in which galactosides act as inducers, the results were completely negative. Nothing whatever indicated that the inductive interaction is accompanied by a chemical change, however transient, or by any kind of covalent reaction in the inducer itself. The kinetics of induction, elaborated on in the elegant work of Kepes (35, 36), also revealed that the inductive interaction is extremely rapid and completely reversible (Fig. 5).

This is quite a remarkable phenomenon, if one thinks of it, since this noncovalent, reversible stereospecific interaction—an interaction that in all probability involves only a few molecules and can involve only a very small amount of energy—triggers the complex transcription mechanism of the operon, the reading of the message, and the synthesis of three proteins, leading to the formation of several thousand peptide links. During this entire process, the inducer acts, it seems, exclusively as a chemical signal,

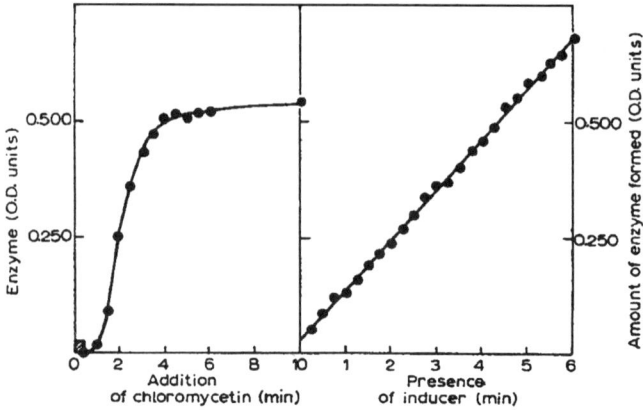

FIG. 5. Kinetics of the synthesis of galactosidase after a short period of induction. Left: Inducer added at time zero. Inducer eliminated after a time corresponding to the width of the cross-hatched rectangle. On the ordinates: accumulation of the enzyme. Right: Total amount of enzyme formed (asymptote of the curve at the left) as a function of the duration of the presence of the inducer. The linear relation obtained indicates that the inductive interaction is practically immediate and reversible (35).

recognized by the repressor, but without directly participating in any of the reactions which it initiates.

One would be inclined to consider such an interpretation of the inductive interaction as highly unlikely if one did not know today of numerous examples in which similar mechanisms participate in the regulation of the activity as well as the synthesis of certain enzymes. It was as a possible model of inductive interactions that Jacob, Changeux, and I first became interested in regulatory enzymes (37). The first example of such an enzyme was undoubtedly phosphorylase *b* from rabbit muscle; as Coris (38) and his group (39) showed, this enzyme is activated specifically by adenosine 5'-phosphate, although the nucleotide does not participate in the reaction in any way. We are indebted to Novick and Szilard (40), to Pardee (41), and to Umbarger (42) for their discovery of feedback inhibition, which regulates the metabolism of biosynthesis—their discovery led to a renewal of studies and demonstrated the extreme importance of these phenomena.

In a review that we devoted to these phenomena (43), a systematic comparison and analysis of the properties of some of the regulatory enzymes led us to conclude that, in most if not all cases, the observed effects were due to *indirect* interactions between distinct stereospecific receptors

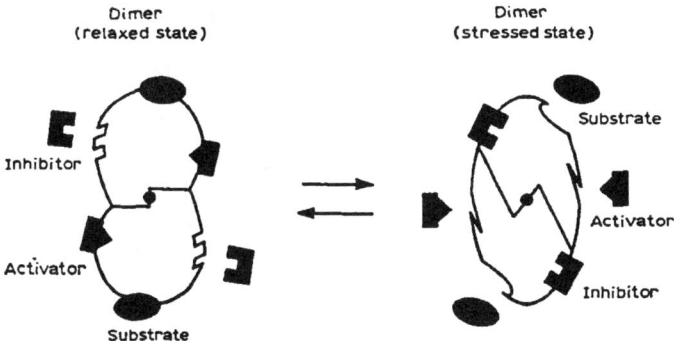

Inhibitor

Activator

Substrate

Substrate

Activator

Inhibitor

FIG. 6. Model of allosteric transition produced in a symmetrical dimer. In one of the two conformations, the protein can attach itself to the substrate as well as to the activating bond. In the other conformation, it can attach itself to the inhibiting bond.

on the surface of the protein molecule, these interactions being in all likelihood transmitted by means of conformational modifications induced or stabilized at the time of the formation of a complex between the enzyme and the specific agent—hence the name "allosteric effects," by which we proposed to distinguish this particular class of interactions, and the term "allosteric transitions," used to designate the modification undergone by the protein (Fig. 6).

By virtue of being indirect, the allosteric interactions do not depend on the structure or the particular chemical reactivity of the ligands themselves, but entirely on the structure of the protein, which acts as a relay. This is what confers upon these effects their profound significance. The metabolism, growth, and division of a cell require, obviously, not only the operation of the principal metabolic pathways—those through which pass the necessary energy and chemical materials—but also that the activity of the various metabolic pathways be closely and precisely coordinated by a network of appropriate specific interactions. The creation and development of such networks during the course of evolution obviously would have been impossible if only direct interactions at the surface of the protein had been used; such interactions would have been severely limited by chemical structure, the reactivity or lack of reactivity of metabolites among which the existence of an interaction could have been physiologically beneficial. The "invention" of indirect allosteric interactions, depending exclusively on the structure of the protein itself, that is on the genetic code, would have freed molecular evolution from this limitation (43).

The disadvantage of this concept is precisely that its ability to explain is so great that it excludes nothing, or nearly nothing; there is no physiological phenomenon so complex and mysterious that it cannot be disposed of, at least on paper, by means of a few allosteric transitions. I was very much in agreement with my friend Boris Magasanik, who remarked to me several years ago that this theory was the most decadent in biology.

It was all the more decadent because there was no *a priori* reason to suppose that allosteric transitions for different proteins need be of the same nature and obey the same rules. One might think that each allosteric system constituted a specific and unique solution to a given problem of regulation. However, as experimental data accumulated on various allosteric enzymes, surprising analogies were found among systems that had apparently nothing in common. In this respect, the comparison of independent observations by Gerhart and Pardee (44) on aspartate transcarbamylase and by Changeux (45) on threonine deaminase of *Escherichia coli* was especially impressive. By their very complexity, the interactions in these two systems presented unusual kinetic characteristics, almost paradoxical and yet quite analogous. Therefore it could not be doubted that the same basic solution to the problem of allosteric interactions had been found during evolution in both cases; it remained only for the researcher to try to discover it in his turn.

Among the properties common to these two systems, as well as to the great majority of known allosteric enzymes, the most significant seemed to us to be the fact that their saturation functions are not linear (as is the case for "classic" enzymes) but multimolecular. An example of such a pattern of saturation has been known for a long time: it is that of hemoglobin by oxygen (Fig. 7). Jeffries Wyman had noted several years earlier (46) that the symmetry of the saturation curves of hemoglobin by oxygen seemed to suggest the existence of a structural symmetry within the protein molecule itself; this idea was brilliantly confirmed by the work of Perutz (47).

These indications encouraged us—Wyman, Changeux, and myself—to look for a physical interpretation of the allosteric interactions in terms of molecular structure. This exploration led us to study the properties of a model defined in the main by the following postulates:

(1) An allosteric protein is made up of several identical subunits (protomers).
(2) The protomers are arranged in such a way that none can be distinguished from the others; this implies that there are one or more axes of molecular symmetry.

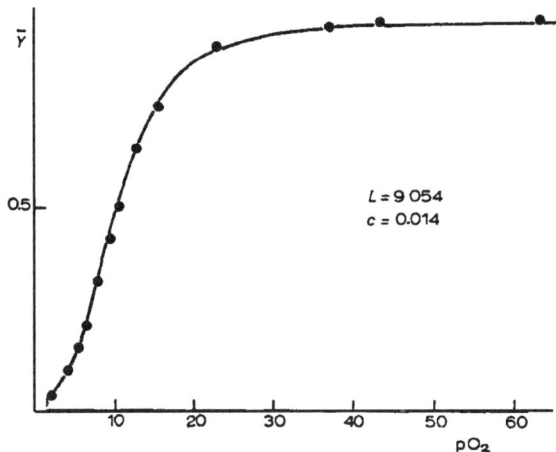

FIG. 7. Saturation of hemoglobin with oxygen. Abscissa: Partial pressure of O_2. Ordinate: Saturated fraction. The points correspond to experimental points (51). The interpolation curve was calculated from a theoretical model essentially similar to that of Fig. 6.

(3) Two (or more) conformational states are accessible to this protein.

(4) These conformational transitions tend to preserve the molecular symmetry, or, more generally, the equivalence of the protomers (48).

We were pleasantly surprised to find that this very simple model made it possible to explain, classify, and predict most of the kinetic properties, sometimes very complex in appearance, of many allosteric systems (Figs. 7 and 8). Obviously, this model represents only a first approximation in the description of real systems. It is not likely, moreover, that it represents the only solution to the problem of regulative interactions found during evolution; certain systems seem to function according to quite different principles (49), which will also need to be clarified.

However, the ambition of molecular biology is to interpret the essential properties of organisms in terms of molecular structures. This objective has already been achieved for DNA, and it is in sight for RNA, but it still seems very remote for the proteins. The model that we have studied is interesting primarily because it proposes a functional correlation between certain elements of the molecular structure of proteins and certain of their physiologic properties, specifically those that are significant at the level of integration, of dynamic organization, of metabolism. If the proposed correlation is experimentally verified, I would see an additional reason for having confidence in the development of our discipline which,

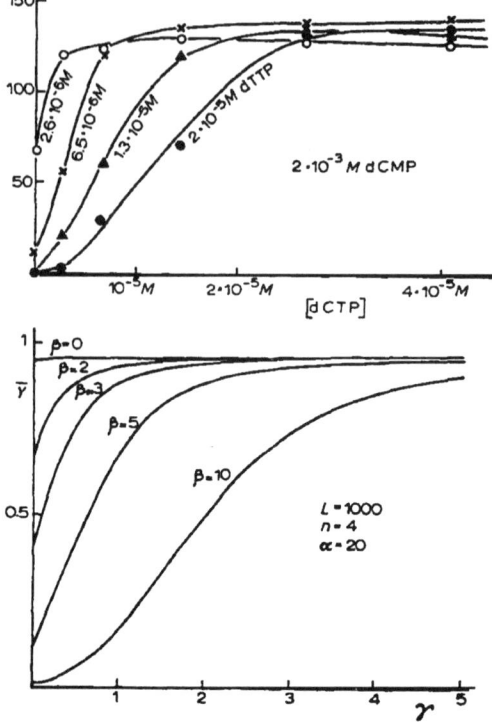

FIG. 8. Activity of deoxycytidine deaminase as a function of the concentration of the substrate (dCMP), of the activator (dCTP), and of the inhibitor (dTTP). (Top) Experimental results (from Scarano; see ref. 48). (Bottom) Theoretical curve calculated for a similar case according to the model of Monod, Wyman, and Changeux (48).

transcending its original domain, the chemistry of heredity, today is oriented toward the analysis of the more complex biological phenomena: the development of higher organisms and the operation of their networks of functional coordinations.

Acknowledgment

The research by my collaborators and myself since 1945 has been carried out entirely at the Pasteur Institute. This work has received decisive assistance from numerous institutions, in particular the Centre National de la Recherche Scientifique, the Rockefeller Foundation of New York, the National Science Foundation and the National Institutes of Health of the United States, the Jane Coffin Childs Memorial Fund, the Commissariat à l'Energie Atomique, and the Délégation Générale à la Recherche Scientifique et Technique. A donation by Mesdames Edouard de Rothschild and Bethsabée de Rothschild permitted, in large part, the establishment in 1954 of the Department of Cellular Biochemistry at the Pasteur Institute.

References

1. **Monod, J.** 1947. *Growth.* **11:**223.
2. **Magasanik, B.** 1965. *Mécanismes de régulation des activités cellulaires chez les microorganismes.* Centre National de la Recherche Scientifique, Paris, p. 179.
3. **Monod, J.** 1944. *Ann. Inst. Pasteur.* **70:**381.
4. **Monod, J. and Andureau, A.** 1944. *Ann. Inst. Pasteur.* **70:**381.
5. **Luria, S. E. and Delbrück, M.** 1943. *Genetics.* **28:**491.
6. **Avery, O. T., MacLeod, C. M., and McCarty, M.** 1944. *J. Exptl. Med.* **79:**409.
7. **Lederberg, J. and Tatum, E. L.** 1946. *Cold Spring Harbor Symp. Quant. Biol.* **11:**113.
8. **Monod, J., Torriani, A. M., and Gribetz, J.** 1948. *Compt. Rend.* **227:**315. J. Monod, *Intern. Congr. Biochem., 1st, Cambridge, 1949,* Abs. Commun., p. 308; *Unités biologiques douées de continuité génétique,* Centre National de la Recherche Scientifique, Paris, 1949, p. 181.
9. **Monod, J. and Cohn, M.** 1951. *Biochim. Biophys. Acta.* **7:**153.
10. **Cohn, M. and Torriani, A. M.** 1952. *J. Immunol.* **69:**471.
11. **Monod, J., Pappenheimer, A. M., and Cohen-Bazire, G.** 1952. *Biochim. Biophys. Acta.* **9:**648.
12. **Hogness, D. S., Cohn, M., and Monod, J.** 1955. *Biochim. Biophys. Acta.* **16:**99. J. Monod and M. Cohn, *Intern. Congr. Microbiol., 6th, Rome, 1953, Symp. Microbial Metabolism,* p. 42.
13. **Schoenheimer, R.** 1942. *The Dynamic State of Body Constituents.* Harvard Univ. Press, Cambridge.
14. **Monod, J., Cohen-Bazire, G., and Cohn, M.** 1951. *Biochim. Biophys. Acta.* **7:**585.
15. **Monod, J. and Cohn, M.** 1952. *Advan. Enzymol.* **13:**67.
16. **Pollock, M. R.** 1950. *Brit. J. Exptl. Pathol.* **31:**739.
17. **Monod, J. and Cohen-Bazire, G.** 1953. *Compt. Rend.* **236:**530. M. Cohn and J. Monod, *Adaptation in Microorganisms,* Cambridge Univ. Press, Cambridge, 1953, p. 132.
18. **Vogel, H. J. and Davis, B. D.** 1952. *Federation Proc.* **11:**485.
19. **Cohen, G. N.** 1965. *Ann. Rev. Microbiol.* **19:**105.
20. **Monod, J.** 1957. *Enzymes: Units of Biological Structure and Function.* Academic Press, New York, 1956, p. 7; G. N. Cohen and J. Monod, *Bacteriol. Rev.* **21:**169.
21. **Rickenberg, H. V., Cohen, G. N., Buttin, G., and Monod, J.** 1956. *Ann. Inst. Pasteur.* **91:**829.

22. **Fox, C. F. and Kennedy, E. P.** 1965. *Proc. Natl. Acad. Sci. (U.S.)* **54:**891.

23. **Zabin, I., Kepes, A., and Monod, J.** 1959. *Biochem. Biophys. Res. Commun.* **1:**289. 1962. *J. Biol. Chem.* **237:**253.

24. **Pauling, L., Itano, H. A., Singer, S. J., and Wells, I. C.** 1950. *Nature.* **166:**677.

25. **Watson, J. D.** 1964. The involvement of RNA in the synthesis of proteins; F. H. C. Crick, On the genetic code, in *Nobel Lectures, Physiology or Medicine, 1942–1962*, Elsevier, Amsterdam, 1964, pp. 785, 811.

26. **Gamow, G.** 1954. *Nature.* **173:**318.

27. **Jacob, F.** 1972. Genetics of the bacterial cell, in *Nobel Lectures, Physiology or Medicine, 1963–1970*, Elsevier, Amsterdam, p. 148.

28. **Pardee, A. B., Jacob, F., and Monod, J.** 1958. *Compt. Rend.* **246:**3125. **A. B. Pardee, F. Jacob, and J. Monod,** 1959. *J. Mol. Biol.* **1:**165. **F. Jacob and J. Monod,** 1959. *Compt. Rend.* **249:**1282.

29. **Perrin, D., Jacob, F., and Monod, J.** 1960. *Compt. Rend.* **250:**155.

30. **Willson, C., Perrin, D., Cohn, M., Jacob, F., and Monod, J.** 1964. *J. Mol. Biol.* **8:**582.

31. **Bourgeois, S., Cohn, M., and Orgel, L.** 1965. *J. Mol. Biol.* **14:**300.

32. **Benzer, S. and Charupe, S. P.** 1962. *Proc. Natl. Acad. Sci. (U.S.)* **48:**1114.

33. **Brenner, S., Stretton, A. O. W., and Kaplan, S.** 1965. *Nature.* **206:**994.

34. **Weigert, M. G. and Garen, A.** 1065. *Nature.* **206:**992.

35. **Kepes, A.** 1960. *Biochim. Biophys. Acta.* **40:**70.

36. **Kepes, A.** 1963. *Biochim. Biophys. Acta.* **76:**293. 1963. *Cold Spring Harbor Symp. Quant. Biol.* **28:**325.

37. **Monod, J. and Jacob, F.** 1961. *Cold Spring Harbor Symp. Quant. Biol.* **26:**389. J. P. Changeux, 1961. *Cold Spring Harbor Symp. Quant. Biol.* **26:**313.

38. **Cori, C. F.** et al., see references in C. F. Cori and G. T. Cori, Polysaccharide phosphorylase, in *Nobel Lectures, Physiology or Medicine, 1942–1962*, Elsevier, Amsterdam, 1964, p. 186.

39. **Hehnreich, E. and Cori, C. F.** 1964. *Proc. Natl. Acad. Sci. (U.S.)* **51:**131.

40. **Novick, A. and Szilard, L.** 1954. *Dynamics of Growth Process.* Princeton Univ. Press, Princeton, N.J., p. 21.

41. **Yates, R. A. and Pardee, A. B.** 1956. *J. Biol. Chem.* **221:**757.

42. **Umbarger, H. E.** 1956. *Science.* **123:**848.

43. **Monod, J., Changeux, J. P., and Jacob, F.** 1963. *J. Mol. Biol.* **6:**306.

44. **Gerhart, J. C. and Pardee, A. B.** 1961. *Federation Proc.* **20:**224. 1962. *J. Biol. Chem.* **237:**891. 1963. *Cold Spring Harbor Symp. Quant. Biol.* **28:**495. 1964. *Federation Proc.* **23:**727.

45. **Changeux, J. P.** 1961. *Cold Spring Harbor Symp. Quant. Biol.* **26**:303. 1962. *J. Mol. Biol.* **4**:220. 1964. *Bull. Soc. Chim. Biol.* **46**:927, 947, 1151; (1965) **47**: 115, 267, 281.

46. **Allen, D. W., Guthe, K. F., and Wyman, J.** 1950. *J. Biol. Chem.* **187**:393.

47. **Perutz, M. F.** X-ray analysis of haemoglobin, in *Nobel Lectures, Chemistry, 1942–1962*, Elsevier, Amsterdam, 1964, p. 653.

48. **Monod, J., Wyman, J., and Changeux, J. P.** 1965. *J. Mol. Biol.* **12**:88.

49. **Woolfolk, C. A. and Stadman, E. R.** 1964. *Biochem. Biophys. Res. Commun.* **17**:313.

50. **Monod, J.** 1941. *Recherches sur la croissance des cultures bactériennes*, Hermann, Paris.

51. **Lyster,** Unpublished results.

Scientific Publications by Jacques Monod*

1931

1. La formation de l'ébauche buccale postérieure chez les Ciliés en division et ses relations de continuité topographique et génétique avec la bouche antérieure. E. CHATTON, A. LWOFF, M. LWOFF and J.-L. MONOD. *C. R. Soc. Biol.*, 1931, **108**, 540–544.

2. Sur la topographie, la structure et la continuité génétique du système ciliaire de l'infusoire *Chilodon uncinatus*. E. CHATTON, A. LWOFF, M. LWOFF and J.-L. MONOD. *Bull. Soc. Zool. de France*, 1931, **66**, 367–374.

1933

3. Mise en évidence du gradient axial chez les infusoires ciliés par photolyse à l'aide de rayons ultraviolets. J. MONOD. *C. R. Acad. Sci.*, 1933, **196**, 212–214.

4. Données quantitatives sur le galvanotropisme des infusoires ciliés. J. MONOD. *Bull. Biol. Fr. et Belg.*, 1933, **67**, 474–479.

1934

5. Indépendance du galvanotropisme et de la densité du courant chez les infusoires ciliés. J. MONOD. *C. R. Acad. Sci.*, 1934, **198**, 122–124.

6. Sur le rôle des chlorelles symbiotiques dans la nutrition de *Paramecium bursaria*. I. GOLDBERG and J. MONOD. *C. R. Acad. Sci.*, 1934, **198**, 1183–1185.

7. Galvanotropisme et âge physiologique. J. MONOD. *C. R. Acad. Sci.*, 1934, **198**, 1882–1883.

* Reprinted with the kind permission of the Royal Society from the *Biographical Memoirs of Fellows of the Royal Society* (vol. 23, p. 385–412, 1977).

1935

8. Rapport préliminaire sur les observations d'histoire naturelle et de géographie physique. P. DRACH and J. MONOD. *Ann. Hydrographiques*, 1935, 3–11.

9. Le taux de croissance en fonction de la concentration de l'aliment dans une population de *Glaucoma piriformis* en culture pure. J. MONOD. *C. R. Acad. Sci.*, 1935, **201**, 1513–1515.

1936

10. La concentration de l'aliment, facteur quantitatif de l'accroissement des populations d'infusoires. J. MONOD and G. TEISSIER. *C. R. Acad. Sci.*, 1936, **202**, 162–164.

1937

11. Specific reactions of the ovary to interspecific transplantation among members of the *Melanogaster* group of *Drosophila*. J. MONOD and D. F. POULSON. *Genetics*, 1937, **22**, 257–263.

12. Ration d'entretien et ration de croissance dans les populations bactériennes. J. MONOD. *C. R. Acad. Sci.*,1937, **205**, 1456–1457.

1938

13. Extraction et dosage du pigment de l'oeil de la Drosophile. J. L. MONOD and Y. NEEFS. *C. R. Acad. Sci.*, 1938, **206**, 1677–1679.

1941

14. Croissance des populations bactériennes en fonction de la concentration de l'aliment hydrocarboné. J. MONOD. *C. R. Acad. Sci.*, 1941, **212**, 771–773.

15. Sur un phénomène nouveau de croissance complexe dans les cultures bactériennes. J. MONOD. *C. R. Acad. Sci.*, 1941, **212**, 934–936.

16. Recherches sur la croissance des cultures bactériennes. J. MONOD. Thèse Doctorat ès Sciences, Paris, 1941. Hermann Edit., Paris, 1942 (Deuxième Edition, 1958).

1942

17. Influence de l'amide de l'acide nicotinique de l'aneurine et de l'acide ascorbique sur la croissance des cultures de *B. coli*. J. MONOD. *Ann. Inst. Pasteur*, 1942, **68**, 435–438.

18. Sur un phénomène de lyse lié à l'inanition carbonée. J. MONOD. *Ann. Inst. Pasteur*, 1942, **68**, 444–451.

19. Diauxie et respiration au cours de la croissance des cultures de *Bacillus coli*. J. MONOD. *Ann. Inst. Pasteur*, 1942, **68**, 548–550.
20. Sur l'expression analytique de la croissance des populations bactériennes. F. MORIN and J. MONOD. *Rev. Scientifique*, 1942, **5**, 227–229.

1943

21. Influence de la concentration des substrats sur la rapidité d'adaptation chez le *Bacillus coli*. J. MONOD. *Ann. Inst. Pasteur*, 1943, **69**, 179–181.

1944

22. Sur la non-additivité d'action de certains enzymes bactériens. J. MONOD. *Ann. Inst. Pasteur*, 1944, **70**, 57–59.
23. Remarques sur le problème de la spécificité des enzymes bactériens. J. MONOD. *Ann. Inst. Pasteur*, 1944, **70**, 60–61.
24. Inhibition de l'adaptation enzymatique chez *Bacillus coli* en présence de 2,4 dinitrophénol. J. MONOD. *Ann. Inst. Pasteur*, 1944, **70**, 381–384.

1945

25. Sur la nature du phénomène de diauxie. J. MONOD. *Ann. Inst. Pasteur*, 1945, **71**, 37–40.

1946

26. Sur l'utilisation du saccharose par *Proteus vulgaris*. M. MOREL and J. MONOD. *Ann. Inst. Pasteur*, 1946, **72**, 647–651.
27. L'anhydride carbonique considéré comme substance indispensable aux microorganismes. La biosynthèse des acides dicarboxyliques. A. LWOFF and J. MONOD. *C. R. Acad. Sci.*, 1946, **222**, 696–697.
28. Mutation et adaptation enzymatique chez *Escherichia coli*. J. MONOD and A. AUDUREAU. *Ann. Inst. Pasteur*, 1946, **72**, 868–878.
29. Sur une mutation spontanée affectant le pouvoir de synthèse de la méthionine chez une bactérie coliforme. J. MONOD. *Ann. Inst. Pasteur*, 1946, **72**, 879–890.
30. Remarques à propos du rapport présenté par S. SPIEGELMAN. J. MONOD. *Cold Spring Harbor Symposia* **XII**, 1946, 274–275.

1947

31. Essai d'analyse du rôle de l'anhydride carbonique dans la croissance microbienne. A. LWOFF and J. MONOD. *Ann. Inst. Pasteur*, 1947, **73**, 323–347.
32. Inhibition de l'adaptation enzymatique chez une bactérie (*Escherichia coli*) infectée par un bactériophage. J. MONOD and E. WOLLMAN. *C. R. Acad. Sci.*, 1947, **224**, 417–419.

33. Excitation par l'acide succinique du centre respiratoire du Chien en apnée d'hyperventilation. D. BOVET, J. MONOD and A. LWOFF. *C. R. Acad. Sci.*, 1947, **224,** 1844–1846.

34. L'inhibition de la croissance et de l'adaptation enzymatique chez les bactéries infectées par le bactériophage. J. MONOD and E. WOLLMAN. *Ann. Inst. Pasteur*, 1947, **73,** 937–956.

35. Aminoacids in the physiology of bacteriophage. J. MONOD. *J. Gen. Microbiology*, 1947, **2,** vii–viii.

36. The phenomenon of enzymatic adaptation and its bearing on problems of genetics and cellular differentiation. J. MONOD. *Growth Symposium*, 1947, **XI,** 223–289.

37. Le rôle de l'anhydride carbonique dans le métabolisme bactérien. A. LWOFF and J. MONOD. IVème Congrès International de Microbiologie, Copenhague, 1947, p. 150–151 du *Compte Rendu*.

1948

38. Synthèse d'un polysaccharide du type amidon aux dépens du maltose, en présence d'un extrait enzymatique d'origine bactérienne. J. MONOD and A. M. TORRIANI (avec la collaboration technique de M. VUILLET). *C. R. Acad. Sci.*, 1948, **227,** 240–242.

39. Sur une lactase extraite d'une souche d'*Escherichia coli* mutabile. J. MONOD, A. M. TORRIANI and J. GRIBETZ. *C. R. Acad. Sci.*, 1948, **227,** 315–316.

1949

40. Sur les propriétés de l'amylomaltase. A. M. TORRIANI and J. MONOD. *First Intern. Congr. Biochem.*, Cambridge, England, 1949, p. 303–304 *Abst of Comm.*

41. Sur la réversibilité de la réaction catalysée par l'amylomaltase. A. M. TORRIANI and J. MONOD. *C. R. Acad. Sci.*, 1949, **228,** 718–720.

42. Facteurs génétiques et facteurs chimiques spécifiques dans la synthèse des enzymes bactériens. J. MONOD. "Unités biologiques douées de continuité genétique," C.N.R.S., éd. Paris, 1949, 181–200.

43. The problem of heterocarboxylic metabolites. A. LWOFF and J. MONOD. *Arch. Biochem.*, 1949, **22,** 482–483.

44. Sur la réactivation de bactéries sterilisées par le rayonnement UV. J. MONOD, A. M. TORRIANI and M. JOLIT. *C. R. Acad. Sci.*, 1949, **229,** 557–559.

45. The growth of bacterial cultures. J. MONOD. *Annual Rev. Microb.*, 1949, **3,** 371–394.

1950

46. De l'amylomaltase d'*Escherichia coli*. J. MONOD and A. M. TORRIANI. *Ann. Inst. Pasteur*, 1950, **78,** 65–77.

47. Adaptation, mutation and segregation in the formation of bacterial enzymes. J. MONOD. *Biochem. Soc. Symp.*, 1950, **4,** 51–58.

48. La technique de culture continue. Théorie et applications. J. MONOD. *Ann. Inst. Pasteur*, 1950, **79,** 390–410.

1951

49. La compétition entre les ions hydrogène et sodium dans l'activation de la β-D-galactosidase d'*Escherichia coli* et la notion d'antagonisme ionique. G. COHEN-BAZIRE and J. MONOD. *C. R. Acad. Sci.*, 1951, **232,** 1515–1517.

50. Purification et propriétés de la β-galactosidase (lactase) d'*Escherichia coli*. M. COHN and J. MONOD. *Biochim. Biophys. Acta*, 1951, **7,** 153–174.

51. Sur la biosynthèse de la β-galactosidase (lactase) chez *Escherichia coli*. La spécificité de l'induction. J. MONOD, G. COHEN-BAZIRE and M. COHN. *Biochim. Biophys. Acta*, 1951, **7,** 585–599.

52. L'effet du rayonnement ultraviolet sur la biosynthèse de la β-galactosidase et sur la multiplication du bacteriophage T2 chez *Escherichia coli*. F. JACOB, A. M. TORRIANI and J. MONOD. *C. R. Acad. Sci.*, 1951, **233,** 1230–1232.

1952

53. La synthèse de la β-galactosidase chez les Entérobactériacées. Facteurs génétiques et facteurs chimiques. J. MONOD. *Revue Suisse de Path. gén. et de Bact.*, 1952, **15,** 407–417.

54. Inducteurs et inhibiteurs spécifiques dans la biosynthèse d'un enzyme. La β-galactosidase d'*Escherichia coli*. J. MONOD. *Bull. Org. Mond. Santé*, 1952, **6,** 59–64.

55. La biosynthèse induite des enzymes (adaptation enzymatique). J. MONOD and M. COHN. *Adv. Enzymol.*, 1952, **13,** 67–119.

56. Le role des inducteurs spécifiques dans la biosynthèse des enzymes. J. MONOD. *Symposium sur la Biogénèse des Protéines*, IIème Congrès Intern. Biochimie, Paris, 1952, 75–84.

57. La cinétique de la biosynthèse de la β-galactosidase chez *Escherichia coli* considérée comme fonction de la croissance. J. MONOD, A. M. PAPPENHEIMER Jr. and G. COHEN-BAZIRE. *Biochim. Biophys. Acta*, 1952, **9,** 648–660.

1953

58. L'effet inhibiteur spécifique des β-galactosides dans la biosynthèse 'constitutive' de la β-galactosidase chez *Escherichia coli*. J. MONOD and G. COHEN-BAZIRE. *C. R. Acad. Sci.*, 1953, **236**, 417–419.

59. L'effet d'inhibition spécifique dans la biosynthèse de la tryptophane-desmase chez *Aerobacter aerogenes*. J. MONOD and G. COHEN-BAZIRE. *C. R. Acad. Sci.*, 1953, **236**, 530–532.

60. Specific inhibition and induction of enzyme biosynthesis. M. COHN and J. MONOD. *"Adaptation in microorganisms,"* London Symposium, 14–15 April 1953. Cambridge Univ. Press, 1953, 132–149.

61. L'effet inhibiteur spécifique de la méthionine dans la formation de la méthionine-synthase chez *Escherichia coli*. M. COHN, G. N. COHEN and J. MONOD. *C. R. Acad. Sci.*, 1953, **236**, 746–748.

62. Sur le mécanisme de la synthèse d'une protéine bactérienne. J. MONOD and M. COHN. *Symposium on Microbial Metabolism*, VIth Intern. Congr. of Microbiol., Rome, 7–11 September 1953, 42–62.

63. Terminology of enzyme formation. M. COHN, J. MONOD, M. R. POLLOCK, S. SPIEGELMAN and R.Y. STANIER. *Nature*, 1953, **172**, 1096.

1954

64. Les facteurs de la biosynthèse des enzymes. J. MONOD. *Cahiers de Physique*, 1954, Cahier no. **47**, 70–84.

1955

65. Studies on the induced synthesis of β-galactosidase in *Escherichia coli*: the kinetics and mechanism of sulfur incorporation. D.S. HOGNESS, M. COHN and J. MONOD. *Biochim. Biophys. Acta*, 1955, **16**, 99–116.

66. Données nouvelles sur la biosynthèse des enzymes. Adaptation enzymatique. J. MONOD. *Exposés Annuels Biochim. Méd.*, Masson Edit., Paris, 1955, **série XVII**, 195–211.

1956

67. Remarks on the mechanism of enzyme induction. J. MONOD. In *Enzymes: Units of Biological Structure and Function*, Henry Ford Hosp. Intern. Symposium (1 November 1955). Acad. Press Inc., Publ., New York, 1956, 7–28.

68. La galactoside-perméase d'*Escherichia coli*. H. W. RICKENBERG, G. N. COHEN, G. BUTTIN and J. MONOD. *Ann. Inst. Pasteur*, 1956, **91**, 829–857.

69. Etude du fonctionnement de la galactoside-perméase d'*Escherichia coli*. A. KEPES and J. MONOD. *C. R. Acad. Sci.*, 1957, **244**, 809–811.

70. Bacterial permeases. G. N. COHEN and J. MONOD. *Bact. Reviews*, 1957, **21**, 169–194.

1958

71. An outline of enzyme induction. J. MONOD. *Recueil Trav. Chim. des Pays-Bas*, 1958, **77**, 569–585.

72. Sur l'expression et le rôle des allèles "inductible" et "constitutif" dans la synthèse de la β-galactosidase chez des zygotes d'*Escherichia coli*. A. B. PARDEE, F. JACOB and J. MONOD. *C. R. Acad. Sci.*, 1958, **246**, 3125–3128.

1959

73. Antibodies and induced enzymes. J. MONOD. In *Cellular and Humoral Aspects of Hypersensitive States* (Symposium of the Section of Microbiology, The New York Academy of Medicine, 1957), H. Sherwood Lawrence Edit., New York, 1959, 628–645 (Discussion 645–650).

74. The genetic control and cytoplasmic expression of "inducibility" in the synthesis of β-galactosidase by *Escherichia coli*. A. B. PARDEE, F. JACOB and J. MONOD. *J. Mol. Biol.*, 1959, **1**, 165–178.

75. Information, induction, répression dans la biosynthèse d'un enzyme. J. MONOD. *Colloquium der Gesellschaft fur physiologische Chemie*, 9–12 April 1959, Mosbach/Baden, 120–178.

76. Biosynthese eines Enzyms. Information, Induktion, Repression. J. MONOD. *Angew. Chemie*, 1959, **71**, 685–691.

77. Sur la présence de protéines apparentées à la β-galactosidase chez certains mutants d'*Escherichia coli*. D. PERRIN, A. BUSSARD and J. MONOD. *C. R. Acad. Sci.*, 1959, **249**, 778–780.

78. Gènes de structure et gènes de régulation dans la biosynthèse des protéines. F. JACOB and J. MONOD. *C. R. Acad. Sci.*, 1959, **249**, 1282–1284.

79. On the enzymic acetylation of isopropyl-β-D-thiogalactoside and its association with galactoside-permease. I. ZABIN, A. KEPES and J. MONOD. *Biochem. and Biophys. Res. Comm.*, 1959, **1**, 289–292.

1960

80. Biosynthèse induite d'une protéine génétiquement modifiée, ne presentant pas d'affinité pour l'inducteur. D. PERRIN, F. JACOB and J. MONOD. *C. R. Acad. Sci.*, 1960, **250**, 155–157.

81. L'opéron: groupe de gènes à expression coordonnée par un operateur. F. JACOB, D. PERRIN, C. SANCHEZ and J. MONOD. *C. R. Acad. Sci.*, 1960, **250**, 1727–1729.

82. Synthèse constitutive de galactokinase consécutive au développement des bactériophages lambda chez *Escherichia coli* K 12. G. BUTTIN, F. JACOB and J. MONOD. *C. R. Acad. Sci.*, 1960, **250**, 2471–2473.

83. Effets d'un analogue de l'uracile sur les propriétés d'une protéine enzymatique synthetisée en sa presence. A. BUSSARD, S. NAONO, F. GROS and J. MONOD. *C. R. Acad. Sci.*, 1960, 250, 4049–4051.

84. Galactose transport in *Escherichia coli*. I. General properties as studied in a galactokinaseless mutant. B. L. HORECKER, J. THOMAS and J. MONOD. *J. Biol. Chem.*, 1960, **235**, 1580–1585.

85. Galactose transport in *Escherichia coli*. II. Characteristics of the exit process. B. L. HORECKER, J. THOMAS and J. MONOD. *J. Biol. Chem.*, 1960, **235**, 1586–1590.

86. On the expression of a structural gene. M. RILEY, A. B. PARDEE, F. JACOB and J. MONOD. *J. Mol. Biol.*, 1960, **2**, 216–225.

1961

87. Genetic regulatory mechanisms in the synthesis of proteins. F. JACOB and J. MONOD. *J. Mol. Biol.*, 1961, **3**, 318–356.

88. Structural and rate-determining factors in the biosynthesis of adaptive enzymes. J. MONOD, F. JACOB and F. GROS. *Biochem. Soc. Symposium*, 1962, **No.21**, 104–132.

89. Carbon source repression of β-galactosidase in *Escherichia coli*. D. BROWN and J. MONOD. *Fed. Proc.*, 1961, **20**, 222f.

90. On the regulation of gene activity. F. JACOB and J. MONOD. *Cold Spring Harbor Symp. Quant. Biol.*, 1961, **26**, 193–211.

91. General conclusions: Teleonomic mechanisms in cellular metabolism, growth, and differentiation. J. MONOD and F. JACOB. *Cold Spring Harbor Symp. Quant. Biol.*, 1961, **26**, 389–401.

92. Sur la régulation et sur le mode d'action des gènes. F. JACOB, F. GROS and J. MONOD. *J. Chim. Phys.*, 1961, **58**, 1100–1102.

1962

93. Sur le mode d'action des gènes et leur regulation. F. JACOB and J. MONOD. In *Semaine d'étude sur le problème des macromolécules d'intérêt biologique*. Acad. Pontificale des Sciences, Rome, 1962, 85–95.

94. Thiogalactoside transacetylase. I. ZABIN, A. KEPES and J. MONOD. *J. Biol. Chem.*, 1962, **237,** 253–257.

95. Summary of the Columbia Symposium. J. MONOD. In *Basic Problems in Neoplastic Disease* (Columbia University Symposium, March 1962), Columbia Univ. Press, 1962, 218–237.

96. Sur la nature du répresseur assurant l'immunité des bactéries lysogènes. F. JACOB, R. SUSSMAN and J. MONOD. *C. R. Acad. Sci.*, 1962, **254,** 4214–4216.

1963

97. Determinisme et régulation spécifique de la synthèse des protéines. F. JACOB and J. MONOD. *Proc. Vth Intern. Congr. Biochem.* (Moscow, 1961), Pergamon Press, Oxford-London-New York-Paris, 1963, vol. 1, 132–154.

98. Genetic repression, allosteric inhibition, and cellular differentiation. F. JACOB and J. MONOD. In *Cytodifferential and Macromolecular Synthesis*, Acad. Press Inc., New York, N.Y., 1963, 30–64.

99. Elements of regulatory circuits in bacteria. F. JACOB and J. MONOD. In *Biological Organization at the Cellular and Supercellular Level*, Acad. Press Inc., New York and London, 1963, 1–24.

100. Allosteric proteins and cellular control systems. J. MONOD, J. P. CHANGEUX and F. JACOB. *J. Mol. Biol.*, 1963, **6,** 306–329.

101. On the reversibility by treatment with urea of the thermal inactivation of *Escherichia coli* β-galactosidase. D. PERRIN and J. MONOD. *Biochem. Biophys. Res. Comm.*, 1963, **12,** 425–428.

1964

102. Le promoteur, élément génétique nécessaire a l'expression d'un opéron. F. JACOB, A. ULLMANN and J. MONOD. *C. R. Acad. Sci.*, 1964, **258,** 3125–3128.

103. Mécanismes bichimiques et génétiques de la régulation dans la cellule bactérienne. F. JACOB and J. MONOD. *Bull. Soc. Chim. Biol.*, 1964, **46,** 1499–1532.

103a. Id., Russian translation.

104. Non-inducible mutants of the regulator gene in the "Lactose" system of *Escherichia coli*. C. WILLSON, D. PERRIN, M. COHN, F. JACOB and J. MONOD. *J. Mol. Biol.*, 1964, **8,** 582–592.

105. The effect of 5'adenylic acid upon the association between bromoth-ymol blue and muscle phosphorylase b. A. ULLMANN, P. R. VAGE-LOS and J. MONOD. *Biochem. Biophys. Res. Comm.*, 1964, **17**, 86–92.

1965

106. On the nature of allosteric transitions: a plausible model. J. MONOD, J. WYMAN and J. P. CHANGEUX. *J. Mol. Biol.*, 1965, **12**, 88–118.

106a. Id. in *Theoretical physics and biology*, Compte rendu de la conference intitulée *Some considerations of the functional importance of molecular symmetry of some biological macromolecules*, given by J. MONOD for l'Institut de la Vie at Versailles (26–30 June 1967), at the first interna-tional conference on *Theoretical Physics and Biology*.

107. Identification par complémentation *in vitro* et purification d'un seg-ment peptidique de la β-galactosidase d'*Escherichia coli*. A. ULL-MANN, D. PERRIN, F. JACOB and J. MONOD. *J. Mol. Biol.*, 1965, **12**, 918–923.

108. Un modèle plausible de la transition allostérique. J. P. CHANGEUX, A. ULLMANN and J. MONOD. In *Mécanismes de régulation des ac-tivités cellulaires chez les microorganismes*, Colloque international du CNRS No. 124 (Marseille, 23–27 July 1963), 1965, 285–295.

109. Rôle du lactose et de ses produits métaboliques dans l'induction de l'operon lactose chez *Escherichia coli*. C. BURSTEIN, M. COHN, A. KEPES and J. MONOD. *Biochim. Biophys. Acta*, 1965, **95**, 634–639.

110. Genetic mapping of the elements of the lactose region in *Escherichia coli*. F. JACOB and J. MONOD. *Biochem. Biophys. Res. Comm.*, 1965, **18**, 693–701.

111. Quelques réflexions sur les relations entre structures et fonctions dans les protéines globulaires. J. MONOD. *Année Biologique*, 1965, **4**, 231–240.

112. Délétions fusionnant l'opéron lactose et un opéron purine chez *Es-cherichia coli*. F. JACOB, A. ULLMANN and J. MONOD. *J. Mol. Biol.*, 1965, **13**, 704–719.

1966

113. On the mechanism of molecular interactions in the control of cellular metabolism. J. MONOD. *Endocrinology*, 1966, **78**, 412–425.

114. De l'adaptation enzymatique aux transitions allostériques. J. MONOD. Conférence Nobel, In *Les Prix Nobel en 1965*, Imprimerie Royale P.A. Norstedt and Soner, Stockholm, 1966, p. 244–263.

114a. From enzymatic adaptation to allosteric transitions. J. MONOD. En-glish translation, Nobel Conference: *Science*, 1966, **154**, 475–483.

114b. Von der enzymatischen Adaptation zur allosterischen Umlagerung. J. MONOD. German translation, Nobel Conference: *Angew. Chem.*, 1966, **14**, 694–703.

1967

115. The operon: a unit of coordinated gene action. G. BUTTIN, F. JACOB and J. MONOD. In *Heritage from Mendel*, R.A Brink Edit., Univ. Wisconsin Press, 1967, 155–177.

116. Characterization by *in vitro* complementation of a peptide corresponding to an operator-proximal segment of the β-galactosidase structural gene of *Escherichia coli*. A. ULLMANN, F. JACOB and J. MONOD. *J. Mol. Biol.*, 1967, **24**, 339–343.

117. Leçon Inaugurale (Chaire de Biologie moléculaire), Collège de France, 3 November 1967. J. MONOD. Publication du Collège de France, 1968, No. 47, 31 pages.

117a. De la biologie moléculaire a l'ethique de la science. J. MONOD. Leçon Inaugurale, In *L'Age de la Science*. Dunod Edit., No. 1, 1968, 3–18.

117b. From molecular biology to the ethics of knowledge. J. MONOD. Inaugural Lesson, in *The Human Context*, The Chaucer Publ. Co. Ltd. (17, Platt's Lane, London N.W. 3, England), 1968 (?), 325–336.

117c. From biology to ethics. J. MONOD. Inaugural Lesson, in *The Salk Institute for Biological Studies Occasional Papers* (San Diego, California), 1969, **1**, 22 pages.

1968

118. Kinetics of the allosteric interactions of phosphofructokinase from *Escherichia coli*. D. BLANGY, H. BUC and J. MONOD. *J. Mol. Biol.*, 1968, **31**, 13–35.

119. On the determination of molecular weight of proteins and protein subunits in the presence of 6 M guanidine hydrochloride. A. ULLMANN, M. E. GOLDBERG, D. PERRIN and J. MONOD. *Biochemistry*, 1968, **7**, 261–265.

120. Sur certaines implications de l'hypothèse d'une équivalence stricte entre les protomères des protéines oligomériques. P. CLAVERIE, M. HOFNUNG and J. MONOD. *C. R. Acad. Sci.*, 1968, **266**, 1616–1618.

121. On the subunit structure of wild-type versus complemented β-galactosidase of *Escherichia coli*. A ULLMANN, F. JACOB and J. MONOD. *J. Mol. Biol.*, 1968, **32**, 1–13.

122. On symmetry and function in biological systems. J. MONOD. In *Symmetry and function of biological systems at the macromolecular level*, Nobel

Symposium No. 11, Edited by Arne Engstrom and Bror Strandberg, Almqvist and Wiksell, Stockholm, 1968, 15–27.

123. Cyclic AMP as an antagonist of catabolite repression in *Escherichia coli*. A. ULLMANN and J. MONOD. *FEBS Letters*, 1968, **2 no. 1,** 57–60.

1969

124. On the effect of divalent cations and protein concentration upon re-naturation of β-galactosidase from *Escherichia coli*. A. ULLMANN and J. MONOD. *Biochem. Biophys. Res. Comm.*, 1969, **35,** 35–42.

125. On values in the age of science. J. MONOD. In *The place of value in a world of facts*. Nobel Symposium No. 14, Edited by Arne Tiselius and Sam Nilsson, Almqvist and Wiksell, Stockholm, 1969, 19–27.

1970

126. Introduction. F. JACOB and J. MONOD. In *The Lactose Operon*, Cold Spring Harb. Lab., J. R. Beckwith and D. Zipser Edit., 1970, 1–14.

127. On the stoichiometry and kinetics of ω complementation of *Escherichia coli* β-galactosidase. A. ULLMANN and J. MONOD. In *The Lactose Operon*, Cold Spring Harb. Lab., J. R. Beckwith and D. Zipser Edit., 1970, 265–272.

128. On the mechanism of catabolite repression. G. CONTESSE, M. CREPIN, F. GROS, A. ULLMANN and J. MONOD. In *The Lactose Operon*, Cold Spring Harb. Lab., J. R. Beckwith and D. Zipser Edit., 1970, 401–415.

129. *In vitro* studies of the Lac operon regulatory system. S. BOURGEOIS and J. MONOD. In *Ciba Foundation Symposium on Control Processes in Multicellular Organisms*. G.E.W. Wolstenholme and Julie Knight Edit., J. and A. Churchill Ltd (104 Gloucester Place, London Publ.), 1970, 3–21.

130. Cyclic AMP and catabolite repression in *Escherichia coli*. A. ULLMANN, G. CONTESSE, M. CREPIN, F. GROS and J. MONOD. In *Fogarty International Center, Proceedings No. 4*, U.S. Dept. of Health, Education, and Welfare, NIH, 1970, 215–231.

131. *Le hasard et la nécessité. Essai sur la philosophie naturelle de la biologie moderne.* J. MONOD.

    1 - Editions *"Le Seuil,"* Paris, 1970, 197 pp.

    2 - Editions *"Le Seuil."* Collection *"Points - Sciences,"* 1973, 244 pp.

    3 - Editions *"La Bibliothèque du XXe siècle"* (preface by Henri Laborit), France Loisirs (123, boulevard de Grenelle, Paris), 1989, 237 pp.

Foreign translations

Italy: *Il caso e la Necessità.* Mondadori Edit., November 1970 (plus pocket edition).

United States: *Chance and Necessity,* Knopf, October 1971; Vintage Books (pocket edition), October 1972.

Brazil: *O acaso e a necessidade,* Editora Vozes, November 1971.

Greece: Lambis Rappas Edit., November 1971.

Portugal: *O acaso e a necessidade,* Europa-America Edit., February 1972.

The Netherlands: *Toeval en onvermijdelijkheid,* Bruna et Zoon Edit., March 1972.

United Kingdom: *Chance and Necessity,* Collins Edit., April 1972.

Norway: *Tilfeldigheten og nodvendigheten,* Gyldendal Norsk Edit., May 1972.

Sweden: Albert Bonniers Edit., 1972.

Finland: *Sattuma ja välttämätömyys.* Werner Soderstrom Osakeyhtio Edit., 1973.

Germany: *Zufall und Notwendigkeit,* D.T.V. Edit. (pocket edition), 1975; Piper Edit., 1983.

Denmark: *Tilfaeldigheden og nodvendigheden.* Fremad Edit., November 1975.

Spain: *El azar y la necesidad,* Tusquets Edit., December 1981.

1971

132. Suggestion for a study on the implications of various chemicals and non-disease specific drugs. P. BEACONFIELD, R. RAINSBURY, J. HUXLEY, R. PETERS, J. TREFOUEL, J. MONOD, R. PAUL and H. THEORELL *Experientia,* 1971, **27,** 715–717.

133. Du microbe à l'homme. J. MONOD. In *Of Microbes and Life* (Tribute to A. Lwoff). J. Monod and E. Borek Edit., Columbia Univ. Press, New York, London, 1971, 1–12.

1973

134. On the molecular theory of evolution. J. MONOD, Herbert Spencer Lecture, Oxford, Nov. 2th, 1973. In *Herbert Spencer Lectures Book 1973,* p. 11–24.

1974

135. An immunological study of complementary fragments of β-galactosidase. F. CELADA, A. ULLMANN, and J. MONOD. *Biochemistry,* 1974, **13,** 5543–5547.

1976

136. Catabolite modulator factor: a possible mediator of catabolite repression in bacteria. A. ULLMANN, F. TILLIER and J. MONOD. *Proc. Natl. Acad. Sci. USA*, 1976, **73,** 3476–3479.

1977

137. L'evolution microscopique. J. MONOD. (Translation from the 4th Hartmann-Miller Conference, given by J. Monod at the University of Zurich, 14 February 1975). *Theoria to Theory*, Gordon and Breach Science Publ. Ltd., 1977, **10,** 303–311.

Index